The Science of Imaging

An Introduction

SECOND EDITION

The Science of Imaging
An Introduction
SECOND EDITION

Graham Saxby

CRC Press
Taylor & Francis Group
Boca Raton   London   New York

CRC Press is an imprint of the
Taylor & Francis Group, an **informa** business

A TAYLOR & FRANCIS BOOK

Taylor & Francis
6000 Broken Sound Parkway NW, Suite 300
Boca Raton, FL 33487-2742

© 2011 by Taylor and Francis Group, LLC
Taylor & Francis is an Informa business

No claim to original U.S. Government works

Printed in the United States of America on acid-free paper
10 9 8 7 6 5 4 3 2 1

International Standard Book Number: 978-1-4398-1286-0 (Hardback)

Library of Congress Cataloging-in-Publication Data

Saxby, Graham.
    The Science of imaging : an introduction. / Graham Saxby. -- 2nd ed.
        p. cm.
    Summary: "Edited and expanded to keep pace with the digital revolution, this highly popular and critically acclaimed work continues to provide students with a comprehensive exploration of imaging science. Brilliantly written and extensively illustrated, it covers fundamental laws of physics as well as the cutting-edge techniques that are defining the current direction of the field. It adds a new section on astronomical imaging, as well as major revisions in the areas of digital imaging and modern technology. All references have been updated and now include a significant number of URLs leading to both teaching materials and expanded information. A fundamental introduction to the subject, this volume takes readers on a grand tour of imaging. Starting with the fundamentals of light and basic cameras, the author journeys through television and holography to advanced scientific and medical imaging. Topics such as digital recording of images, the photographic process, and film development are dealt with in an informative and entertaining manner. The book keeps mathematics to a minimum and offers copious examples."-- Provided by publisher.
    Includes bibliographical references and index.
    ISBN 978-1-4398-1286-0 (hardback)
    1. Images, Photographic. 2. Photography--Processing. 3. Image processing. 4. Photography--Digital techniques. I. Title.

TR222.S295 2010
621.36'7--dc22   2010032659

Visit the Taylor & Francis Web site at
http://www.taylorandfrancis.com

and the CRC Press Web site at
http://www.crcpress.com

To the memory of Michael Langford,
a gifted and dedicated teacher, and a much valued friend

Contents

Preface to the First Edition

When my publisher suggested that I should write a book on the science of imaging, my first thought was that this must have been done already a dozen times or more. However, I roughed out a plan, and then, in order to see what the competition was, did a literature search. To my surprise it turned up only six books. Four were out of print. The other two were nothing like what I had in mind: they were advanced textbooks, and both omitted certain areas I felt were essential in a comprehensive book. There was nothing for beginners.

Over the years I have had to read a great many textbooks, and I have usually been cautious when coming across one with the word "Introduction" in its title. All too often the first chapter turns out to be a recap on the principles of tensor calculus, or a brief review of quantum electrodynamics, after which it gets down to the serious stuff. To rub salt in, the blurb usually twitters about its being "suitable for first-year students."

This book is different: it really is an introduction. I wasn't able to follow Stephen Hawking's example and write a whole book with only one equation in it (not that that makes *A Brief History of Time* any easier reading), but I have tried to keep the equations to the irreducible minimum, and mostly they are quarantined in boxes that you can ignore (if equations make you nervous) without losing the thread. Chapter 4 does have a few, though: these are unavoidable if you are to understand how a lens works.

It has always been my firm belief that a textbook should be the next best thing to a live tutor at one's elbow. In this respect the Open University's texts are impressive. I haven't hesitated to follow their example both in their personal approach to the reader (so often absent from traditional textbooks), and in their practice of using wide margins to accommodate notes, small diagrams, and parenthetic remarks. I hate footnotes as much as do printers, and comments that are placed within the text break its continuity. So my own asides have gone into the margins, and if you are not interested in the fact that Charles Wheatstone invented the concertina, or that Joseph Fourier was appointed Governor of Egypt by Napoleon, you can ignore the marginal note and read on.

Lists of references can be useful if the discussion of a particular topic has whetted your appetite for more information, but in practice such lists rarely give an indication of the quality or scope (or even reliability) of the sources. In this respect I have tried to be more helpful, by including at the end of each chapter a section titled "Digging Deeper." Rather than giving an exhaustive and perhaps less than discriminant bibliography, I have chosen the books that seem to me the best ones to lead on from what I have written in each chapter. These days there is a tendency among publishers to allow a book to go out of print as soon as it is selling fewer than a hundred or so copies a year. This has been the fate of many texts I personally would have regarded as indispensable. Fortunately, it is always possible to borrow a copy through the public library system, or through a university or other academic library. There are also book dealers, often with their own Web sites, who specialize in hunting down out-of-print books. With that in mind I haven't hesitated to recommend one or two seminal books written many years ago, which still deserve a place in any enthusiast's personal reference library.

G.S.

Preface to the Second Edition

When I began the first edition of this book, imaging science was in the middle of a revolution. It still is. In 2002 digital imaging techniques were steadily taking over from analog techniques. Digital sound recording had already ousted analog entirely, and digital imaging was in the process of doing the same with visual records. Digital imaging techniques have now accomplished this almost completely. As its name proclaims, this book is primarily concerned with science; it has been some consolation to me while preparing a new edition of it that in spite of all the technological advances of the last seven years nobody has succeeded in changing the laws of physics (and I hope they don't). So you will find the first few chapters unaltered, except where a few syntactic and semantic infelicities have been eliminated. Later chapters have been updated where necessary, in some cases substantially, and some of their contents have been rearranged in a more logical fashion. As much of the content of the three chapters on the photographic process in the first edition has now become redundant in the context of present-day photographic technology, this material has been reduced to a single chapter. On the other hand, the scientific discipline of astrophysics has grown in importance to such an extent that its imaging principles and techniques can no longer be left out. So there is a whole new chapter on astronomical imaging. The final chapters in the first edition have also been rearranged and tidied up. I have adopted American spelling because British readers are less likely to have difficuly with this than American readers might have with the quirks of British spelling. Where appropriate I have updated the reading lists at the end of chapters, though it is surprising how much of the older material is still valid. But it grieves me to have had to add the adjective "late" to the names of some of the authors in these lists, in particular to my fellow author and good friend Ron Graham.

Since the first edition of this book there has been a veritable explosion of scientific and technical material on the Internet. At one time it might have taken weeks for an author to obtain or confirm a particular piece of information: now this can often be achieved with a few clicks of a mouse. There is plenty of good teaching material on the Net too, and I have not hesitated to recommend this where it has seemed appropriate. This all bodes well for the scientific enlightenment of the general public. May the Internet continue to flourish!

Acknowledgments

It would be dishonest to pretend that I had written the whole of this book off the top of my head. Certainly, much of it represents my own experience. But whereas in the fifteenth century it was just about possible for one individual to possess all the knowledge there was in the world (this was said of Leonardo da Vinci, and, much later, with somewhat less justification, of Johann Wolfgang Goethe), it is now impossible for anyone to claim a comprehensive knowledge of even a comparatively narrow subject. And imaging is certainly not a narrow subject.

You can get an idea of some of the sources I have consulted from the reading lists at the end of each chapter. I owe a debt of gratitude to each of these authors. I also owe a large lunch to those image makers who have helped me by providing illustrations I could not have made myself and material I could not have obtained from reference sources. I must mention in particular Sidney Ray, David Burder, Jon Tarrant, Adrian Davies, Bob Gibson, Justin Quinnell, Hans Bjelkhagen, Ric Parker, and Ron Graham, as well as the various publishers who gave their permission to use some of their illustrative material.

About the Author

Graham Saxby served in the Royal Air Force (RAF) for 27 years, the first 19 in the trade of photographer, where he undertook almost every possible form of assignment. After being commissioned into the Education Branch, he was officer commanding photographic science flight at the RAF School of Photography at Cosford for seven years. On leaving the RAF he joined the staff of what is now the University of Wolverhampton as senior lecturer in educational technology, later moving to the Department of Applied Sciences to teach modern optics. His research into display holographic techniques has earned him an international reputation, and his books have won several prestigious awards. Now formally retired, he works as a freelance editor and reviewer of technical books and as a consultant in optical and photographic matters. He is a fellow of the Institute of Physics and an honorary fellow of the Royal Photographic Society.

Chapter 1 The Nature of Light

Models for the Behavior of Light

For thousands of years people have wondered what light really is, and have tried to construct models predicting its behavior. In the seventeenth century Sir Isaac Newton put forward the concept of "energy," and used it as a fundamental property of objects in motion. He also ascribed it to such things as unwinding springs, burning gases, sound and—important in our context—light. Not everyone agreed with him at the time. But today, when we operate switches, drill teeth, and even weld metal with light beams, there is no longer any doubt.

Newton himself believed that light consisted of particles like tiny bullets, travelling at enormous speed. This model does predict much of the more obvious behavior of light such as reflection and the formation of shadows, and refraction too, if one makes some dodgy assumptions. Indeed, the entire system of what we now call geometrical optics is based on the idea of the rectilinear propagation of light.

Newton's contemporary Christiaan Huyghens suggested that the behavior of light, particularly with regard to refraction, could be accounted for better if light consisted of waves like sound waves. Newton strongly opposed this theory, and this disagreement led to a permanent antagonism between the two men.

When the polarization of light was discovered, it became necessary to modify Huyghens's model of longitudinal waves (which can't be polarized) to transverse waves (which can) (Figure 1.1).

The wave model provided a reasonably good account of diffraction and interference, as the principle assumed that each point on a wavefront was a source of "wavelets," the envelope of which would form the new wavefront (Figure 1.2).

By the middle of the nineteenth century the speed of light had been established to within a few percent, but there still seemed to be no logical reason for this particular speed, until James Clerk Maxwell postulated a connection between electricity and magnetism, and light. By combining two fundamental constants of electricity and magnetism he obtained an expression that gave a value to the speed of electromagnetic propagation (see Box). A few years later Heinrich Hertz's experiments with radio waves showed that there was a whole electromagnetic spectrum, of which light was a part.

Newton seems to have been good at antagonisms. He was even more vehement about Robert Hooke, who had anticipated some of Newton's most important ideas. It has been suggested that Newton's famous remark about having stood on the shoulders of giants was partly a jibe at these two men, who were both short in stature. Christiaan Huyghens (1629–1695) was a Dutch physicist and astronomer. Apart from his work on optics he invented the pendulum clock, virtually founded the discipline of mechanics, and identified the nature of Saturn's rings, as well as discovering its moon, Titan.

James Clerk Maxwell was a brilliant theoretical physicist who proposed the kinetic theory of gases, established the structure of Saturn's rings, and made notable contributions to thermodynamics and to the theory of color perception (he produced the first-ever color photograph). In the formulation of his electromagnetic equations, perhaps his greatest contribution to modern science, he was the first person to apply the newly minted mathematics of vector calculus. This paved the way for Heinrich Rudolf Hertz (1857–1894) to discover and identify radio waves, and to show that they behaved in a manner similar to light. Both men died tragically young.

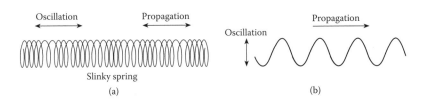

Figure 1.1 Longitudinal and transverse waves: (a) Slinky spring; (b) rubber rope.

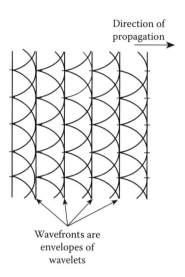

Figure 1.2 Propagation by wavelets (Huyghens principle).

MAXWELL AND ELECTROMAGNETISM

Maxwell found that the speed of electromagnetic propagation was related to the permittivity ε_0 and permeability μ_0 of free space by the simple relationship

$$c = 1/\sqrt{(\varepsilon_0 \mu_0)}$$

Maxwell's work led to four famous equations that describe the behavior of electromagnetic radiation. To appreciate them mathematically you need a fair understanding of vector calculus, but when translated into simple prose they go something like this:

1. The distribution of electric charges creates electric fields;
2. Magnetic fields that change in time also produce electric fields;
3. Magnetic fields are continuous, with no beginning and no end;
4. Both electric currents and electric fields that change in time can produce magnetic fields.

I am indebted to Milo Sholt of the Open University for this insight.

It was for his work on the photoelectric effect and the principle of photon emission and absorption that Einstein (1879–1955) was awarded the Nobel Prize for Physics in 1921, and not, as many people think, for his much more famous relativity theories. Max Planck (1858–1947) discovered that energy consists of fundamental indivisible units, which he called *quanta*. For this discovery he was awarded the Nobel Prize for Physics in 1918.

When the photoelectric effect was discovered and quantified by Albert Einstein, it became clear that a continuous electromagnetic wave, in Maxwell's terms, whose energy would depend only on its intensity could not possess the appropriate energy. In order to obtain the right answers Einstein suggested that light was not a continuous wave, but was emitted in the form of tiny pulses of light energy (which he called *photons*), and that the energy carried by a photon was directly proportional to its frequency. This concept fitted Max Planck's quantum theory, and the photon took its place among fundamental particles. A photon was considered as having zero rest mass, and could therefore travel "at the speed of light" without violating the laws of special relativity.

Electromagnetic Radiation

Electromagnetic radiation is so much part of our daily lives that we scarcely ever think about it. But now we need to look at its nature a little more closely. Why "electromagnetic"?

The first person to appreciate the connection between electricity and magnetism was Michael Faraday. One of his experiments is very simple to repeat:

If you pass an electric current through a piece of copper wire it will set up a magnetic field, and if you take a small compass needle and hold it with its spindle parallel to the wire it will align itself with the direction of the field. If you reverse the direction of the current through the wire the needle will reverse its direction too (Figure 1.3).

If the wire is carrying a.c. mains current, this changes direction back and forwards again 50 times a second (60 in the United States). This represents a frequency of 50 hertz (Hz). If the compass needle were small and light enough it would follow these reversals too. It is electromagnetic energy that moves the needle. The strength of the field becomes less as you move away from the wire, but the effect is still present, and would still be there even if you were to go millions of miles away. Now, in basic electrical theory we tend to think of a magnetic field as being set up instantaneously everywhere, but Maxwell showed that it is not. If your compass needle happened to be 300,000 kilometers from the wire, it would sense the existence of the field one whole second later. If our a.c. were to be 1 Hz rather than 50 Hz, such a compass needle would always be exactly one cycle behind. A needle at half this distance would be half a cycle behind, and so on. If you could freeze the system at a particular instant it would look like Figure 1.4.

As the distance 300,000 km contains one complete cycle, we say that the *wavelength* is 300,000 km (3×10^8 m). The number of cycles per second is the *frequency*; we can say, therefore, that

Velocity = Wavelength × Frequency

Michael Faraday (1791–1867) was undoubtedly the greatest experimental physicist of the nineteenth century. But as well as discovering the relationship between magnetism and electricity and designing the first electrical generator and motor, he made a large number of practical discoveries in organic chemistry (he was the first person to isolate benzene). He was a bookbinder by training, and had no mathematical ability: all his successes were the result of intuitions.

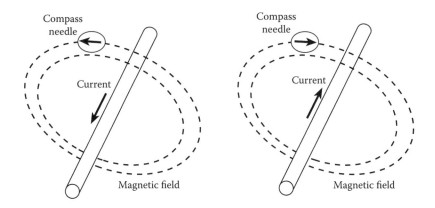

Figure 1.3 Reversing the current reverses the direction of the magnetic field.

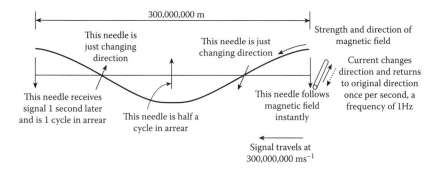

Figure 1.4 Instantaneous state of the magnetic field.

The Electromagnetic Spectrum

The electromagnetic field set up by a 50 Hz a.c. is not very strong. If you place a compass needle under an electricity pylon it will barely quiver. But as the frequency increases, so does the energy. To obtain high frequencies you need to use an electronic device called an oscillator, and if you adjust this to give a frequency of, say, 198,000 Hz (198 kHz) the effects can be detected many miles away.

In fact, if you superimpose a microphone signal on it you will be broadcasting (illegally) on the same frequency as Radio 4 (long wave).

This is almost the lowest frequency that can be used for radio broadcasting. The radio spectrum continues up to a frequency of around 1,000,000,000 Hz (10^9 Hz or 1 GHz), which represents the (surface-based) TV band. Here the radiation takes on some of the properties shown by light: it can be reflected, and it travels in straight lines; large objects such as tower blocks and hills can obstruct and reflect the beam. Above this region, corresponding to wavelengths of tens of millimeters, are microwaves, which, of course, possess enough energy to cook potatoes. They are used in radar and in satellite communications, as well as in ground-based links, and when used for such purposes require parabolic reflectors to concentrate the beam on the receiving antenna. The next step up is the band of wavelengths running from around 1 mm to 10 micrometers (μm). These are known as terahertz waves (1 terahertz [THz] = 10^{12} Hz), or T-rays. It has recently become possible to generate terahertz radiation and to produce images using these wavelengths. Beyond this is the beginning of the infrared (IR) region, which extends from some 10 μm to the beginning of the visible wavelengths.

Electronic devices can't generate IR wavelengths. But lasers exist that can. Lasers produce single frequencies, just like oscillators, at wavelengths that range right through the IR and visible spectrum into the ultraviolet (UV), and there is even an X-ray laser operating at less than 10 nanometers (1 nm = 10^{-9} m). At present we can't generate shorter wavelengths such as gamma radiation at single frequencies, though the electron laser has already become a fact. Figure 1.5 summarizes the electromagnetic spectrum. You will notice that on this gigantic keyboard the visible spectrum spans less than a single octave.

Calling it a keyboard isn't mere whimsy. Each octave on a (musical) keyboard represents a doubling of sound frequency. In Figure 1.5 each interval represents a multiplication of electromagnetic frequency by 10. Both scales are *logarithmic*. There is more about logarithms in Appendix 1; the concept underlies much of the nature of visual perception (see Chapter 3).

We don't use single frequencies much in making images. Only radar and holography demand such highly disciplined radiation. For most imaging purposes we need to use the whole visible spectrum as generated by fairly ordinary light sources, including daylight.

Polarization

An important characteristic of light that is predicted by the transverse wave model is *polarization*. If you were to take up the end of the rubber rope of Figure 1.1b and shake it in random directions, waves would still travel along it, but the plane of vibration would fluctuate wildly. This is an *unpolarized* wave. If you shake the rope in one direction only, the vibration will be in a single plane, and this is a *linearly polarized* wave. Ordinary water waves are linearly polarized: although the energy spreads out from the source in all directions; at any one point the water is only going up and down. To produce polarized light from an ordinary source you need a polarizing filter, which is described in Chapter 2. Laser light is polarized, and polarized light also occurs naturally in skylight and in light reflected from a shiny nonmetallic surface. Certain materials, such as the minerals mica, tourmaline, and calcite, have the property of *double refraction* or *birefringence*. Light passing into such materials is split into two beams, which are linearly polarized at right angles to one another.

Many minerals are birefringent, or become so when stressed, and birefringence is important in the study of mineralogy. Oddly enough, even ice is birefringent when grown as a single crystal.

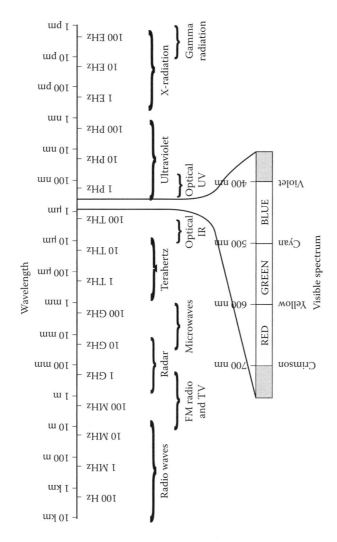

Figure 1.5 The electromagnetic spectrum.

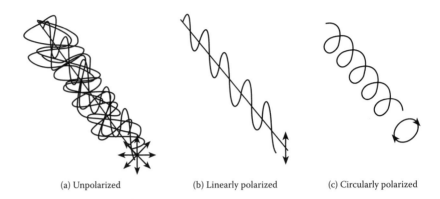

(a) Unpolarized   (b) Linearly polarized   (c) Circularly polarized

Figure 1.6 Polarization of light: (a) unpolarized; (b) linearly polarized; (c) circularly polarized.

If you take your rubber rope and give it a rolling motion by moving your hand in a circle, you will still produce a wave, but of a different kind. It is certainly not random. Is it another form of polarization? Yes, it is: it is called *circular polarization*. The direction of polarization rotates through 360° for every wavelength. This type of polarization doesn't seem to occur in nature, but it does have a place in certain physics experiments, and even in photography (Figure 1.6).

Interference

Two waves of the same wavelength passing through the same space in the same direction are said to undergo *interference*. (Note: this has nothing to do with the sort of interference that sometimes mars analog TV and shortwave radio.) Their amplitudes (i.e., the heights of their crests above the median) will add algebraically. If the crests coincide, the resultant wave has an amplitude that is the arithmetic sum of the two amplitudes. If the crest of one wave coincides with the trough of the other, the resultant will be the arithmetic difference between the two amplitudes. If these are equal they will cancel, and in space there will be no disturbance at all (Figure 1.7).

These two conditions are described respectively as "in phase" and "in antiphase." In between ("out of phase"), the algebraic sum of the two amplitudes may be more or less than the average of the two amplitudes, depending on whether the phase

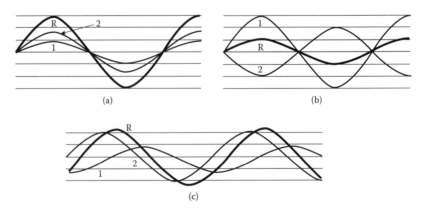

(a)   (b)

(c)

Figure 1.7 Interference between two transverse waves (1 and 2), and resultant (R): (a) in phase; (b) in antiphase; (c) out of phase.

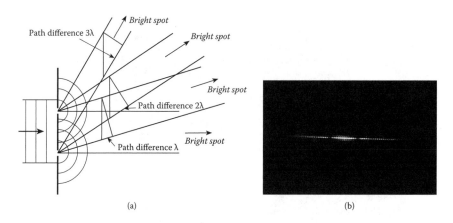

Figure 1.8 (a) Formation of bright spots where optical path differences are integral multiples of wavelength; (b) photograph of fringes.

relationship is nearer to in-phase or antiphase. The position of the crest of the resultant will be somewhere in between the other two.

The term "interference" is qualified as "constructive" or "destructive" depending on whether the resultant is greater or less in amplitude than its components.

A striking demonstration of interference occurs when a beam of monochromatic (single wavelength) light passes through two closely spaced slits onto a screen. This is known as the *Young's slits* experiment after its originator, Thomas Young.

Where the two beams overlap, the screen shows a series of evenly spaced spots or *fringes*. These mark the angles at which the difference in distance from the two slits to that point on the screen is either a whole number of wavelengths (0, 1, 2, 3 etc.) for a bright fringe, or an odd number of half-wavelengths (½, 1½, 2½, 3½ etc.) for a dark one. This is illustrated in Figure 1.8.

The best light source for this demonstration is a small laser such as a diode laser pointer, as the beam is very bright (the laser being still 160 years in the future, Young had to make do with sodium light). You can replicate the demonstration if you make a double slit by scratching two fine lines, as close together as you can, with a needle on a piece of fogged black photographic film, or better, make a high-contrast photographic negative of two parallel lines drawn 2-mm apart, from a distance of around 1 m. When you shine your laser pointer through these slits at a white wall in a darkened room you will see Young's fringes clearly. You can replicate the experiment on a large scale using water waves in a bath. Make the two wave sources by dipping your fingers in the water several times a second, with your fingers 10–20 cm apart. You will be able to see the interference pattern radiating outwards from the two sources in alternate lines of still and disturbed water.

You may have noticed if you did the bath experiment that there is a relationship between the separation of the sources and the separation of the spots. It is an inverse relationship: halve the slit separation and you double the spot separation. There is also a direct relation between the spot separation and the wavelength of the light for any given separation. So if you illuminate the slits with blue light the spots have a separation smaller than if you illuminate them with red light. In fact, if you illuminate the slits with white light (which contains all wavelengths from 400 to 700 nm) you get a drawn-out spectrum instead of the spots (Figure 1.9b, color plate).

Note that "out of phase" doesn't mean the same as "in antiphase." Many textbooks fail to make this distinction.

Thomas Young (1773–1829) not only confirmed the wave nature of light, but explained polarization and additive color synthesis (see Chapter 7). He was also largely responsible for bringing the Rosetta Stone to England and deciphering its Egyptian hieroglyphics.

(a)

(b)

Figure 1.9 (See color insert following page 154.) Spectra formed by (a) a prism; (b) a diffraction grating.

Interference occurs in many natural situations. The iridescent colors of beetles, tropical butterflies' wings, oil films, and soap bubbles are all the result of interference. The type involved here is known as *thin-film interference*, and it happens when the space between two adjacent transparent surfaces is of the order of half a wavelength of light. The light is reflected more or less equally from the first and second surfaces of the scale, chitin layer, or liquid film, and light with a wavelength equal to twice their spacing will have both these reflected beams in phase and will be reflected strongly, whereas light of other wavelengths will be out of phase and will be weakened or suppressed (Figure 1.10).

One effect of interference that can be particularly irritating to photographers is called *Newton's rings* (or, more correctly, *Newton's fringes*, as they are not always circular). Newton was the first to describe the effect, though he was unable to suggest a satisfactory explanation. If you place a surface of low curvature, such as a very thin lens, in contact with a polished glass plate on a white sheet, and illuminate it from above with a small light source, you will see a set of concentric colored fringes. With monochromatic light these will be better defined (Figure 1.11).

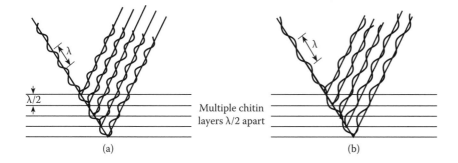

Figure 1.10 Bragg reflection: (a) light of wavelength twice the spacing is reflected in phase and reinforced; (b) other wavelengths are reflected out of phase and are suppressed.

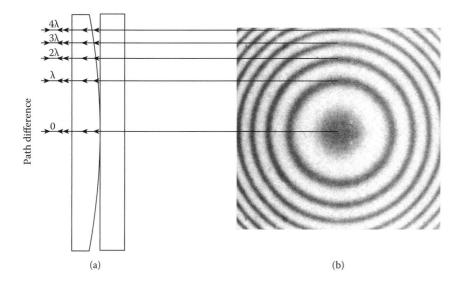

Figure 1.11 (a) Formation of Newton's rings (monochromatic light). (b) The central spot is dark because there is a 180° (π) phase reversal at the lower air–glass surface, so that destructive interference occurs when the path difference for the two reflected rays is 0, λ, 2λ, 3λ, etc.

These irregular colored fringes are often called *Fizeau's fringes*. Armand Hippolyte Louis Fizeau (1819–1896) worked with Louis Foucault, and first applied the Doppler principle to the measurement of the motion of stars. He also made the first reasonably accurate determination of the speed of light in 1849, and contributed to the development of astronomical photography.

Jean-Baptiste Joseph Fourier (1768–1830) was a French mathematical physicist whose main research was in the field of what we now call thermodynamics. He accompanied Napoleon on his Egyptian campaign, where he conducted a number of diplomatic affairs, which resulted in his being appointed Governor. In his researches into heat transfer he developed the theory of Fourier series to describe complex periodic functions. The Fourier principle has been shown to underlie much of modern optics.

The innermost of these represent the position where the spacing between the adjacent surfaces is one half-wavelength; the second represents a spacing of 1½ wavelengths; and so on. With white light the fringes quickly become muddled together, but if you use monochromatic light you can see many more fringes, Lens makers use these fringes as a guide to accuracy when testing a lens against a pattern, but to the photographer they are simply a nuisance, occurring as irregular patches in slides mounted between glasses, or on photographic enlargements made with the negative held in a glass stage. The colors you see in soap bubbles and in oil films on water are also Newton's fringes, this time between internal rather than external surfaces. Interference also plays a large part in the antireflection coating of lenses (Chapter 4), and it is fundamental to the making of holograms (Chapter 16).

Diffraction

Diffraction is a wave phenomenon closely related to interference. It describes the way a beam of light behaves when it passes through an aperture, or past an obstruction. There are several models that describe diffraction with varying degrees of success. The best is the Fourier model, which deserves, and has received more than once, a whole book to itself. There is a short account of it in Appendix 3, which may whet your appetite. However, the Huyghens wave principle is simpler, and is adequate for our present purposes.

If you consider a wavefront passing through a very small aperture only one or two wavelengths wide, then beyond it the waves will spread out in all directions, as the aperture acts like a point source, in effect a single wavelet. However, if the aperture is a slit that is somewhat wider, say, a few tens of wavelengths, a screen placed on the exit side will show an interference pattern, with spacing similar to that given by a double slit. How does this happen? Well, to explain it you need to do a little constructive thinking. Imagine the single slit aperture to be made up of a large number of mini-slits, all touching. Now take the mini-slit at the extreme left, and match it to the one next to the midline of the aperture. Think of these two mini-slits as a pair of Young's slits. They will form a double-slit interference pattern. Now consider the adjacent pair of mini-slits. They will also form an identical pattern, and so on until you reach the right-hand edge of the main slit. So you will have a pattern like a Young's slits pattern, but with a fringe spacing corresponding to that given by a pair of Young's slits spaced at half the width of the single aperture; and so it proves (Figure 1.12).

THE GRATING CONDITION

The relationship between the grating spacing (the *pitch* or *spatial period*) and the spot spacing follows directly from Figure 1.8. For the bright fringes the angles at which the beams emerge is given by

$\sin \theta = n\lambda/d$

where θ is the angle to the normal, λ the wavelength, d the period of the grating, and n = 0. 1. 2. 3. etc. If the screen is a distance x from the grating, the first-order fringe will be formed at $x \sin \theta_1 = \lambda x/d$, the second-order fringe at $x \sin \theta_2 = 2\lambda x/d$, and so on. For a single slit of width d, the fringes will form at $2\lambda x/d$, $4\lambda x/d$, etc.

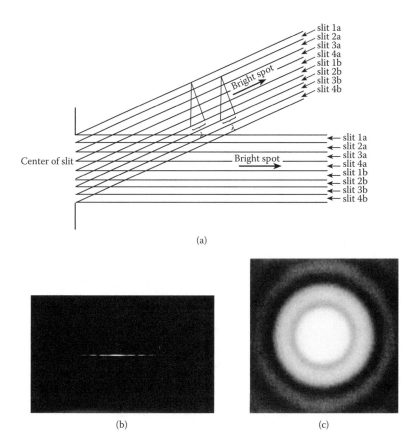

(a)

(b)   (c)

Figure 1.12 (a) The diffraction pattern formed by a single slit aperture can be explained by considering the aperture to be a number of pairs of Young's slits separated by half the aperture width; (b) diffraction pattern for a single slit; (c) Airy diffraction pattern for a circular aperture.

The Airy Diffraction Pattern

It's somewhat more complicated to work out the diffraction pattern for a two-dimensional figure.

One such figure is very important in imaging: a circular aperture. Its diffraction pattern was first described by Sir George Airy in connection with the images of stars, and it is named after him.

The central disc (the *Airy disc*) contains more than 80 percent of the light, surrounded by rings. Its diameter D to the first zero is given by

$$D = 2.44\lambda x/d$$

where d is the diameter of the Airy disc, x the aperture diameter, and d the screen distance.

In a telescope or camera system, d represents the focal length and x/d is simply the f-number of the lens or objective. (For an explanation of "f-number" see Chapter 4.)

George Biddell Airy (1801–1892) was Astronomer Royal for 46 years. He was also Director of Cambridge Observatory. He introduced many innovations into astronomy. He also established the border between Canada and the United States, assisted with the method of laying of the first transatlantic cable, and helped with the design of the chimes of Big Ben.

Reflection and Refraction

In oblique reflection each part of the wavefront arrives at the reflecting surface at a different time, so that the envelope of the wavelets sets off in a new direction that is symmetrical with respect to the surface. In practice we don't use the surface as our reference, because it may be (indeed, usually is) curved. Instead, we use the perpendicular at the reference point, called the *normal*. We can use a simple geometrical model here. Instead of the tedious business of drawing wavefronts we can simply draw a line showing the direction in which they are travelling, a *ray*. Reflection is then a straightforward matter. The angle the entering (incident) ray makes with the normal is called the *angle of incidence*, and the angle the emerging (reflected) ray makes with it is called the *angle of reflection* (Figure 1.13).

The two laws of reflection are:

1. The incident and reflected rays are in the same plane as the normal.
2. The angle of incidence is equal to the angle of reflection.

These laws also apply to curved surfaces, and here it is much easier to follow the path of a light beam by using rays rather than wavefronts.

The term *refraction* refers to the change in direction of a light beam entering a substance having optical qualities different from the one it has left. The speed of light discussed earlier was actually the speed of light in empty space. In air this is very slightly less, and in liquids and transparent solids it is very much less. When a beam of light enters a glass block obliquely, the wavefront is slowed down, and successive crests become closer together. There is thus a change in direction of the wavefront towards the normal. On emergence from a parallel surface the process is reversed, and the beam resumes its original direction (Figure 1.14). The change in direction is called *refraction*. Again, we can simplify the geometry by using the ray model.

As in reflection, the incident ray, the refracted ray, and the normal are all in the same plane.

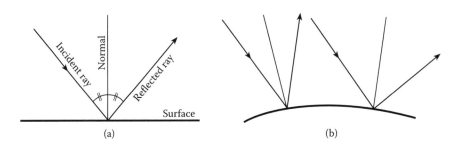

(a)   (b)

Figure 1.13 Reflection (a) at a flat surface; (b) at a curved surface.

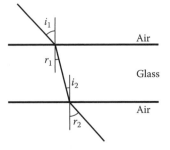

Figure 1.14 Refraction in a parallel-sided glass block: $i_1 = r_2$ and $i_2 = r_1$.

For small angles the angle of incidence bears a constant relationship to the angle of refraction (i.e., the angle the refracted ray makes with the normal). It is the same as the ratio of the speeds of light in the two media, and is called the *refractive index*. This constant is of immense importance in lens design. For angles of incidence larger than a few degrees we have to use a more precise relationship, and this is shown in the Box.

SNELL'S LAW

Snell's law follows directly from the observation that light changes its speed when it passes from one optical medium to another of different optical density (this term refers to its light-slowing powers, not its physical density). It states that where the refractive indices of the first and second media are n_1 and n_2, respectively, and i and r are the angles of incidence and of refraction, respectively, then

$$n_1 \sin i = n_2 \sin r, \text{ or}$$

$$\sin i / \sin r = n_2 / n_1$$

If the first medium is air, n_1 is approximately 1. (The refractive index of empty space is exactly 1.) The refractive index of an optical medium with respect to air is often represented by the Greek letter μ (Figure 1.15).

Snell's law leads directly to the formula for the critical angle. At this angle $\sin r = 1$, so that

$$n_2 / n_1 = \sin i$$

and for glass to air, $\sin i = 1/n_{\text{glass}}$.

Willebrord Snell (1580–1626) was a trained lawyer, and became Professor of Mathematics at Leyden University, where in addition to his work in optics he developed the method of survey by triangulation. He also made an accurate determination of the radius of the Earth.

Total Internal Reflection (TIR)

If you pass a beam of light obliquely through a glass block into air, the angle of refraction will be greater than the angle of incidence, and will increase faster than the angle of incidence. So at a certain angle, the *critical angle*, the emergent beam will have an angle of refraction of 90°, and will travel along the surface (Figure 1.16).

If you increase the angle of incidence beyond the critical angle the beam can't escape. Instead, it is reflected. This effect is known as *total internal reflection* (*TIR*). It is more efficient than a conventional mirror, which can manage at best no more than about 96 percent reflectance. As the critical angle for glass is less than 45°, TIR is used in prismatic reflectors (*roof or Porro prisms*) for binoculars, and for retroreflectors (corner cubes).

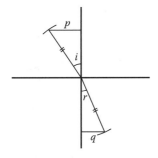

Figure 1.15 Construction for Snell's law: $p/q = \mu = \sin i / \sin r$.

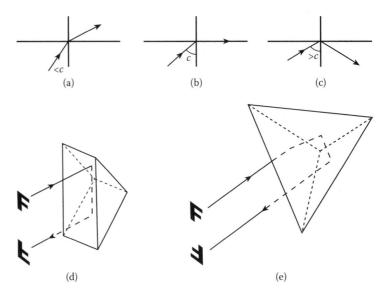

Figure 1.16 Total internal reflection: (a) glass to air; (b) at the critical angle; (c) total internal reflection (TIR); (d) Porro prism for image inversion; (e) corner cube reflector.

Prisms

When a beam of light passes through a parallel-sided glass block the emergent beam is parallel to the incident beam, but is displaced from it a little. However, if the second surface is not parallel to the first, as in a prism, the beam undergoes a second deviation (Figure 1.17).

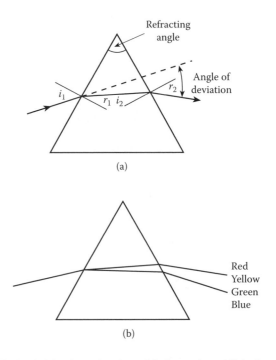

Figure 1.17 (a) Path of light through prism; (b) dispersion of light by prism (see Figure 1.9a).

The combined angle of deviation depends on the angle between the prism's faces (the *refracting angle*) and the angle of incidence at the first surface. The deviation is at a minimum when the incident and emergent beams are symmetrical to the prism. This is another property that is useful in optical design.

A prism has another important effect on white light. The refractive indices of transparent media are greater for shorter wavelengths than for longer ones, and a prism, like a diffraction grating, spreads white light into a spectrum (Figure 1.9a, color plate). Notice that the spectrum produced by a prism is not linear in its spread, and runs in the opposite direction to that produced by a diffraction grating. Newton was the first person to make a proper examination of the spectrum produced by a prism. Perhaps because of the mystical importance of the number seven, he assigned seven hues to the spectrum, namely red, orange, yellow, green, blue, indigo, and violet. This may seem somewhat confusing today, as the dye indigo (used to color denim jeans) is a desaturated color not in the spectrum at all. Modern colorimetry (see Chapter 2) suggests that the spectrum of white light can be better described as consisting of five basic hues: red, yellow, green, cyan, and blue-violet (usually called simply "blue").

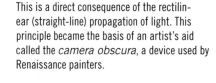

In Newton's day "indigo" meant a color more like royal blue, and what he termed "blue" was more greenish than the way we visualize it today.

The Pinhole Camera

As early as the second century AD, it was noted by Arab philosophers that if a small hole was made in the outside wall of a dark chamber, an inverted image of the outside scene would appear on the opposite wall (Figure 1.18).

This is a direct consequence of the rectilinear (straight-line) propagation of light. This principle became the basis of an artist's aid called the *camera obscura*, a device used by Renaissance painters.

Figure 1.18 The principle of the pinhole camera was known in the Middle East before AD 200.

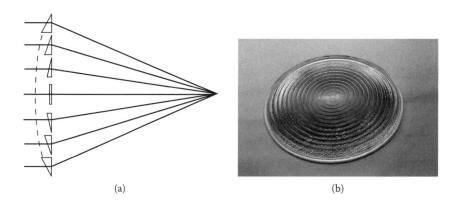

(a)   (b)

Figure 1.19 (a) A lens as a series of concentric ring prisms; (b) a Fresnel lens as used in a spotlight.

Today we call this device a *pinhole camera*. The image is in full color, but is not very bright, nor is it very sharp, owing to the finite size of the pinhole. If you reduce the pinhole diameter, at a certain point diffraction takes over, and the image becomes even less sharp. There is thus an optimum pinhole size. There is more about pinhole cameras in Chapter 5. (See Figure 5.18, color plate.)

Development of a Lens

In the form of magnifiers and burning glasses, lenses have been known for a very long time. In the fourteenth century some unknown person was smart enough to couple a lens with a camera obscura aperture and obtain a bright, sharp image. In order to see how a lens focuses, we may look on it initially as a series of prisms (Figure 1.19).

You need to think of these prisms as cross-sections of a ring of glass, of course, because a lens is circular. In fact, lighthouse lenses are traditionally made in this way, to save weight. They are called *Fresnel lenses*. This type of stepped lens is also used in spotlights, and as field brighteners in overhead projectors and camera viewfinders. As a rule lenses have smooth rather than stepped profiles, but you can think of a lens as being a very large number of concentric ring prisms, each giving just the right amount of deviation to bring light to a focus at a single point.

When you use a lens to burn a hole in a piece of paper, you are focusing an image of the sun on the paper, and the distance between the lens and the paper when the image is tiniest is called the *focal length*. If you move the lens a little farther from the paper you can bring into focus nearer objects such as trees. Chapter 3 discusses focusing relationships.

A good shape for a lens surface is approximately part of a sphere, and as this is also the easiest curved surface to shape, most lenses have spherical surfaces. However, a spherical shape isn't exactly correct, and later in Chapter 3 we shall see some of the consequences of this discrepancy, and how lens designers minimize them.

A curved mirror behaves similarly to a lens in most respects, and it has certain advantages that make it particularly suitable for generating some types of optical image. Again, a spherical shape is very nearly (but not quite) ideal.

In the Archaeological Museum in Iraklion, Crete, there is a beautiful rock crystal magnifying lens dating from Minoan times, i.e., before 1500 BC (Figure 1.20). It would have been used by a goldsmith for fine work.

Augustin Fresnel (1788–1827) was a civil engineer working for the French government, and when Napoleon returned from Elba, Fresnel was put under house arrest. With nothing useful to do, he sat and worked out the transverse wave theory of light and the mathematics of diffraction. He invoked polarization to explain the double refraction of certain crystals. After he was reinstated he won a government award for his lighthouse lens design.

The German word for focal length is *Brennweite* ("burning distance")!

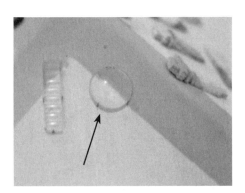

Figure 1.20 Minoan rock crystal lens, ca 1500 BC (Museum of Archaeology, Iraklion, Crete).

ANOMALOUS REFRACTION

Two odd possibilities for modifying the behavior of light waves have cropped up. The first is so-called *slow light*. Under certain conditions light photons can be absorbed by atoms of a substance, and released in the same direction after a brief delay. The light is thus effectively slowed down considerably. This phenomenon is at present the subject of a good deal of active research.

The second odd possibility is *negative refractive index*. Refractive index (RI) depends on two electromagnetic constants, namely permeability and permittivity (see Box, Maxwell and Electromagnetism). If both these quantities are negative (and they can be), the refractive index must be negative too. This means that a ray entering a substance possessing these properties will be bent *away* from the normal. All kinds of optical anomalies follow: for instance, it becomes possible for a flat plate to act as a focusing lens. Negative RI has already been achieved for microwaves, using complicated three-dimensional meshes, but at the time of writing has not been achieved for visible light.

Digging Deeper

When it comes to the physics of light, you can dig as deep as you like; but the deeper you dig the tougher it gets. There are many standard textbooks up to degree level. *Light*, by Michael Sobel (University of Chicago Press, 1989) is an excellent non-mathematical survey of the whole field of light and optics. *Insight into Optics*, by Oliver Heavens and Robert Ditchburn (John Wiley, 1991), is a good introduction to all branches of optics, with fairly rigorous but not too difficult mathematical backgrounds. *Modern Optics*, by Robert Guenther (John Wiley, 1990), is a full degree-level course. The standard work on optics is *Principles of Optics*, by Max Born and Emil Wolf. The last edition these two outstanding teachers actually wrote was the third edition in 1965, but it has been regularly updated since then. The latest edition is the ninth, published in 1999 by Cambridge University Press. It is not an easy read. But my number one choice and all-time favorite is *Optics*, fourth edition, by Eugene Hecht (Addison-Wesley, 2002). Nothing is left out; everything is patiently explained, all the mathematics is there; there are excellent problems at the end of each section, with answers where these are important, and even a table of sinc [$(\sin x)/x$] functions, which are relevant to resolution theory.

Chapter 2 Photometry, Lighting, and Light Filters

Photometric Units

The formal name for the measurement of light is *photometry*. It is part of the wider system of measurement of electromagnetic radiation, which is called *radiometry*, but as it is concerned with visible radiation its units have to be associated with the visual process. In all aspects of imaging technology we need terms to define such matters as light sources, quantity of illumination, reflection, transmission, and so on, in terms that can readily be related to one another and to units of mechanical and electrical energy. In the past there were three different systems—FPS, CGS, and MKS—resulting in a proliferation of named units, to the confusion of students and the exasperation of their tutors. Unfortunately, some academic texts still employ these outdated systems. In this book I use only SI units, with the occasional centimeter or inch where either is obviously called for.

FPS was foot, pound, second, the old Imperial system. CGS was centimeter, gram, second, the old metric system. MKS was meter, kilogram, second, the forerunner of SI (Système International d'Unités).

Luminous Intensity

The SI unit of *luminous intensity* is the candela (cd). It is a fundamental SI unit, the others being the meter, kilogram, second, ampere, kelvin, and mole. You will meet them all in due course. The candela is a unit of power, and thus has the same physical dimensions as mechanical or electrical power, namely energy emitted (or work done) per second. The SI unit of power is the watt, and you might expect the candela to be interconvertible with it, but this is in fact possible only for

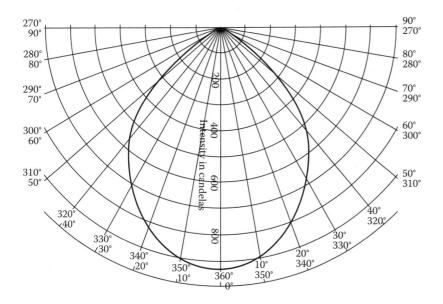

Figure 2.1 Typical polar diagram for the intensity of a photographic floodlamp.

This was indeed a candle, the Imperial Standard Candle, no less, manufactured and burned under specific conditions. Later a specified filament lamp became the standard, and later still, platinum at its melting point.

Whereas linear angles are measured in radians (rad), solid angles are measured in steradians (sr). One steradian is represented by a portion of a sphere of 1 meter radius and 1 square meter area. There are 4π steradians in a sphere. To find a solid angle you take the projected area and divide it by the square of the radius (Figure 2.2).

Flux is an old-fashioned word dating from Newton's time, and meaning "flow." I have no idea how or why it has survived.

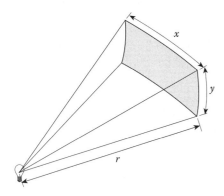

Figure 2.2 Solid angle: here the solid angle is xy/r^2.

Johann Heinrich Lambert (1728–1777) was a Swiss polymath who wrote successful books on philosophy and cosmology, put the science of photometry on a sound basis, and proved that the numbers π and e are irrational as well as propounding a number of theorems in what became non-Euclidean geometry.

monochromatic (single wavelength) light. So the candela was originally defined in terms of white light, using a standard light source.

However, the modern definition of the candela is the luminous intensity in a given direction of a monochromatic source of frequency 540×10^{12} Hz whose radiant intensity is $1/683$ W sr^{-1} (watts per steradian). In practice we more often call this frequency by its corresponding wavelength of 555 nm. This particular wavelength is chosen as the one to which the human eye is most sensitive.

If you want to draw up a complete specification for a light source such as a photographic spotlight you have to do so by means of a polar diagram (Figure 2.1). But you can't gauge the overall power of a floodlamp, a flash head, or even a car headlamp simply by studying a polar diagram. You have to integrate all the values over a complete sphere to do this. You can, however, specify *mean beam intensity* over a specified angular area, say a $10° \times 20°$ rectangle. No source (except possibly an aerial flare) is uniform over a complete sphere, but we use this hypothetical uniform source to relate the candela to a more generally useful unit, the lumen.

Luminous Flux

Except for special purposes we are not interested in measuring light output in just a single direction. We're more interested in the aggregate output of a light source designed to cover a particular solid angle. The lumen is linked directly to the candela, 1 lumen (lm) being the luminous flux emitted by a uniform source of 1 candela into unit solid angle. Thus 1 watt of radiation at 555 nm produces a flux of 683 lumens.

As solid angle has no physical dimension, both the candela and the lumen have the same dimensions, i.e., power. Thus a *uniform* source of 1 candela is equal to 4π lumens.

PLANCK'S EQUATION AND RETINAL SENSITIVITY

The radiant energy emitted by a hot, totally nonreflective surface (usually dubbed a *black body*) has the form of a continuous spectrum. When energy is plotted against wavelength for different temperatures the result is a family of curves (Figure 2.3), broadly similar in shape, their maximum value both increasing and moving towards shorter wavelengths as the temperature increases. Max Planck found an equation that generated the family of curves, as well as its theoretical justification. It is

$M_c = c_1\lambda^{-5}/[\exp(c_2/\lambda T) - 1]$ W m^{-3} (watts per square meter per wavelength in meters)

where $c_1 = 3.74183 \times 10^{-16}$ W m^2 and $c_2 = 1.4388 \times 10^{-2}$ m K (T is in kelvins).

From this and the sensitivity versus wavelength curve for the human retina (see Chapter 3) we can relate lumens to watts for any visible wavelength, the maximum value being 683 lm W^{-1} at 555 nm.

Illuminance

A hypothetical uniform source of 1 candela emits 1 lumen per square meter (lm m^{-2}). When this light falls perpendicularly on a surface, the amount of illumination, called the *illuminance*, is 1 lux (lx). If the light beam is oblique, with an angle of incidence θ (theta), the illuminance is multiplied by $\cos\theta$ (Lambert's law, Figure 2.4).

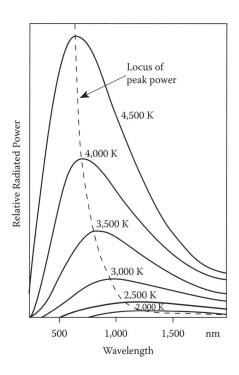

Figure 2.3 Spectral power distribution for a black body radiator.

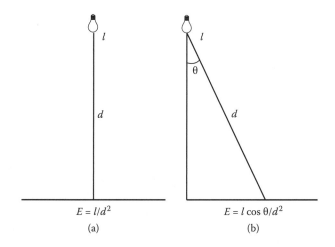

Figure 2.4 Lambert's law: (a) perpendicular illumination; (b) oblique illumination.

Inverse Square Law

The illuminance on a surface illuminated by a small source is proportional to the intensity of the source and inversely proportional to the square of its distance from the source. This is because when you double the distance you quadruple the area covered, and when you triple it you multiply the area ninefold (Figure 2.5). So the full expression for calculating the illuminance E at a distance d from a source of intensity I at an angle of incidence θ is given by

$E = (I \cos \theta)/d^2$

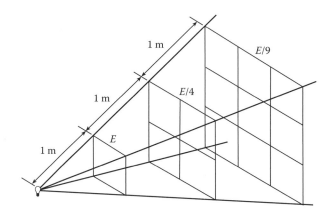

Figure 2.5 Illustration of the inverse square law.

Luminance

If you look at a 200-candela filament lamp it seems very bright. If you look at a 200-candela fluorescent light it seems much less bright. The same amount of light is being emitted, but from a much larger surface area. The quantitative term for brightness is *luminance*, and you obtain it by dividing the luminous intensity by the projected area of the luminous surface. As with luminous intensity, you have to specify the direction. The unit of luminance is the candela per square meter (cd m^{-2}).

Reflectance

You can specify the luminance of an illuminated surface in the same way. You can also calculate this, if you know both the illuminance and the *reflectance* of the surface in a particular direction. There are two kinds of reflection:

> *Specular reflection.* "Speculum" is Latin for "mirror," and specular reflection is direct reflection from a mirror-like surface, obeying the laws of reflection. The incident light doesn't penetrate the surface, so the reflected beam possesses the same color as the incident beam (Figure 2.6a).
>
> *Diffuse reflection.* In diffuse reflection the light penetrates the surface and is modified spectrally by any pigment present in the material. On emergence it is partially scattered. You can represent the pattern of scattering by a polar diagram (Figure 2.6b). If the material is perfectly diffusing, the reflected light obeys Lambert's law (Figure 2.6c), and the reflectance is said to be *Lambertian*.

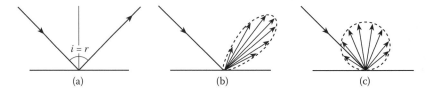

Figure 2.6 (a) Reflection at a surface; (b) specular; (c) diffuse. Lengths of arrows show relative intensities.

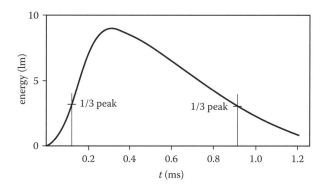

Figure 2.7 Typical intensity-time graph for an electronic flash.

Luminous Energy

The luminous energy emitted by a photoflash or other transient source is measured in lumen seconds (lm s). As the intensity varies throughout the duration of the flash, we obtain the total light energy by measuring the area under the time-intensity curve (Figure 2.7). In practice the measurement is usually made between the rising and falling one-third peaks, and the time interval between them is taken as the effective flash duration. (It usually contains around 90 percent of the total light energy.)

Luminous Efficacy

For a given light source the luminous energy emitted is a fixed proportion of the electrical energy fed into it. If we were in a position to make a direct comparison, i.e., watts out ÷ watts in, the answer would be simply the luminous (or rather radiant) efficiency. You can do this for monochromatic sources such as lasers, and state the output efficiency as a simple percentage. You can convert watts to lumens too for single wavelengths, based on the sensitivity curve for photopic (bright light) vision (see Chapter 3). At 555 nm (yellow-green), 1 watt corresponds to 683 lumens. For other wavelengths the figure falls as the wavelength moves farther away from this central point. For light sources with a broad spectrum there is no single conversion figure. Instead we quote the *luminous efficacy* in lumens per watt (lm W^{-1}). The luminous efficacy of a tungsten filament lamp is around 20 lm W^{-1} and of a xenon discharge tube about 40 lm W^{-1}. The luminous efficacy of a light-emitting diode (LED) can be much higher, as little energy is wasted as heat.

Photographic flash heads are rated in joules (J). A joule is a fairly large amount of energy (a 10-joule pulse laser can blow paint off a wall!), so it seems odd that a studio flash should be rated at 200 joules or more. In fact, it is the electrical *input* that is being quoted. And this has to be multiplied by the luminous efficacy to find the output in lumen seconds.

Spectral Energy Distribution

The simplest way to depict this is by percentage figures for red, green, and blue content, or as a simple column chart (Figure 2.8a). A more accurate description is a histogram (Figure 2.8b). The third and most accurate method is to show a full plot of energy versus wavelength (Figure 2.8c).

The curves of Figure 2.8c are of the kind produced by a spectrophotometer, a device that scans the energy via a monochromator (i.e., wavelength by wavelength) and plots the values as a graph.

23

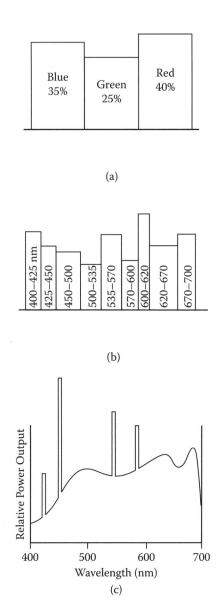

Figure 2.8 Depiction of spectral power distribution: (a) primary hue content; (b) histogram; (c) power output versus wavelength for fluorescent lamp. The bars are the superimposed line spectrum.

An incandescent light source has a spectral energy distribution close to that of a black body at approximately the same temperature, following the Planckian energy distribution system (Figure 2.9).

The Kelvin (K) is the SI unit of thermodynamic temperature. It has the same magnitude as the degree Celsius, but 0K corresponds to absolute zero (–273 degrees C), the point at which all molecular activity has ceased.

Color Temperature

As the Planckian distribution provides a precise description of the spectrum of an incandescent light source, we can specify the spectral energy distribution of any source of this type by stating the equivalent temperature in kelvins (see marginal note). This figure is called the *color temperature*. A halogen lamp as used in overhead projectors has a color temperature of 3400 K. The true temperature of the

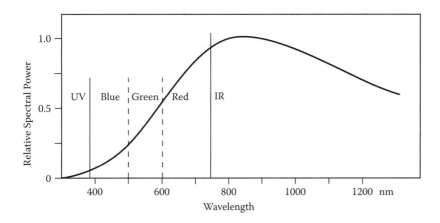

Figure 2.9 Spectral power distribution of a filament lamp (3400 K).

William Thompson, first Baron Kelvin (1824–1907), was professor of natural philosophy at Glasgow University for 53 years. His researches made thermodynamics a respectable branch of physics and he was knighted for his part in the design and laying of the first Atlantic cable.

filament is a little lower, which is just as well, as 3400 K is uncomfortably close to the melting point of tungsten. The sun, too, is an incandescent body: standard (or "mean noon") sunlight is established as 5400 K, a figure obtained from a series of measurements made in Chicago in the 1920s.

This figure is somewhat low, and modern color films for use in daylight are balanced for 5500 K.

The Mirek Scale

Equal changes in the perceived hue of an incandescent body are not matched by equal changes in color temperature. With its reciprocal there is a much closer match. As this figure is very small, in order to bring it up to whole numbers we multiply it by a million. The resultant figure is in units called *mireks*, from *mi*cro *re*ciprocal *k*elvins (10^6 K^{-1}).

Because hue shift expressed in mireks is more or less constant within the visible spectrum, color filters to convert the color temperature of various black body sources to match daylight or tungsten light emulsions are specified in mireks, negative for a blue shift and positive for a red shift. For example, a blue filter designed to match a 3200 K tungsten floodlamp (312 mireks) to 5500 K daylight film (182 mireks) has a value of (182 − 312) = 130 mireks. Figure 2.10 relates color temperature to its mirek equivalent.

Full marks if you spotted that it should be mega, not micro. In the days when kelvins were degrees Kelvin, mireks were mireds, and many textbooks still refer to them as such. "Hue" is a more exact term than "color" here, because "color" also involves lightness and saturation (see Chapter 7).

Note: Some types of light source that are not incandescent, and therefore have an irregular or discontinuous spectrum (e.g., electronic flash, color-balanced fluorescent tubes) have an overall spectrum within the visible range that is close enough to a Planckian curve to be allotted an honorary value, known as a *correlated color temperature*. Two sources that have a different spectrum but possess the same hue to the eye are said to be *metameric*.

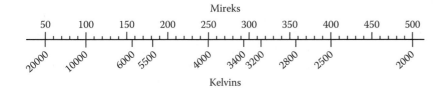

Figure 2.10 Color temperature equivalents in mirek and Kelvin scales.

Types of Light Sources

Light sources for imaging can be continuous, as in daylight and floodlamps, or transient, as in flash, and their spectra can be quasi-black body or irregular. The main light sources used in photography are given below.

Daylight

This is a very variable source. Around sunrise and sunset the color temperature is as low as 3000 K, but it rises to more than 5000 K at midday, and in summer the presence of blue sky can raise it to 6000 K and more. On an overcast day the color temperature is higher still, up to 8000 K. On a clear day the color temperature of the sky itself (which is the illuminant for shadows) may be as high as 16,000 K.

Tungsten Filament Lamps

Photographic floodlamps have a color temperature of 3200 K; the halogen types have a color temperature of 3400 K (see Figure 2.9). Household filament lamps are 2800–3000 K.

Fluorescent Lamps

These have an irregular spectrum, which is the sum of the spectra of the various phosphors used in the tube, plus the line spectrum of low-pressure mercury vapor broadened into bands (see Figure 2.8c). Lamps intended for color matching are balanced close to daylight. Ordinary fluorescent lamps don't render colors correctly in a color transparency, and show a greenish cast; film manufacturers recommend combinations of weak magenta and yellow CC (color correction) filters. Fluorescent lamps have a 100-Hz flicker (120 Hz in the United States), which may cause strobing effects with TV or digital cameras. So-called energy-saving lamps are compact fluorescent lamps that emit light in three or four broad overlapping wavelength bands. The control circuitry in the base consists of an electronic ballast that operates the lamp at high frequency, thus avoiding the flicker associated with the older type of fluorescent tube.

Discharge Lamps

Pulsed xenon tubes are popular for cine and video work. Their overall hue is fairly close to that of daylight; the better examples incorporate various metal halides, which produce a near-exact match. Any flicker effect is avoided by converting the applied voltage to a very high-frequency a.c.

Flashbulbs

Now seldom seen, these were at one time the mainstay of press photographers. They operate by the electrical ignition of a bunch of very fine zirconium or magnesium ribbon in an atmosphere of oxygen at reduced pressure. Their inherent color temperature is 3800 K, but the protective coating usually contains a blue dye, which raises the color temperature to 5500 K. The duration of a flash may be between 5 and 50 milliseconds (ms) according to the type of bulb. The shape of the intensity curve is much the same as that of Figure 2.7, but with a much-expanded time scale.

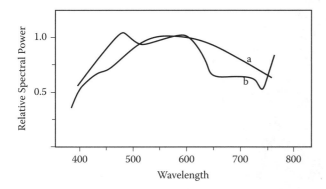

Figure 2.11 Spectral power distributions of (a) sunlight; (b) electronic flash.

Electronic Flash

Xenon gas in a tube at low pressure is ionized and made conducting by a trigger pulse of high voltage. This allows a bank of capacitors to discharge very rapidly through the tube, generating an intense flash of white light, typically of 1 ms duration, but which can be quenched to give durations as short as 20 microseconds (μs). Older flash heads produce a spectrum that is slightly deficient in green, and may have a pale yellow-green tint to the protective screen to compensate. Modern flash tubes contain small amounts of other gases added to produce a spectrum that is a close match to daylight (Figure 2.11, curve b).

Photographic Light Filters

When photographers talk of filters they almost always mean light filters. A light filter is an optical device that modifies the color, quality or intensity of the light it transmits. There are nine categories of light filter, each with a different purpose.

Emulsion Balancing Filters

These are now chiefly of historical interest, though older photographers will remember the yellow filter they used in order to make the color rendering of orthochromatic (red-blind) emulsions more believable, especially in the rendering of skies and flesh tones. More recently, a pale yellow-green filter has been used to balance the tone rendering of high-speed panchromatic films possessing enhanced red and blue sensitivity. Dufaycolor films (see Chapter 7) used to include a gelatin emulsion-balancing filter with each film. An oddity among these filters was the panchromatic vision (PV) filter, which was a deep purple and was intended to be looked through to give a visual impression of the way the monochrome image would appear. Nowadays this category of filters applies chiefly to the color correction (CC) filters necessary when making very long exposures on color transparency film, and to so-called haze-cutting filters, which are UV-absorbing filters with a very slight pinkish-yellow tinge.

Contrast Filters

These belong specifically to black-and-white photography. They function by lightening parts of the image that are the same color as (or similar to) that of the filter, and darkening parts that are of colors different from its own, particularly the complementary (opposite) color (see Figure 2.12). Thus a red filter lightens red, and to a lesser extent yellows and greens, but darkens cyan, and to a lesser extent blues and

27

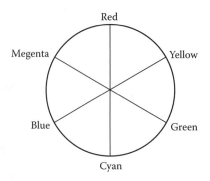

Figure 2.12 The color wheel: a color filter will lighten the tone of its own and adjacent colors and darken those diametrically opposite to it.

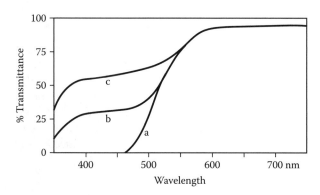

Figure 2.13 Spectral transmittance of three grades of yellow filter: (a) deep yellow (minus blue); (b) medium yellow; (c) light yellow.

To say that a filter lightens its own color is an oversimplification. A filter can only absorb light, so it can't actually make a color lighter. In practice you achieve the effect by increasing the exposure. This introduces the concept of a filter factor. This is the factor by which you have to multiply the exposure in order to obtain the same rendering of a neutral grey with the filter on the lens as you would without the filter. Typically, a medium yellow filter has a factor of ×2, and a deep red a factor of ×5. Figure 2.13 shows the spectral transmittances of three grades of yellow filter.

greens. It is the effect on tone quality that is important. If you were photographing, say, a red mahogany surface and you wanted to bring out the detail of the grain, you could use a green filter to increase the contrast of the reds. This works because a pale red reflects a good deal of green light, whereas a more intense (saturated) red reflects scarcely any. The tone scale of reds is thus expanded. A yellow or orange filter darkens the rendering of blue sky, in the same manner. On the other hand, if you were photographing a green landscape in bright contrasty sunlight, a green or yellow filter would lower the contrast to give a more delicate range of tones. Most of these effects can be obtained with digital imagery by changing the RGB balance when printing for monochrome.

Color Separation Filters

The production of some types of color print, and most color photomechanical processes, require the making of *color separations*.

The principles of color separation are described more fully in Chapter 6.

Briefly, you need to make three negatives (or the equivalent in other imaging techniques), recording, respectively, the red, green, and blue content of the subject matter. The transmittances of the three filters (often termed a *tricolor set*) are shown in Figure 2.14a. They are such that the combined filters cover the whole visible spectrum, with the overlaps of their transmittances arranged such that the aggregate transmittance is uniform.

The separation filters used in the printing process (in processes where these are used) have a different set of transmittances. They are described as *narrow-band filters*, and there is little or no overlap (Figure 2.14b). You have to use this type of filter when you want to make color separations from transparencies of printed matter, otherwise the result will be desaturated.

Lighting Balancing Filters

These apply to color films, and are of two general types. Both types of filter are calibrated in mireks. The first group compensates for the spectral mismatch between daylight-balanced films and tungsten lighting and the converse. The second group has a series of small color balance shifts (5–15 mireks), and pictorial photographers use them for making subtle alterations in color temperature to enhance the lighting quality of a scene in a color transparency. Many amateur digital cameras have settings for lighting balance for daylight, tungsten, and fluorescent illumination, and professional cameras allow the photographer to adjust the color balance as appropriate to the lighting and subject matter.

Color Print (CP) Filters

Owing to the inevitable small variations in color balance in the manufacture of both negative and print materials, you may have to use these in the enlarger light path to obtain a satisfactory color balance in the print. They are available in graded densities in cyan, magenta, and yellow, and in the better enlargers are built into the head and adjusted by rotating knobs. Their application is dealt with more fully in Chapter 7. Similarly, when working with an inkjet printer, it is almost always necessary to make small adjustments to the color balance after trying out a test print.

Neutral Density (ND) Filters

These reduce the intensity of light reaching the emulsion (or other sensitive medium) without affecting its spectral distribution. In practical photography you need ND filters only when you need to cut down the exposure and there is no possibility of adjusting either the shutter or the lens aperture. ND filters are used mainly in laboratory research and in sensitometric testing (see Chapter 10). With a digital camera the need for an ND filter can be avoided altogether, by simply lowering the ISO speed setting.

Infrared (IR) and Ultraviolet (UV) Filters

IR films are also sensitive to red and blue light, so an IR film has to be shielded from blue light and at least some of the red. IR filters typically pass a range of wavelengths from 680 to 900 nm. Some earlier digital camera sensors are able to record IR images with one of these filters, but require comparatively long exposures. More recent equipment has all IR filtered out, and such cameras require modification in order to produce IR images. UV filter material transmits radiation between 300 and 400 nm, but also allows some red light through, so UV filters also contain blue dye. To photograph fluorescence you need to have the UV filter on the illuminating lamp. Most modern camera lenses are effectively opaque to UV radiation, but it is still advisable to fit a UV-absorbing filter on the camera lens, so that only the visible fluorescence is recorded.

Special Effects Filters (SFX)

These include graduated density filters, multiple-image filters, soft-focus and starburst filters, and filters giving various diffraction effects. These do not lie within the remit of this book.

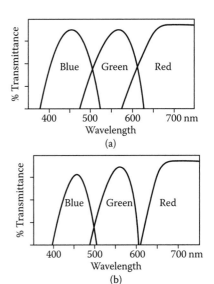

Figure 2.14 Tricolor separation filters: (a) broad band; (b) narrow band.

29

The earliest polarizing filters were called *Nicol prisms*. William Nicol (1768–1851) was a Scottish geologist who used polarized light in the microscopic examination of rock specimens. A Nicol prism is a single crystal of the doubly refracting material calcite, cut so that one of the rays is suppressed (the rays are orthogonally polarized). Modern polarizing filters are sheets of plastic material containing a layer of transparent polymer with its molecular axes all aligned.

Sir David Brewster (1781–1868) was awarded the Rumford Medal by the Royal Society for his discoveries in polarization. He is better known for his invention of the kaleidoscope. He also had a hand in the development of the stereoscope. He was a friend and mentor to Henry Talbot, and introduced David Octavius Hill to Robert Adamson.

Polarizing Filters

A polarizing filter transmits only linearly polarized light. Ordinary unpolarized light consists of waves vibrating in all planes, and of those vibrations the polarizing filter transmits only the vector components that are parallel to its polarizing axis. This means theoretically that only 50 percent of the light incident on the filter actually gets through it. So the filter factor ought to be ×2. In fact it is nearer ×3½, as the filter material is not completely transparent. Two polarizing filters with their axes orthogonal will block off virtually all light (Figure 2.15).

APPLICATIONS OF POLARIZING FILTERS

There are two main areas of application for polarizing filters. The first is in the suppression of specular reflections, as in the photography of objects behind glass or under water. Light reflected from a shiny nonmetallic surface at an angle is partially polarized; at angles of reflection around 56°–57° (the *Brewster* angle) it is totally polarized, in a plane that is perpendicular to the plane containing the incident and reflected rays and the normal. This is called *s*-polarization ("s" is the initial of the German word *senkrecht*, which means "perpendicular"). Polarization parallel to this plane (i.e., at right angles to the surface) is called *p*-polarization (*parallel* ["*p*"] is the same in both German and English). Setting a polarizing filter on the camera lens with its axis perpendicular to the surface will help to eliminate surface reflections. The Brewster angle for water is about 53°.

The Brewster angle is defined as the angle of incidence for which the reflected and refracted rays are orthogonal (Figure 2.16). At this angle the refracted ray is partially *p*-polarized and the reflected ray is totally *s*-polarized. The Brewster angle is the angle whose tangent is equal to the refractive index of the material.

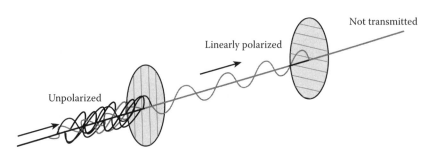

Figure 2.15 Action of a polarizing filter.

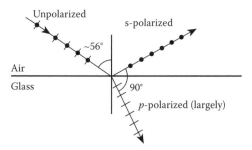

Figure 2.16 Reflection at the Brewster angle.

An extension of the principle is to illuminate the subject with light that is already polarized, and to use a polarizing filter on the camera lens with its axis crossed with respect to that of the illuminant. Specular (mirror-like) highlight reflections are suppressed, but diffuse reflected light, which becomes depolarized when diffused, passes through the second filter.

The second use for a polarizing filter is in landscape photography. Sky light away from the sun is polarized to varying extents.

If you fit a polarizing filter with its axis directed at the sun you will darken the sky without changing its hue. If you use the filter with its axis vertical, the effect of reflected skylight will be minimized, and the result will be an improvement in the color saturation of fields and horizontal spaces.

The maximum degree of polarization occurs along an arc at right angles to the axis joining the sun to the camera.

Circular Polarization

As I explained in Chapter 1, circular polarization is the condition where the polarization vector rotates one complete revolution for each cycle of the light wave. A circularly polarizing filter (also called a quarter-wave plate) is a sheet of doubly refracting material such as mica, of a thickness such that one of the two rays is delayed by one-quarter of a wavelength. This is the requirement for producing circular polarization.

Why do we need circular polarization? There is a problem in most camera metering systems when the light entering the camera is linearly polarized. A polarizing beamsplitter directs part of the incoming light to the exposure and autofocus sensors, and if you use a polarizing filter, rotating it will vary the division of the light, resulting in incorrect settings and possible loss of focus. This problem is overcome by mounting a quarter-wave plate on the camera-facing side of the polarizing filter. It has no effect on the behavior of the filter in eliminating reflections, etc., but it allows the sensors to operate correctly for any orientation of the filter axis. In one type of stereoscopic projection the two images are projected through circular polarizing filters polarized in opposite senses and viewed through matching glasses.

The term "vector" is used in mathematics and physics to define a quantity that has both magnitude and direction.

Digging Deeper

Many textbooks contain chapters on photometry that are less than reliable. Some otherwise excellent texts, especially from the United States, use obsolete and confusing units such as the erg (CGS) and the foot-lambert (FPS), sometimes together with metric units. One thoroughly reliable source is the *Manual of Photography* (10th edition, Focal Press, 2010). The most comprehensive coverage of the subject is *Lamps and Lighting* (4th edition, eds. J. R. Coaton and A. M. Marsden, Arnold, 1996), which is updated at regular intervals. It consists of sections on light, vision, color, and measurement; lamps and control equipment; and lighting fittings and techniques. Eastman Kodak publishes *Filters Handbook*, with a list of all the filters available from Kodak, their characteristics, and their Wratten numbers. The Kodak Workshop series includes *Using Filters* (1995). If your interests are largely pictorial, Joseph Meehan's *The Photographer's Guide to Using Filters* (Watson-Guptill, 1998) is excellent, and although it is now long out of print, Clyde Reynolds's *Focalguide to Filters* (Focal Press, 1974) packs a lot of useful information into a small space. Your local library should be able to find a copy. Surprisingly, Hecht's *Optics* (see Digging Deeper for Chapter 1), while otherwise admirably comprehensive, contains nothing about photometry.

Chapter 3 Visual Perception

The Eye and Evolution

The eye is such a complicated organ that creationists have used it for many years as evidence against the theory of evolution, saying, "What use to an animal would an only half-developed eye be?" In his book *The Blind Watchmaker*, Richard Dawkins demolishes such arguments, showing that at every stage of the building of the mammalian eye, from a simple light-sensitive patch to its present complexity, the organ became steadily more useful to its owner, and moreover that every stage in that development is still extant, in creatures ranging from the earthworm to the eagle.

Optics of the Eye

Older textbooks on light and optics always contained a section on the eye, and it was often the weakest section in the book. The authors usually treated the eye as a kind of camera, and got the optics wrong too, showing all the refraction taking place at the lens—in fact, most of the refraction occurs at the cornea. Discussion of the visual process has now largely moved to textbooks on the psychology of perception, where the eye is treated as having a resemblance to a computer input. Both these views nevertheless contain a measure of truth, and both may be helpful in an appreciation of what really goes on in the process of visual perception. Certainly, the eye is an optical device that focuses an image on a light-sensitive surface, so that is probably the best place to begin.

You can demonstrate the optical workings of the eye quite easily, though unless you have some rather old-fashioned equipment such as a large goldfish bowl and the condenser lens from an old photographic enlarger (or a large and powerful magnifying glass) it will have to be a thought experiment. If you fill the goldfish bowl with water you will have made a lens—a very thick one, but a lens nonetheless. If you shine a beam of collimated (parallel) light, such as sunlight or the beam from a focusing torch, through it, it will bring the light to a focus about one-third of its diameter behind its rear surface. (In the 1930s many a table runner was set on fire this way.) Water has a refractive index of 1.33, and glass is around 1.5, so if you suspend your lens in the water near the front of the bowl, it will shorten the focus so that it is more or less at the rear surface of the bowl. If you use a large glass ball (like a fortune-teller's crystal) instead of the lens, it acts as a more powerful lens, and by moving it backwards or forwards in the water you can bring nearer or farther objects into focus. In fact, this is the way the eyes of most fishes work. You can see this focusing effect if you hold a thin sheet of paper against the rear surface of the bowl. But to simulate the human eye more exactly you would need a flexible lens, say, one made of gelatin. You could then change the focus not by moving it, but by squeezing or stretching it to make it fatter or thinner. But the image still won't be very sharp. You can sharpen up the focus by placing a sheet of opaque material with a medium-sized hole in the middle between the front of the bowl and the lens. You have now created a model of the optical system of the human eye. Figure 3.1 shows the anatomy of the eye itself.

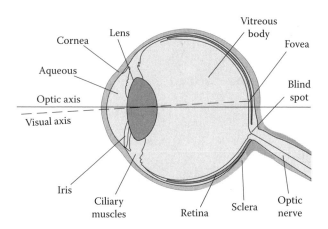

Figure 3.1 Structure of the human eye (right eye from above).

The optical functions of the parts that make up the eye are simple. As in the case of the goldfish bowl, the front surface (known as the *cornea*) does most of the refracting. The medium in front of the lens is called the *aqueous humor* (or, more often, just the *aqueous*; "humor" is an antiquated word meaning simply "fluid"). Its refractive index is 1.34, the same as seawater—which may say something about our fishy ancestry. Behind the lens is the jelly-like *vitreous body*, also with a refractive index of 1.34. The lens is built in layers like a rather flat shallot, and has a refractive index that varies from 1.38 on the outside to 1.41 at its center.

The lens is kept in tension by the ciliary muscles, which relax for close focusing, allowing its curvature to increase. As we get older the lens becomes less flexible, so that it becomes difficult to focus on close objects. Sometimes the lens slowly turns yellowish and cloudy—a pickled shallot. Its refractive index increases, the eye becomes shortsighted, and the visual image becomes progressively hazier. This condition is called cataract, and it can be dealt with by a simple surgical procedure, which breaks up and removes the lens and replaces it with a soft silicone implant.

The *iris* is an opaque muscular diaphragm with a circular aperture, the *pupil*. The iris controls the level of illumination that reaches the light-sensitive tissue lining the back of the eye, called the *retina*; when you move from bright light into dimmer surroundings the iris opens wide, until your retinal sensitivity has adjusted to the new conditions. Its range of apertures, in photographic terms, is from about *f*/3 to *f*/16—a ratio of about 28 to 1. (*f*-numbers are explained in Chapter 4.)

Graded refractive index lenses have been around for only about 20 years. Their use solves many of the problems associated with lens aberrations (see Chapter 4). It seems that in this respect, as in many others, nature got there first.

Some animals do much better than this. Domestic cats have a slit pupil that can vary its aperture between *f*/1.5 and around *f*/100, a useful capability in an animal that hunts by both day and night.

Note that this description of ocular astigmatism has nothing whatever to do with the lens aberration that is also called astigmatism. It is somewhat unfortunate that two totally different optical phenomena happen to have been given the same name—one by the medical profession and the other by physicists. Chapter 4 discusses lens aberrations.

Short and Long Sight

Other eye defects not associated with aging are short and long sight. In short sight the eyeball is too long or the curvature of the cornea too great, resulting in the image being focused in front of the retina; in long sight the eyeball is too short or the curvature of the cornea too little, so that the image is focused behind the retina (Figure 3.2). Astigmatism is a combination of these defects: the cornea has too great a curvature in one direction and too little in the other (i.e., instead of being part of a sphere, like a soccer ball, it is like a rugby ball). The eye is thus shortsighted for verticals and longsighted for horizontals (or some other pair of orthogonal axes).

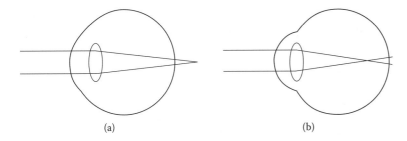

Figure 3.2 (a) Hypermetropia (long sight); (b) myopia (short sight).

The Retina

The embryonic eye doesn't grow as an entity. The cornea and the frontal part of the eye, including the eyeball itself, are outgrowths of the fetal skin, but the retina is an outgrowth of the brain. The way the retina grows in mammals has one unfortunate consequence: it is effectively inside out. The layer of light-sensitive cells is overlaid by relay cells, nerves, and blood vessels. There is also a sizeable blind spot (some 3° across) about 15° out from the center of your vision; this is the spot where the individual nerve fibers disappear into the optic nerve. You can't usually see the nerves and blood vessels, as they are stationary with respect to the retina, but when an optician examines your eyes with a bright light it will sometimes surprise your retina into perceiving the shadows of this tree-like network. Figure 3.3 shows a cross-section of the retina, showing the way it is constructed. The optical image produced at the back of the eye isn't a good one by photographic standards, though it is rather better than you might imagine from the description above.

Only a tiny area at the center of vision, the *fovea*, with a field of about 1°, picks up a really sharp image. The visual receptor cells become fatigued very quickly, and turn off altogether after a few seconds unless there is a change in the stimulus.

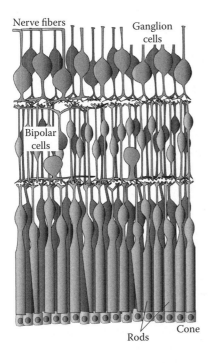

Figure 3.3 Structure of the retina.

These two factors are alleviated by constant small involuntary oscillations of the eye, which allow the stimulation to continue, and spread the effective angle of sharp vision to several degrees.

Rods and Cones

The primary light receptors of the retina are of two types: *rods*, which function only in dim light, and *cones*, which respond only to bright light. There is a twilight level (literally) where both types are operating, though neither operates very efficiently here, as anyone who has to drive a car around dusk can confirm. Cones operate efficiently at light levels above 10 cd m^{-2}, and cease functioning altogether at about 0.1 cd m^{-2}. Rods, on the other hand, continue to function down almost to the level of single photons. They contain a purple dye called *rhodopsin* that is destroyed by light, and this provides the operating mechanism. It takes up to half an hour to build up a full complement of rhodopsin, and this is why it takes so long to become fully dark-adapted.

Cones operate in a similar manner to rods, but the light-sensitive substance is different. It is a dye called *iodopsin*, and comes in three different forms. These have sensitivity peaks in the red-orange, green, and blue regions of the spectrum, so the cone system can detect color, much in the manner of the three-color additive process of television.

The blue-sensitive cones are much less numerous than the others, but make up for this by being more sensitive. There are also many more cones than rods in the central region of the retina. About 6000 cones are packed very tightly in the fovea, where there are no rods at all.

Cone vision, called *photopic vision*, has a maximum sensitivity at about 555 nm. This represents a yellow-green hue, and it closely matches the peak spectral output of the sun. Rod vision, called *scotopic vision*, has a sensitivity peak around 507 nm, well into the bluish-green region, and is totally insensitive to red light. As there is only one type of rod, there is no color discrimination with scotopic vision: by moonlight everything looks grey. The shift in maximum spectral sensitivity from yellow-green to blue-green is known as the *Purkinje shift*.

Because of the huge number of cones in the foveal area, humans have excellent visual acuity in bright light. In dim light, however, the resolution of the human retina (though not its sensitivity) is poor. Rod cells are connected together in parallel, sometimes in quite large numbers, which increases their efficiency at detecting very low light levels. At these levels the foveal area, having no rod cells, is not operating at all. The iris is fully open, increasing the effect of optical aberrations. So although everything *looks* bright and sharp on a moonlit night, you still can't read a newspaper, apart from the largest headlines. About 45° out from the fovea the cones peter out altogether: there is no color vision in your peripheral visual field. Near the edges of the retina the rods are also much larger, and can only detect movement.

Sensitivity Range

The sensitivity range of the human eye is enormous. From full sunlight to starlight represents an illuminance ratio of more than 10 million to 1. It's just as well, then, that we don't perceive equal increments of luminance as equal increments of brightness. Instead we perceive them on something close to a logarithmic scale, that is, equal

Additive color systems are discussed in more detail in Chapter 7.

You can test this by watching an electric fire in the dark, when you switch it off. The element goes a deeper red as it cools, but just before it goes out it turns grey. As rods are insensitive to red, you can dark-adapt your eyes in full sunlight by wearing red goggles, something I discovered many years ago when working as a photographer in the Far East.

Jan Evangelista Purkinje (properly spelled Purkyňe) (1787–1869) was a Czech histologist of considerable distinction. Apart from his work on the structure and functions of the eye, he investigated neural structures in the brain and heart and in embryos (several structures in each of these bear his name); he discovered the sweat glands in skin, founded the techniques of fingerprint identification, and was the first person to use a microtome in preparing specimens for microscopy. He also translated the works of Goethe and Schiller.

This is why, on a starry night, you can see dim stars better if you look slightly away from them.

multiples of luminance (1, 2, 4, 8, 16, etc.) are perceived as equal increments of brightness (1, 2, 3, 4, 5, etc.). This is an important characteristic of visual perception, and it appears more than once in later chapters. This principle also applies to other senses, and is equally important in hearing. It is called the *Weber-Fechner law*.

You can find out all about logarithms from Appendix 1.

THE WEBER-FECHNER LAW

This is in fact a conflation of two laws. The earlier one, due to the German physiologist Ernst Heinrich Weber (1793–1878), states that the "just noticeable differences" in stimuli are proportional to the magnitudes of the original stimuli. For example, if you can just distinguish between two weights of 10 grams and 11 grams, then you will just be able to distinguish between weights of 100 grams and 110 grams, not 101 grams, as you might have supposed. The German psychophysicist Gustav Theodor Fechner (1801–1887) carried this principle further. His law states that the intensity of subjective sensation increases as the logarithm of the stimulus intensity. Thus it is Fechner's share of the law that we are concerned with here.

Visual Pathways

Most of the operations of the right side of the body are processed and controlled by the left hemisphere of the brain, and vice versa. This also applies to visual processes, though the eyes of humans are wired up differently from those of many other animals. The terminus of the neural pathway from the retina is at the outermost part of the rear of the brain, and is known as the *visual cortex*. In most mammals all the nerves from the whole of the right retinal field go to the left visual cortex and those from the left retina go to the right visual cortex. However, in humans and some other mammals (particularly primates) the left *halves* of the two retinal fields both go to the left visual cortex and the right halves to the right visual cortex. The foveal areas cross over entire. The important thing about this is not that

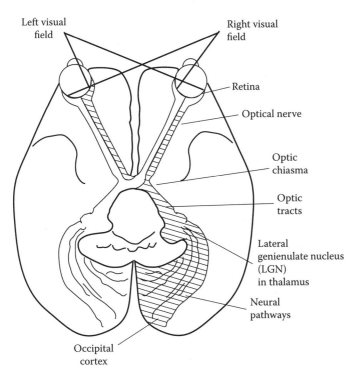

Figure 3.4 Neural pathways for the visual process, from above.

the field is split down the middle (there is some overlap), but that both retinas send their messages to the cortex as it were superimposed. (We shall see the importance of this later.) This sorting out of the visual pathways takes place at a ganglion called the *optic chiasma*. From the optic chiasma the nerve bundles travel to two locations right and left within the midbrain called the *lateral geniculate nucleus* (LGN). From there they fan out to the visual cortex (Figure 3.4).

NEURAL PROCESSING OF THE VISUAL SIGNAL

The details of neural processing in the visual pathways are complicated. Most of the pioneering work on this was done by Hubel and Wiesel. The following is a much-simplified version of their findings.

There are more than 100 million light receptors in each retina, but the two optic nerves, with diameters less than that of a pencil, contain only about 1 million fibers, so plainly a good deal of processing has to have gone on in the retina itself. When a receptor cell in the retina fires, its signal is passed first to a so-called *bipolar cell*, then to a further cell called a *ganglion cell*, which is connected to a large number of adjacent bipolar cells. This passes a (modified) signal down the optic nerve to the LGN. The connections between retinal cells are intricate, and appear to collect and modify the stimulus in such a way as to compensate for optical deficiencies in the eye. For example, where there is a stimulus such as a point of light, the signals from the cells surrounding the area are inhibited. This makes the image clearer and sharper, in much the same way as unsharp masking (Chapter 10).

The optic nerve can carry some 10 million bits of information per second. It passes this information directly to the LGN, which acts as a kind of relay station, sorting out types of signal such as color information, shapes, movements, contrast, etc., and routing them to the appropriate areas of the visual cortex. Here, different areas register lines and edges at various angles, movement in various directions, differences between the two retinal images, color information, and so on, and assemble everything into a coherent visual appreciation by matching the perceived stimuli to those of known objects, and performing an interpretation.

David Hubel (b. 1926) was until his retirement professor of neurobiology at Harvard University, where he met Torsten Wiesel (b. 1924). For many years they worked together on the correlation of the anatomical structure of the visual cortex with the physiological responses to visual stimuli. In 1981 they were awarded the Nobel Prize for Physiology for their work.

Visual Fields and Binocular Vision

Vertebrates have evolved with broadly bilateral symmetry, so that in general our organs are either symmetrical or duplicated. In the process of evolution our auditory and visual processes have taken advantage of this duplication, in the former case awarding us binaural hearing and the ability to locate sounds accurately. Binocular perception seems to have gone in two different directions, depending on whether the species in question belongs to the hunted (pigeons, antelopes) or the hunters (falcons, leopards). The hunted have their eyes at the sides and take in an all-round view, often nearly a full sphere (useful for spotting something sneaking up behind you or about to dive on you), with each eye contributing half the visual field, and with a small overlap at the front (helpful in avoiding hitting a tree). The hunters have both eyes at the front, giving them excellent binocular judgment of distance over a wide field of view. Humans, too, appear to have stemmed from a line of predators and tree-dwellers, and also have a wide field of binocular vision. This field is about 125° vertically and about 120° horizontally (this depends to some extent on the size of your nose). Each eye sees a slightly different view, and part of the visual cortex is dedicated to recording this difference and deducing the three-dimensional nature of what we are observing. Outside this area we have some 40° each side of poorish monocular vision devoted mainly to the detection of movement (our ancestors, after all, were themselves hunted sometimes). This is illustrated in Figure 3.5.

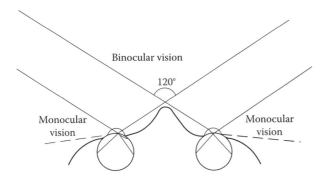

Figure 3.5 Horizontal limits of binocular vision in humans.

Stereoscopic perception, as it is called, is a consequence of the differences between the left-eye and right-eye images because of their different viewpoints; the bifurcated signal at the optic chiasma may be a sign of the evolutionary importance of the rapid and accurate distance judgment afforded by efficient stereoscopic perception. Be that as it may, most of us scarcely notice its presence, and it is only the surprise on viewing a stereoscopic pair of photographs for the first time, and suddenly seeing what looks like real depth in a photograph, that makes us aware of it. Stereoscopic perception is not a very powerful process in humans. Indeed, something like one person in four has poor stereoscopic vision and finds difficulty in "fusing" traditional stereoscopic pairs. Roughly one person in 20 has no stereoscopic perception at all. Other, largely unconscious, clues to depth are collectively more useful: apparent size of objects, obscuration of one object by another, parallax (change of appearance of a scene when you move your head), bluishness and lowered contrast in distant objects (called "aerial perspective" by photographers), and the need for eye convergence and accommodation (change of focus) for near objects. Chapter 15 is concerned with three-dimensional images, and treats the subject more thoroughly.

Color Perception

Few animals seem to possess as fully developed a perception of color as we do, though some birds, fish, insects, and even advanced molluscs such as octopuses and cuttlefish, undoubtedly run us close. The cones of the human eye, as we saw earlier, are of three types, with sensitivity peaks in the red-orange, green, and blue regions of the visible spectrum. The sensitivities overlap, especially those of the red-sensitive and green-sensitive cones. Figure 3.6 shows the (normalized) spectral sensitivity ranges of the three types of cone.

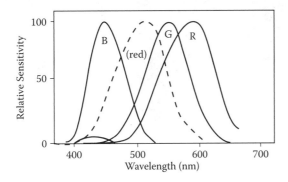

Figure 3.6 Spectral sensitivities of the three types of cone (normalized). Rod sensitivity is shown as a broken line.

Roughly one person in 10 has some deficiency in one type of cone (usually red or green). Such people, who are almost always male, find difficulty in distinguishing certain colors, usually red and green, which they tend to register as various shades of brown. Anomalous color vision can be diagnosed by Ishihara test cards. These are a set of patterns of pale-colored spots, which contain a number printed in a different hue. People with color anomalies see either a different number or no number at all (see Figure 3.7, color plate).

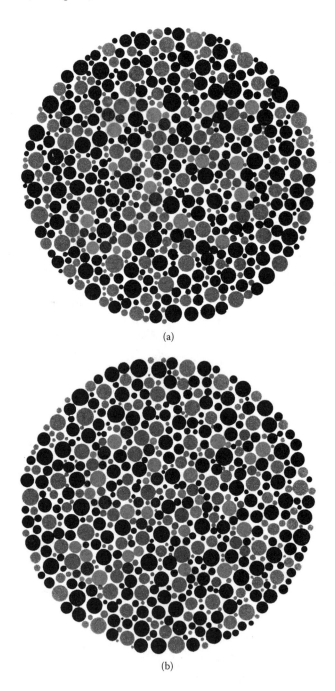

(a)

(b)

Figure 3.7 (See color insert following page 154.) A page from the Ishihara color vision test book. Persons with normal color vision see the number 35. Persons with protanopia (red cone deficiency) will see only the figure 5 and those with deuteranopia (green cone deficiency) will see only the figure 3.

Seeing a Range of Colors

With only three types of cone, how is it that we can manage to see such a huge range of colors? It really is huge: Judd and Wyszecki in the 1970s carried out reliable research, which indicated that people with normal color vision could distinguish 10 million colors. The answer lies in the way a colored surface stimulates the three types of cone differentially. Physiologists refer to the three types by the Greek letters ρ (rho) for red, γ (gamma) for green, and β (beta) for blue. If a certain combination of wavelengths stimulates the ρ and γ cones equally, we perceive the hue halfway along the spectrum between red and green, namely yellow. If the ρ cones are stimulated more than the γ cones, we see orange, and if the γ cones are stimulated more than the ρ cones, we see yellow-green. Similarly, with equal stimulation of the γ and β cones we see blue-green (cyan), and with varying stimulations of the two types of cone we see the spectrum running from green through bluish green, cyan, and greenish blue to full blue. When the β and ρ cones are stimulated together in various proportions we see the range of purples, magentas, and crimsons—hues that are not in the spectrum at all, but are real colors nonetheless. This may seem odd, until you realize that light waves do not themselves possess color: it is their effect on the retinal cells that produces the sensation of color.

It must be clear now that there is more than one way of producing a particular color sensation such as yellow. As long as you stimulate the γ and ρ cones equally you will get yellow, so you can use monochromatic sodium light with a wavelength of 589 nm, or two bands of wavelengths centered round 549 and 649 nm: the result will be the same color sensation.

But what about brown, slate grey, white, or pink? These colors certainly exist. Up until now I have been discussing only saturated (pure) colors, such as those in the spectrum. A desaturated color is like a pure color with grey added. Now, you get grey if you stimulate all three types of cones equally, to a low level. Add a little extra stimulus to the ρ and γ cones and you have brown, which is simply desaturated yellow or orange. Add a little stimulation to the β cones instead, and you see a bluish or slate grey. Increase the stimulus all round and grey becomes white, and take away a little from the β and γ stimulation and you are left with pink, which is simply a light desaturated red.

This description has been a simplified one. For one thing, you can't produce a fully saturated cyan with green and blue light, because the ρ sensitivity curve overlaps at the only point where the γ and β sensitivities are equal. But it is broadly correct, and is the basis of color photography as described in Chapter 7.

On being introduced to the three-color additive principle many people find the summation of red and green to produce yellow hard to accept. Yellow seems to be a primary hue, not related to either red or green. Yet when you project a red light and a green light onto a white screen, superimposed, the result is indisputably yellow (you have already seen why this should be so). Part of the reason may be cultural: yellow had a name long before cyan or magenta. Again, one can visualize a reddish blue and a greenish blue, but a reddish green? Indeed, red can readily be seen as a kind of opposite to green, just as blue seems opposite to yellow. This intuitive concept led Ewald Hering to propose an "opponent-color" theory of color perception as an alternative to Thomas Young's three-color principle. There is now no doubt that the retina analyzes color according to the three-color principle, but it has recently been confirmed that the LGN does indeed operate in a kind of opponent-color mode, with some neurons firing (or inhibiting firing) for red versus green and others for yellow versus blue.

Ewald Hering (1834–1918) was an Austrian physiologist who took an intuitive view of color perception, postulating that in addition to black and white there were four pure colors, namely red, yellow, green, and blue, and that each visual perception was a mixture of these six sensations opposing one another and interacting.

Constancy

Our eyes are continually moving, and are mounted on an unstable platform. Why, then, do we see objects and landscapes as stationary? If you were to take a video camera and move it around in the same manner, the resulting effect would be vertiginous for the viewer. Again, if you stand at the foot of a tall building and turn your gaze upwards to the top, the building doesn't appear to move. But if you do a similar upward pan with a video camera, the verticals converge and the building seems to be toppling over backwards. Yet this doesn't happen in the visual world. The reason for this appears to be that when we are looking at a scene, all the movements of our head and eyes are being recorded too, and the brain, which receives this information in addition to the visual input, integrates the information and consequently discounts the effect of the motion. The result is termed *constancy of position*.

You can disrupt this effect if you move your eye without the cooperation of your eye muscles. With one eye closed, push gently with your finger on the side of the other eye, through your lower eyelid. You will see the whole field of vision moving! Position constancy is essential if we are to be able to make sense of the visual world—and to be able to look upwards without falling over.

Color constancy is nearly as important. As a sunny day draws to a close the color temperature of the sunlight may fall from more than 5000 K to less than 3000 K. But the trees and houses don't seem to change color. Nor do the objects in your room appear to change color when you switch on the light (although the fading daylight in the window seems to have gone blue). As a rule you don't notice much (if any) change in the color of your surroundings when the color of the illumination changes. Yet if you take a photograph on color transparency material balanced for daylight in a room lit by filament lamps the result has a deep orange cast. Photographers who make their own color prints need to check the color balance of test exposures under standard white illumination. They quickly learn not to examine a print for more than a few seconds without looking away. If you gaze at a print that is poorly color-balanced for half a minute or so it will seem to have become much nearer to being correctly balanced.

Constancy of shape and size are also important. When an object moves away from you, you don't perceive it as shrinking. In fact, if it *does* shrink (as with a slowly deflating balloon), it is hard to resist the impression that it is in retreat. Again, a round plate that is tilted doesn't look any less round, even though the image on your retina has become elliptical.

It seems that some birds, particularly chickens and pigeons, have a different way of dealing with the problem. They keep their heads stationary with respect to the ground while taking a step forward, jerking their head forward at the beginning of each step to catch up.

In the early days of High Street cheap'n'cheerful fast color prints, it was a standard ploy to get customers who were unhappy with their results to examine the prints carefully, while they were assured meantime that the color balance was fine. The traders usually got away with it, too.

Before the laws of perspective were properly established, painters invariably painted distant figures too large and tilted plates too round. A naive person, asked to draw a plate presented almost edge-on, will almost invariably draw it insufficiently foreshortened.

Visual Illusions

All of us have at some time been fooled by a visual illusion. Some of these are very powerful. Visual illusions can often reveal much about our visual and perceptual processes. Some illusions plainly originate in the retina, some almost certainly in the LGN, and some manifestly in the perceptual mechanism of the visual cortex. Others appear to be purely cultural, i.e., we have learned to expect a particular visual experience, and the past experience colors our perception.

Illusions Originating in the Retina

The most common retinal illusions are associated with neural fatigue. The best known of these is the color-reversed U.S. flag, with black stars on a yellow ground, and black and cyan stripes (Figure 3.8, color plate). When you stare at this in a good light for about half a minute, then transfer your gaze to a plain white surface, you see a phantom flag in its correct colors—an after-image. The yellow has fatigued your ρ and γ cones, and the cyan bars have fatigued your γ and β cones, while the black stripes and stars have left the cones in these areas of the image fresh. You can do the same trick with a photographic negative, even with a color negative if it is contrasty (this works best with a color transparency film "cross-processed" to give a color negative). Try it with a digital color image in which you have reversed the colors on your monitor. The result can be a surprisingly accurate after-image.

Another striking retinal illusion happens when you study a pattern of closely spaced circles, spokes, or wavy black lines (Figure 3.9). After a few seconds you begin to see dancing patterns. The involuntary movements of your eyes cause the after-images to jump about and construct a set of dancing moiré patterns. The artist Bridget Riley exploited this illusion in her Op Art series of paintings in the 1960s.

Figure 3.8 (See color insert following page 154.) If you stare at the center of this flag in bright light for about half a minute and then look away at a white surface, you will see the U.S. flag as an after-image, in its correct colors.

Figure 3.9 A typical "op art" pattern. As you look at this it begins to flicker, because of the small involuntary movements of your eyes.

Illusions Originating in the LGN

A set of illusions probably associated with the LGN is a family of geometrical patterns that appear to be distorted, though they are not (Figure 3.10). Illusions of this general type can sometimes appear in real situations, such as the bathroom tiles of Figure 3.11.

Illusions associated with motion are also connected with the LGN. If you place a spiral on a record player turntable and watch it turning for about half a minute, then stop the rotation, the spirals will appear to be moving in the opposite direction (e.g., outwards instead of inwards). You may be familiar with the sensation of moving backwards when you have been driving a car at a steady speed and you have to stop suddenly. This is because your sensory system has become accustomed to seeing the scene in front steadily expanding. It has a similar origin to the spiral illusion.

Illusions Originating in the Visual Cortex

Most of the illusions originating in the visual cortex are concerned with either conservation or perspective. The best known is probably the Necker cube and its derivatives,

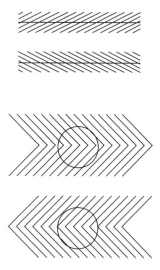

Figure 3.10 Zöllner and Orbison illusions. It is almost impossible to see the two lines as parallel or the circles as true.

Figure 3.11 Bathroom tiles in a Greek hotel. It is hard to believe that they are, in fact, perfectly aligned and regular.

beloved of the artist Maurits Escher (Figure 3.12). You can invert the perspective of this type of drawing just by thinking. The Ponzo "railway lines" illusion and the Müller-Lyer "arrowheads" illusion are examples of misperception of the length of a straight line because of distractions which, though irrelevant, are impossible to ignore (Figure 3.13). The converging lines of the Ponzo illusion create an overwhelming feeling of perspective, so that the upper line is perceived as farther away than the lower line, and therefore larger (conservation of size). The Müller-Lyer illusion is almost certainly cultural in its origins. One figure looks like the outside corner of a building, and the other like the inside corner of a room. This illusion is, it seems, not perceived by people unfamiliar with rectangular buildings, such as Bushmen and some Inuits.

There are large numbers of three-dimensional and moving illusions that, unfortunately, cannot be depicted in a book. One of the most striking (and disturbing) is a hollow cast of a human face. The inside of a Halloween mask will do. If you light it from below, it is almost impossible to believe that it is not convex, like a normal human face, and if you move round it, it seems to rotate in the same direction as you are moving in, but at twice the speed. This is a totally culture-driven illusion: we are so used to seeing human faces as convex, that the belief that this must also be convex overrides any genuine depth perception. You will get used to this illusion if you make transfer holograms, as reversing the master hologram in the illuminating beam reverses the perspective too (see Chapter 16 for more details).

There are many more visual illusions, some of which are dynamic; there are several Web sites where you can see examples. Many are not at all easy to explain. But they are all indications of the vast amount of processing that goes on in our heads in order for us to make sense of the images that fall on our retinas. They also remind us that the process of visual perception is subtle, complex, and altogether remarkable.

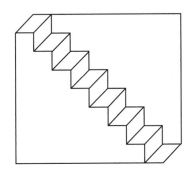

Figure 3.12 Reversing perspective; one of the derivatives of the Neckar cube illusion.

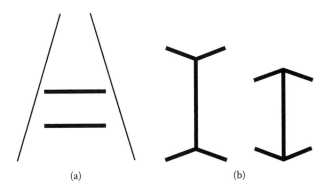

(a) (b)

Figure 3.13 Perspective illusions: (a) Ponzo; (b) Müller-Lyer.

Perception and Imaging

What, you may ask, has all this to do with the science of imaging? The answer is that every image, however technically perfect, has to go through the filter of human visual perception, which, as you have seen, is by no means straightforward. Some of it can be quantified and fitted into the matter dealt with in Chapter 6, but sometimes an image may have to be modified in order to fit what we *think* we see. The ancient Greeks knew this when they built the pillars of the Parthenon and other temples with a slight bulge in the middle so that they appeared to be truly parallel-sided. Modern architects are no less crafty. In later chapters we shall see, from time to time, how the nature of images is affected by the way we look at them.

Digging Deeper

For a book that takes the subject of visual perception about as far as a non-biologist would want to go, I strongly recommend *Eye and Brain* by Richard L Gregory (5th edition, OUP, 1998). It has a comprehensive bibliography citing such research papers as Hubel and Wiesel's original publications. *Perception*, by John Rock (Scientific American Library, 1984), is another excellent introduction, to some extent complementing Gregory's book, but omitting color perception. *The Art and Science of Visual Illusions* by Nicholas Wade (Routledge, 1982) is a comprehensive study, and includes a number of transparent moiré pattern generators to play with, some rather naughty. *Visual Perception*, by Nicholas Wade and Michael Swanston (Routledge, 1991) covers the same ground as this chapter, but in much greater detail. *The Ishihara Tests for Color Blindness* are published by Arnold. They are available in three degrees of complexity, but as the simplest one costs nearly $150 I don't suggest you should get yourself a set. Your local optician or eye hospital would probably be delighted to show you a copy, and perhaps give you a full test. Hubel and Wiesel wrote up their researches in the *Journal of Neurophysiology* in 1965, but the report is strictly for the specialist. David Hubel wrote a much more approachable article for *Scientific American*, November 1963, and this was reprinted in the Open University's reader for their early course *The Biological Bases of Behaviour* (Harper & Row, 1971). This reader also contains an article by Alfred Sherwood Romer on the structure of the eye, and a comparative study of the octopus and mammalian eyes by E. J. W. Barrington, showing how two very different evolutionary lines converged to develop a remarkably similar organ (another nail in the coffin of creationism). This book is out of print, but copies are easy to find second hand or from a public library. If you want a comprehensive physiology of the eye, go to the Web site www.tedmontgomery.com/the_eye.

Chapter 4 Lens Principles

A Model for the Geometry of Camera Lenses

Although there are a number of models describing the behavior of light, the one used by lens designers is the simplest, namely the ray model. Before there were computers, the designing of a new lens was a complicated business involving months of laborious ray tracing. Today, designers can optimize a lens design in a few minutes using an off-the-shelf computer program. The variables to juggle are the refractive indices of the components; the color dispersions of the glasses; the number of components and their curvature, thickness, and separation; and the position of the stop. This all represented a heavy meal for the old-time designer equipped with only a slide-rule and a set of tables of glass types. But to a modern computer they signify no more than a light breakfast. The result still doesn't predict everything about the performance of the final lens; this demands a more sophisticated model, as we will see in Chapter 6. However, a simple ray model is sufficient to describe how a lens forms an image, and to explain why it is necessary to have more than one element in a camera lens. To begin with we will look at the properties of a simple convex lens.

The Simple Lens

The most basic ray model for a lens makes a number of assumptions:

1. The thickness of the lens can be ignored.
2. The lens aperture is small compared with the focal length.
3. The refractive index of the material is the same for all wavelengths.

The geometrical optics of a simple or "thin" lens was established by Sir Isaac Newton, and for this reason is usually known as *Newtonian optics*.

The Lens Laws

The Newtonian lens laws consist of three basic principles concerning a simple lens. The first gives the relationship between the distances of the object and its image from the lens. The positions of the object and the image (which is inverted) are together termed *conjugate foci* (Figure 4.1). They are connected by the *focal length* of the lens, which is the distance of the image from the lens when the object is at infinity. If the distance of the object from the lens is u, the image distance is v, and the focal length is f, the relationship is given by the formula

$1/u + 1/v = 1/f$

The second principle concerns the scale or *magnification* (M) of the image (usually less than 1). This is given by the formula

$M = v/u$

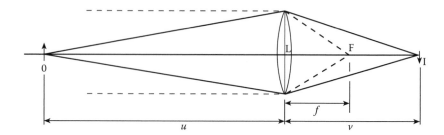

Figure 4.1 Conjugate foci. F is the rear principal focus and *f* the focal length. The object–lens distance OL is *u* and the image–lens distance IL is *v*. As *u* increases, I moves nearer to F.

and this can be substituted into the formula for conjugate foci to give

$$v = f(1 + M)$$

and

$$u = f(1 + 1/M)$$

The third principle is the effect of combining two or more lenses. If two lenses have focal lengths f_1 and f_2 and are mounted close together, their combined focal length *f* is given by

$$1/f = 1/f_1 + 1/f_2$$

or

$$f = f_1 f_2 / (f_1 + f_2)$$

If the lenses are separated by a distance *d*, the formula becomes

$$1/f = 1/f_1 + 1/f_2 - d/f_1 f_2$$

or

$$f = f_1 f_2 / (f_1 + f_2 - d)$$

Thus for two convex lenses, increasing the separation shortens the combined focal length, and for a combination of a convex and a concave lens, increasing the separation increases the combined focal length, as negative (concave) lenses have a negative focal length.

Real and Virtual Images

If you are using the lens formulae, and a distance and/or a magnification comes out negative, the image is a *virtual image*, that is, the rays do not actually pass through it, but instead appear to have come from it. A positive lens can also form a virtual image if *u* is less than *f*. In this case the image is erect (not inverted) and is enlarged. This is the image you see with a magnifying glass (Figure 4.2a). The image formed by a single negative (concave) lens is always virtual, erect, and smaller than the object for all values of *u* (Figure 4.2b). An image that the rays do pass through, i.e., the one we have been discussing above, is called a *real image*, and you can catch it on a screen or on a digital sensor or film.

Throughout this book I am using a sign convention known as "real is positive." This means that all distances to objects and real images are considered to be positive, and all distances to virtual images, as well as their magnifications, are considered negative. The focal length of a negative lens is also considered negative. Some optics textbooks use a convention that puts the lens at the origin of a Cartesian system. This makes *u* negative, and in my view complicates matters unnecessarily.

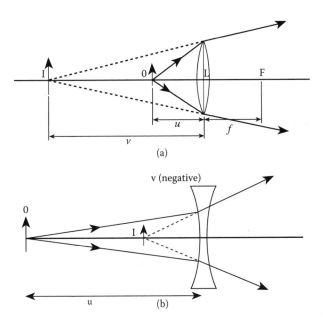

Figure 4.2 (a) Formation of a virtual image I by a positive lens with the object O at a distance less than f from the lens. The distance v is negative, and the rays do not pass through the image, but appear to have originated from it. This is a magnifying-glass configuration: $v/u > 1$; (b) formation of a virtual image by a negative lens.

Depth of Field

In real life the object (i.e., the subject matter) usually has some depth, so that there is a u_{max} and a u_{min}, the farthest and nearest visible points on the subject matter. These give correspondingly different values for v. The question in practice is, how unsharp can we allow the image to be before it becomes unacceptable? This, of course, begs the question of what we mean by "sharp." This is less simple than it appears; it is discussed in more detail in Chapter 6. But for the moment let us go by the most obvious definition: an image is sharp if it *looks* sharp.

Disc of Confusion

Provided the lens aperture is circular, any point on the object that is not at the correct u-distance for the film plane setting will produce an image that is a disc rather than a point. This is called the *disc of confusion*.

If, when you view the final print, you can't distinguish this disc from a true point, then it is sharp (by the definition above). The statistical norm for the resolution of the human eye is taken as the perception of 6 mm detail at a distance of 6 m, hence "6/6 vision" ("20/20" in the United States). So if the disc of confusion on an A4 sized print (which would typically be viewed from a distance of about 40 cm) is no larger than 0.4 mm in diameter, the image is considered acceptably sharp.

Relative Aperture

At this point we need to introduce the concept of *relative aperture*. This is denoted by a number called the f-number (or f/no), and is the focal length of the lens divided by the aperture diameter.

The latter is controlled by an iris diaphragm, an approximately circular annulus of metal leaves which in camera lenses is within the lens and can be adjusted by a ring

Photographers whose grasp of mathematical nomenclature is weak call it a circle of confusion.

Because "6/6" is a statistical mean, half the population has better eyesight than this, and would be able to detect unsharpness in such an image. However, 6 mm at 6 m amounts to 1 milliradian (mrad), which is a conveniently round figure for general purposes, and suffices for pictorial images, though some technical applications require a higher standard.

Strictly, the diameter of the *entrance pupil*, which is the apparent diameter of the image of the aperture seen from the object plane, as explained below.

49

I am using the terms "film" and "film plane" for the sake of brevity. All the material in this chapter applies equally to digital and TV optics. A print intended for direct viewing, or a projected image on a screen, is usually much larger than the original negative or transparency, so the calculated diameter of the acceptable disc of confusion for the final image needs to be divided by the scale of enlargement in order to give the required resolution on the film. This will give a diameter for the disc of confusion of (typically) 0.03 mm for a digital format, 0.04 mm for full-size 35 mm, and 0.15 mm for 4 × 5 in.

If you feel overwhelmed by all these formulae, don't worry. All camera lenses (apart from compacts) are engraved with their own depth of field scales. And if you are operating a large studio camera you can inspect the image on the ground glass screen with a magnifier.

on the lens barrel. At a given f/no the illuminance on the film for a given subject is constant, no matter what the focal length of the lens may be. It follows that for a given f/no, the actual aperture diameter varies in direct proportion to the focal length.

> ## HYPERFOCAL DISTANCE
>
> If a camera lens is focused on infinity (i.e., the film plane is at the principal focus of the lens) the distance to the nearest object that will be rendered acceptably sharp is called the hyperfocal distance (h). It is calculated from the formula
>
> $h = f^2/Nc$
>
> where N is the f/no and c is the diameter of the acceptable disc of confusion.
>
> When the camera is focused on a distance u, the farthest and nearest object distances that produce tolerably sharp images are given, respectively, by
>
> $D_{near} = hu/(h + u)$ and $D_{far} = hu/(h - u)$
>
> From this we can show that the full depth of field D is given by
>
> $D = 2h^2u/(h^2 - u^2)$
>
> If we focus the lens on the hyperfocal distance, D is a maximum, extending from half the hyperfocal distance to infinity. This is the way cheap fixed-focus (so-called focus-free) cameras are constructed.

Depth of Focus

Depth of focus is the tolerance in the film plane. It becomes important in two situations: (a) when you have to use a film holder that is not properly matched to your camera, and (b) when you are making ultraclose-up images and need to focus by adjusting the back of the camera rather than the lens. The formula for depth of focus is comparatively simple. When the magnification is small, the depth of focus t is given by the formula

$t = 2cN$

In passing, you should remember that the optical image is not completely two dimensional: it does have some depth. In fact, at a magnification of 1 it has the same depth as the object. However, the magnification along the axis is the square of the lateral magnification, so except in extreme close-ups (see Chapter 18) this three-dimensional optical image is very much flattened.

Karl Friedrich Gauss (1777–1855) was one of the greatest of all mathematicians. He made important innovations in the theory of almost every branch of science, and even developed a non-Euclidean geometry, which, in the climate of his time, he prudently withheld from publication. He also made useful contributions to practical lens design. One of these is the model for most wide-aperture camera lenses in use today.

Gaussian Optics

One of Gauss's most important contributions to optics was a simplification of the geometry of multi-element ("thick") lenses. He showed that it was possible to apply Newtonian optics to any lens system by specifying six *cardinal points* (also known as *Gauss points*) on the principal axis of the lens. One of these already discussed is the *rear principal focus*. There is also a *front principal focus*, identified by sending parallel rays into the lens in reverse. The four other points are called the *principal* and *nodal points* of entrance and emergence (also called the *front* and *rear nodal points*). In any lens that is wholly in air, the principal and nodal points coincide, though they are defined differently. The planes through these points perpendicular to the principal axis are called the front and rear principal (or nodal) planes.

Gauss's main thesis was that any ray entering the lens, when produced to cut the front principal plane, would, on emergence, appear to have originated at the corresponding point on the rear principal plane. The space in between could be ignored for geometrical purposes. So the u-distance is measured from the front principal (or nodal) plane, and the v-distance and the focal length are measured from the rear principal (or nodal) plane (Figure 4.3).

In general-purpose fixed focal length (so-called *prime*) lenses, the nodal planes are situated about one third of the way inside the lens, but for special purposes they may be designed to be in quite different positions. The important thing is that wherever the rear nodal plane is located, you could in theory substitute an "equivalent thin lens" that would do exactly the same job.

Another important concept is that of the *entrance* and *exit pupils*. The entrance pupil is the image of the aperture stop as seen from an axial point on the object. It is a virtual image, and is usually behind the lens. Its construction is illustrated in Figure 4.4. The exit pupil is the image of the stop as seen from an axial point on the image plane, and is also a virtual image, usually in front of the lens. You will see from the illustration that the true *f*/no of a lens is in fact the focal length divided by the diameter of the entrance pupil, and not that of the aperture stop. In prime lenses the effective aperture diameter (i.e., the diameter of the entrance pupil) is usually larger than the aperture stop, but in retrofocus lenses (see below) it is smaller.

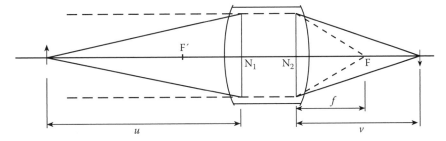

Figure 4.3 Gauss points in a thick (compound) lens. The focal length f and image distance v are measured from the rear nodal point N_2 and the object distance u is measured from the front nodal point N_1. F is the rear and F′ the front principal focus, and F′N_1 = FN_2. Two other points, the principal points, coincide with N_1 and N_2 if the lens is wholly in air.

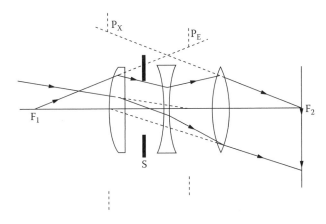

Figure 4.4 Entrance and exit pupils for a simple triplet lens with converging front component. The entrance pupil P_E and the exit pupil P_X are virtual images of the stop S as seen from the front and rear focal points F_1 and F_2, respectively, and in this case are both larger than the stop itself.

Telephoto Lenses

Going back to the lens laws, it is plain that when u, the object distance, is large, v, the image distance, will be approximately equal to f, the focal length. The formula for magnification then simplifies to

$M \approx f/u$

i.e., the image scale is directly proportional to the focal length of the lens. So if you want to be around to be able to impress your friends with a full-frame shot of a charging rhinoceros, you need to use a lens of at least 400-mm focal length (a *long-focus lens*) rather than the usual 35- or 50-mm prime lens. Now, a 400-mm lens of conventional design is a cumbersome object, not the kind of thing one would want to carry on safari. But the focal length is measured, not from the back of the lens, but from the rear nodal plane. If the lens designer can coax this plane out in front of the lens, you will be able to carry around a physically short lens, but with a long focal length—by definition, a *telephoto lens*.

The basic principle of a telephoto lens is that of the Galilean telescope (about which more is discussed in Chapter 17). Figure 4.5a shows how the addition of a negative lens increases the focal length of the combination. If you produce the final convergent rays back until they cut the initial parallel rays entering the lens, you will have marked the position of the equivalent thin lens, the hypothetical simple lens that would have done the same job. The plane of the equivalent thin lens is the rear nodal plane of the combination, and its distance from the film plane is its focal length (strictly, *equivalent focal length*).

Retrofocus Lenses

If you reverse the order of the two lenses (like turning a pair of opera glasses round) you also reverse the position of the rear nodal point. It is now *behind* the lens (Figure 4.5b). This is called a *retrofocus* configuration, and it is most often used in

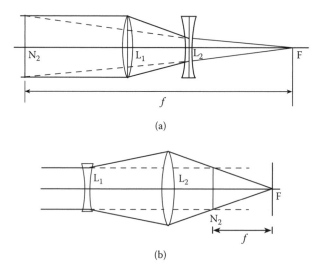

(a)

(b)

Figure 4.5 (a) Telephoto principle: the insertion of a weak negative lens behind the main lens increases the focal length, moving the rear nodal point (i.e., the position of the equivalent thin lens) well in front of the combination. *f* is the equivalent focal length. (b) In a reverse telephoto (retrofocus) configuration the rear nodal point is behind the combination, giving clearance for a reflex mirror.

wide-angle lenses, especially those designed for single-lens reflex (SLR) cameras where there would otherwise be insufficient clearance for the reflex mirror to swing out of the way for the exposure.

Varifocal and Zoom Lenses

Some early lens systems were designed so that they could have their focal length changed by removing the front cell or by changing the separation of the elements, but in general such crude methods gave poor-quality images. A better way of producing a *varifocal lens* is to divide the positive focusing element into two, with a negative element in between (Figure 4.6). The negative element has a focusing power greater than that of either of the outer elements but less than their sum. If you shift this central element from close to the front element towards the rear element, you will move the rear nodal plane from the rear of the combination to the front, changing the configuration smoothly from retrofocus to telephoto.

The system is often employed in slide projector lenses, but you need to refocus each time you move the middle element. By adding a further moving negative element at the front or rear it is possible to hold the focal plane very nearly constant, and this is the basis of the design of the modern *zoom lens*, which does not need refocusing when zoomed. Other factors, some of which we will consider below, complicate the design considerably, and some zoom lenses have 14 or more elements (Figure 4.13).

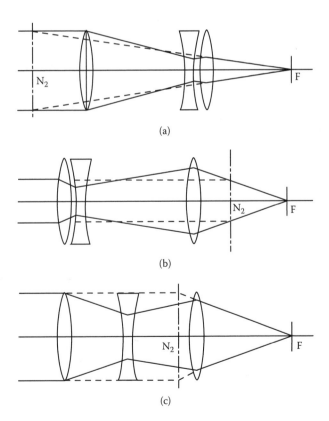

(a)

(b)

(c)

Figure 4.6 Principle of varifocal lens system. The negative element is slightly stronger than either of the positive elements, and when moved close to either of them it gives a negative combination, resulting in either (a) a telephoto or (b) a retrofocus combination. (c) When the negative element is in the halfway position, the rear nodal point is in its conventional position.

As the sensor of the majority of present-day digital cameras is only about seven-tenths the size of a full 35-mm format, it has become customary to describe lenses designed for digital formats in terms of their "35-mm equivalent." Thus the standard 38-mm lens is a "'50-mm equivalent," as its angle of field is roughly the same as that of a 50-mm lens used with a full-frame 35-mm film camera.

Magnifying glasses are specified by their (angular) magnification when used in a standard manner; spectacle lenses are specified in diopters. These are units of focusing power: diopters are the reciprocal of the focal length in meters. Hence a 4-diopter lens has a focal length of 0.25 m (250 mm).

Philip Ludwig von Seidel (1821–1896) was a professor of mathematics and astronomy at Munich University. Apart from his work in optics he made a number of advances in probability theory. A crater on the moon is named after him. By the way, there is nothing arcane about the Seidel equations: they are derived entirely from ray geometry.

Angle of Field

The *angle of field* is the angle subtended at the rear nodal point by the diagonal of the image format, and in a lens giving correct drawing, i.e., an image free from distortion, is the same as the *angle of view*. A "normal" angle of field is around 50°–60°, which is roughly the angle taken in by the human eye when viewing a scene. (This has been confirmed by observations of the distances from which people habitually look at pictures.) Typically, this is about the same as the length of the diagonal of the picture. Now, the diagonal of a 35-mm full-frame camera image is about 43 mm, so a lens of focal length between about 35 and 50 mm is considered "normal-angle." Any lens with a focal length much beyond 50 mm is considered long-focus (narrow-angle), and any lens with a focal length much below 35 mm is considered *wide-angle*. The focal length of the lens used can affect the perceived perspective of a print when it is examined from the normal distance; this is discussed at the end of this chapter. As lenses are designed for specific formats, the *covering power* (disc of sharp imaging) of a lens seldom goes much beyond the corners of the format, except for special "shift" lenses, which are also described later. Thus it is in general unprofitable to try to adapt a lens for a format for which it was not designed.

Lens Aberrations

If you take a simple lens such as a magnifying glass or a 4-diopter spectacle lens, and try to focus an image of a distant light bulb on a wall, as you change the distance of the lens from the wall the image becomes sharper or less sharp. On a close inspection of the sharpest image you can get, you see that the image has a somewhat blurred surround, and that this is colored, too. If you twist the lens out of alignment the image takes on a peculiar shape, which changes when you vary the distance of the lens from the wall. These effects are the manifestations of what are called *lens aberrations*. The name comes from a Latin word meaning "to wander away," which is just what the light rays do. Apart from the color fringes there are five other main aberrations. These were first described mathematically by L. von Seidel, so are called Seidel aberrations, or sometimes "third-order aberrations" from the form of the Seidel equation. They can be described individually in terms of their effects. One of the Seidel aberrations occurs uniformly over the focal field; the others occur only in the outer parts of the field.

1. *Chromatic aberration.* This is not a Seidel aberration, but is the result of dispersion. A lens deviates long wavelengths less than short wavelengths. The focal length of a simple lens is thus greater for red light than for blue. The image of a point of white light on the axis appears fringed with color (longitudinal chromatic aberration), and an image away from the axis appears as a tiny spectrum with the red outermost (lateral chromatic aberration). The effect can be minimized by adding to the focusing element a weak negative lens of a material with a high dispersion (Figure 4.7). Such a combination is termed an *achromatic doublet*.
2. *Spherical aberration.* Lenses are traditionally manufactured in batches by a process that produces surfaces that are parts of spheres, but the curvature of a sphere is not the theoretically correct curvature. It is too great towards the periphery of the lens, so that the outer zones have a shorter focal length than the inner zones (Figure 4.8).

Spherical aberration occurs uniformly over the whole image field. In a single lens it is minimized by sharing the refraction equally between the two surfaces. The result (for parallel light) is very nearly a plano-convex

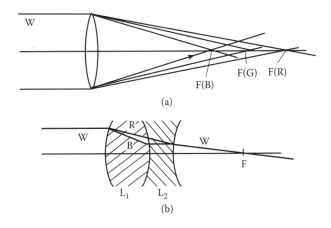

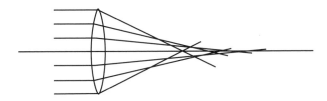

Figure 4.7 (a) A simple lens focuses blue light nearer to the lens than red light; (b) correction by a weak lens with high dispersion. (The thicknesses of the glasses have been exaggerated for clarity.)

Figure 4.8 Spherical aberration: the outer zones of the lens have a shorter focal length than the inner zones.

configuration (i.e., one side of the lens is an optical flat), mounted so the convex surface faces the parallel beam. This accounts for the ubiquity of this type of lens in simple focusing devices such as the condenser lenses in enlargers and slide projectors. Adding a weak negative lens with high (negative) spherical aberration to the focusing lens can further reduce spherical aberration. The negative component of an achromatic doublet can do duty for this.

3. *Coma.* In a simple lens the image of an off-axis point is focused in a different position for each small area on the lens surface. The image from the peripheral zone of the lens is a circle displaced outwards from the true geometrical position (Figure 4.9). The intermediate concentric zones of the lens form circles that are smaller and nearer to the geometrical image. The overlapping of these gives rise to a *coma patch* (*coma* is the Latin for a comet). Coma can be minimized in a single lens by making it a meniscus shape (i.e., like a watch-glass) with the stop at a short distance from its concave side (Figure 4.10). This is a dodge used in cheap cameras with lens apertures of around *f*/11.

4. *Curvature of field.* The "focal plane" of a simple lens is actually part of a sphere with the concave side facing the lens. This means that when the center of the field is in focus, the periphery is out of focus, and vice versa. This aberration can be controlled by closing down the aperture to increase the depth of focus. However, if the lens is split into several components, their respective surface curvatures can be adjusted so that they give a flat field. The reciprocal of the field curvature is called the *Petzval sum*, and a zero Petzval sum indicates a flat field.

Josef Max Petzval (1807–1891) was a Hungarian mathematician who in 1840 designed the first photographic lens to use a rigorous mathematical method in its design. His lens had a relative aperture of *f*/3.4, bringing contemporary exposures down by a factor of at some 20 times, and thereby revolutionizing portrait photography.

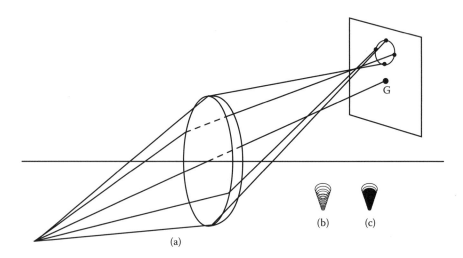

Figure 4.9 Coma: (a) The outermost zone of the lens produces an image of a point object away from the axis as a circle displaced from the geometric position G produced by the primary (central) ray. Intermediate zones produce similar, smaller circles (b), which overlap to produce a comet-shaped image with the tail radially outward (c).

Figure 4.10 Two ways of minimizing coma in a single lens by employing an aperture stop with a meniscus configuration.

5. *Astigmatism.* This term comes from the Greek for "not a point." Even when all the foregoing aberrations have been minimized, the image of an off-axis point is still not a point. In one plane it is a short line radial from the axis (the *radial focus*), and in another it is a short line that is part of a circle drawn round the axis (the *tangential focus*) (Figure 4.11). In a simple lens this effect is convolved with the coma patch and is difficult to see clearly, but in any camera lens made before 1886 it is plain enough.

Two functions are said to be *convolved* when the whole of one function is dealt out to every point on the other function. The corresponding noun is *convolution*.

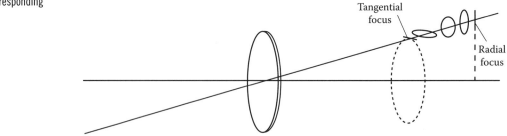

Figure 4.11 Astigmatism: rays from an off-axis point object pass through different thicknesses of glass, and are incident at different angles. The image is not a point but a line that is part of a circle drawn round the optic axis at one distance, and a line radial from the axis at a different distance. In between is a "disc of least confusion," which is reduced in diameter by the use of a small lens aperture.

In that year, the glassmaker Schott, working from theoretical material by Abbe, developed new types of optical glass, which, when incorporated into new lens designs, produced lenses substantially free from astigmatism. These became known as *anastigmats*.

6. *Distortion*. If the aperture stop is positioned in the center of a simple lens (split into two for the purpose) all the rays will pass through the central area, and over the entire field the image points will be in their correct geometrical position, i.e., the lens has correct drawing. However, if the stop is situated in front of or behind the lens, it will select rays that are deviated too much (barrel distortion) or too little (pincushion distortion): the magnification changes with angle of field (Figure 4.12). Distortion can be eliminated most easily by designing lenses to have a quasi-symmetrical configuration of components. By a careful choice of stop position it can be minimized in telephoto and retrofocus lenses too, but it is very difficult to eliminate in zoom lenses.

Yes, "anastigmat" *is* a double negative. Dallmeyer, who produced some of the first anastigmats, pedantically called his designs "Stigmatic," though the name didn't catch on. Ernst Abbe (1840–1905) was responsible for quantifying the dispersion of optical glass in terms of refractive index versus wavelength (a high *Abbe number* indicates low dispersion) as well as designing many optical devices. In particular, he found a rigorous method of defining the resolving power of a microscope, and designed an inverting prism named after him.

Aspheric Surfaces

The introduction of a single aspheric surface can correct a multitude of residual aberrations, even in a single lens. Many simple cameras are now made with a single aspheric lens that is substantially free from both spherical aberration and coma, the most serious aberrations in a normal-angle lens. In more sophisticated lenses such as the multi-element zoom lens of Figure 4.13, a single aspheric surface takes care of almost all residual aberrations.

Aspheric means simply "not a sphere." The simplest aspheric surface is the surface of an ellipsoid, paraboloid, or hyperboloid of revolution. These cannot be made using conventional methods, and must be either diamond-turned (glass) or molded (plastics).

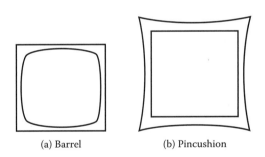

(a) Barrel   (b) Pincushion

Figure 4.12 Distortion: (a) In barrel distortion the magnification decreases with increasing angle of field; (b) in pincushion distortion it increases.

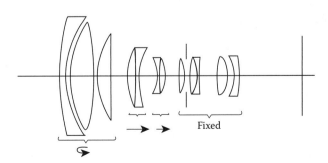

Fixed

Figure 4.13 Arrangement of elements in a typical modern zoom lens (Vivitar). The arrows indicate the direction in which floating groups of elements travel when zooming from 35 to 85 mm.

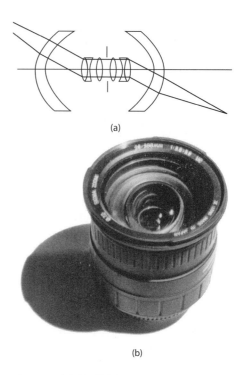

(a)

(b)

Figure 4.14 (a) To reduce cos⁴ fall-off (b) a lens can be designed to deviate the rays to pass less obliquely through the iris diaphragm.

Fall-Off

In any camera, the optical image is sharper at the center of the field than at the corners, as the off-axis aberrations increase as the angle of field increases. Closing down the aperture makes the image sharper up to the point where diffraction takes over, but the required exposure duration can become excessive, and both lateral chromatic aberration and distortion become more obvious at small apertures. The best modern lenses show little fall-off in sharpness over the field they were designed for, but they all show some fall-off in illuminance away from the center of the field. This is called *cos⁴ fall-off*. With a simple lens the fall-off in illuminance is indeed proportional to the fourth power of the cosine of the semi-angle of field. This is for the following reasons:

- The distance of the image plane from the lens is proportional to the cosine of the angle made with the optic axis, so the inverse square fall-off is proportional to $\cos^2\theta$, where θ is the semi-angle of field.
- Owing to the obliquity of the beam, which is spread out in the film plane over an area proportional to $1/\cos\theta$, there is a further fall-off proportional to $\cos\theta$ (Lambert's law).
- The projected area of the aperture itself is reduced proportional to a further $\cos\theta$.

Hence "cos⁴ fall-off." This amounts, for example, to a 75 percent fall-off in illuminance in a 90° wide-angle lens. Lens designers tackle this problem by redirecting the rays in the front lens component so that they pass through the plane of the iris diaphragm more nearly perpendicular to it (Figure 4.14a). In a quasi-symmetrical lens corrected for distortion, this can reduce the fall-off to something like $\cos^2\theta$.

You can observe this effect in a modern wide-angle lens (or a zoom lens set to its shortest focus). If you turn the lens so that you are looking at the front element

Figure 4.15 A fisheye lens gives an increase in peripheral illuminance at the expense of severe barrel distortion.

almost edge on, the iris diaphragm, which you would expect to be reduced to a narrow ellipse, seems to have risen up to face you—a very peculiar effect (Figure 4.14b). This is sometimes termed the *Slussarev effect*, from the designer who originated the idea of introducing coma into the front component and canceling it out with opposite coma in the rear component.

By allowing a large measure of barrel distortion, as in a fisheye lens, fall-off can be almost eliminated, even with a 180° angle of view (Figure 4.15). Fisheye lenses are discussed in Chapter 5.

LENS COATING

Although the basic designs of almost all of our present-day lenses had evolved by the end of the nineteenth century, many potentially good designs could not at that time be used because internal reflections caused flare and "ghost" images of lights and of the iris diaphragm. Losses in a lens with eight glass-air surfaces (such as the Cooke Aviar aerial reconnaissance lens) amounted to 30 percent or more, much of this light reaching the film as contrast-reducing glare.

The proportion of light R reflected directly back from a glass surface at normal incidence is given by Fresnel's law, which states that

$$R = [(n_2 - n_1)/(n_2 + n_1)]^2$$

where n_1 and n_2 are the refractive indices of the first and second medium, respectively. For air and optical glass the figure comes out at about 1/25 or 4 percent. For oblique incidences the amount is somewhat greater. If we coat the glass with a material of intermediate refractive index and do the sums for both interfaces, the total reflectance is lowered to about 2 percent. But we can do better than this. By making the coating just one-quarter of a wavelength thick, the two reflected wavefronts will be one half-wavelength out of phase, and will interfere destructively (Figure 4.16).

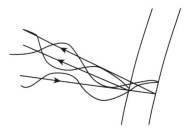

Figure 4.16 When a lens has a low refractive index coating one-quarter of a wavelength thick, the light waves reflected from the two surfaces interfere destructively, suppressing the reflection.

59

Multicoatings matched to three well-chosen wavelengths can reduce reflections to well below 1 percent. This makes it possible to construct lenses with a dozen or more glass-air surfaces, retaining excellent image contrast and freedom from glare. You may still see a row of "ghosts" when a TV camera operator pans too close to the sun. This can happen with simpler lenses, too, though rarely as obtrusively (Figure 4.17, color plate).

Perspective

In general, the term "perspective" refers to the pictorial representation of three-dimensional objects and scenes on a two-dimensional surface. Perspective was well understood by the ancient Greeks and Romans and used in their paintings, but the knowledge was subsequently mislaid, and was only rediscovered in the Renaissance years by Italian painters.

In photography, perspective is best thought of as being concerned with the relative sizes of the images of objects at different distances, and the projected proportions of large, deep objects such as buildings. Although your camera lens records precisely what it sees, and thereby avoids the difficulties facing painters, it can still produce effects that seem somehow wrong. Even professional photography students often don't really understand what is going on in photographic perspective. There is just one simple fact to remember: *in any scene, the perspective depends only on the viewpoint*. It doesn't depend on either the focal length of your lens or on the image format.

Figure 4.17 Almost any lens will produce ghost images if you shoot into the sun.

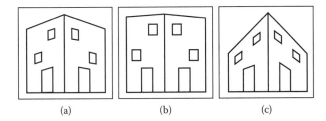

(a)   (b)   (c)

Figure 4.18 Apparent distortion of perspective with change of distance: (a) normal distance; (b) far distance; (c) close up.

Let me give an illustration. Suppose you are photographing a building from a distance of 100 m from its nearest point, and the farthest part visible is 20 m farther away (Figure 4.18). The farthest part will be 120 m away, and from the laws of optics (remember $M = u/f$) it will appear on the image as 100/120 of the height of the front, i.e., 5/6 times the size. If you move your viewpoint back to 1000 m, the proportions will now be 1000/1020 or 50/51, and the angle of the corner will appear flattened. If you change to a long-focus lens to bring the image to the original size, this will not result in any difference in the proportions: the perspective is unchanged. If you now move up close, say 20 m from the nearest point on the building, the proportions will now become 20/40 or 1/2, and the steepness of the corner will be exaggerated. If you substitute a wide-angle lens to get the whole image in, again the perspective stays the same. You can stand in one position with a zoom lens, and make the image any size you like, but you can't change its proportions. Next time you watch a TV play or film, watch what happens when a shot of a person moves in from mid-distance to close-up. If the background enlarges at the same rate as the subject, the camera operator is zooming the lens; if the background goes only slightly larger, and partly disappears behind the subject, the camera is tracking in physically.

Long- and Short-Focus Perspective

There is another important effect that I have already hinted at. It is connected with the way we habitually view a picture. We tend to hold a photographic print so that its diagonal subtends an angle of around 50° to 60° at the eye, which matches what we consciously assimilate visually when we view an actual scene. This represents a distance from the picture roughly equal to its diagonal. A standard (prime) camera lens is designed to cover about the same angle, so that when we look at a print made from the full frame the perspective looks natural.

It is a different matter when the photograph has been taken with a long-focus lens. The objects within the field of view seem to be flattened and piled up one on top of the other (Figure 4.19a). You can notice this especially in TV sports programs where the cameras have to be far away from the action, as in motor racing or cricket, especially in head-on views. It looks as if the cars are very close together, and batsmen seem to be almost running on the spot. That is, if you watch your TV from the usual distance. If you look at it from the bottom of a long garden, the perspective becomes believable.

The opposite applies to short-focus (wide-angle) shots. Here, the perspective seems exaggerated. Near objects seem huge and far objects tiny, and interiors may appear enormous (a dodge often employed by brochures for small holiday hotels with smaller restaurants and even smaller swimming pools). In addition, objects towards the edge of the frame seem stretched out. However, when you view the picture from much closer than usual, the apparent distortions disappear. Photographers call these effects "wide-angle distortion," though the distortion only appears when you view the print from an inappropriate distance (Figure 4.19b).

You can think of it like this: if you were to hold the picture up in front of the scene itself, at your normal viewing distance, every item in the picture would coincide with what you see in the scene. If someone then whipped away the print you would, so to speak, see no difference.

Figure 4.19 The effect of using (a) a very long and (b) a very short focal length lens.

Converging Verticals

A disconcerting effect often seen in amateur photographs of buildings is called "converging verticals." If you have to tilt a camera in order to get the whole of a building into the frame, the top of the building will be farther from the plane of the film than its base, and will appear smaller because its distance from the film plane is greater and the magnification consequently less. Looking at a real building, we don't notice this, because our visual interpretation mechanism takes account of the position of our head, as explained in Chapter 3. In fact, if you take a print with converging verticals and hold it some way above your head to view it, the effect will disappear. Architectural photographers use cameras in which the lens panel can be raised while the lens and film remain upright (Figure 4.20). In this way the film plane remains parallel to the subject, and the verticals don't converge. The same facility has been introduced for 35-mm and medium-format cameras, in the shape of what are called *perspective control* (or simply "shift") lenses, which can be moved in the same way as the lens boards of larger cameras (Figure 4.20a). An example is shown in Figure 4.20b.

Diverging verticals are possible, too. You get them when you take a photograph from the top of a tall building or from the air, with the camera tilted downwards. They are not often seen except in low-level aerial shots of buildings, but we have all seen the unfortunate "lollipop" effect when a society photographer takes an eye-level shot of a celebrity at a party, using a wide-angle lens. The effect can also be seen in horizontals, such as the face of a building to which the camera is not square; this can be compensated in view and studio cameras by a sideways lens or camera back shift. Both types of anomaly can be readily corrected in a digital printing system.

It could be argued that the converging verticals represent correct perspective and that "correcting" by using a rising front actually distorts the perspective; I wouldn't want to enter into an argument about this. What matters is what *looks* right on a print. In practice many professional photographers feel that a small amount of convergence often gives a more realistic image.

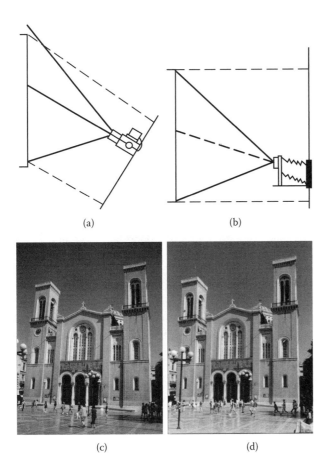

(a) (b)

(c) (d)

Figure 4.20 (a) When a camera is tilted, the top of a building is farther from the film plane than its base. (b) Use of a rising front keeps the camera back parallel to the wall. (c) If you need to tilt the camera to get the top of a building into the picture, the verticals will converge and the building will appear as if tilted backwards. (d) Use of a rising front counteracts this effect, but should not be overdone.

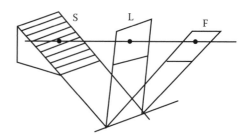

Figure 4.21 The Scheimpflug rule: for overall sharp focus of a tilted subject plane, this plane must meet both the lens plane and the film plane in a single line.

THE SCHEIMPFLUG RULE

Both converging and diverging verticals are part of a phenomenon known as *keystoning*. A keystone is the trapezoidal-shaped stone at the top of an arch, and it matches the shape of the image of a rectangle photographed with the camera axis not perpendicular to the plane of the rectangle. Sometimes an existing image with keystoning needs to be corrected or "rectified." This happens not only with shots of buildings, but also with aerial survey photography, where the camera axis may not be truly vertical for every shot. A rather different problem arises when a professional photographer needs to have a particular item in the image (such as a carpet, or a table laid with food) all in focus simultaneously. The solution to all these problems lies in what is known as the *Scheimpflug rule*, after the person who first worked out its geometry. The rule states that when a camera is used to photograph a tilted plane, that plane will appear in sharp focus only when the planes of the subject, the lens panel, and the film all meet in a single line (Figure 4.21). The proof is an extension of the Newtonian lens laws. Some shift lenses are now made with a tilt facility, enabling the focusing system to satisfy the Scheimpflug rule even on an SLR camera.

Digging Deeper

In this chapter I have been conscious of having had to skim over a great deal of important material. If you feel frustrated by the lack of derivations and proofs for the various formulae, you can find them all (including the Scheimpflug rule) in *The Manual of Photography* (10th edition, Focal Press, 2010), which also contains a fuller account of lens aberrations. Sidney Ray has written a number of books on photographic optics for Focal Press, the most recent being *Applied Photographic Optics* (2nd edition, 1998). Perhaps the most comprehensive book on the subject is the aforementioned *Optics* (4th edition, Eugene Hecht, Addison-Wesley, 2002), which is eminently readable and keeps the maths to the minimum necessary. One of the best books aimed at a more general readership level is *Optics in Photography* by Rudolf Kingslake (SPIE Press, 1992).

Chapter 5 Types of Lenses

The standard lens designed for general use with professional cameras, known as the *prime lens*, has a wide aperture, an angle of view of some 50°–60° to match the width of scene taken in by the average human eye, and all aberrations reduced to a minimum. Most amateur compact cameras are instead equipped with a zoom lens as standard. The telephoto and retrofocus principles were discussed in Chapter 4. However, there are a number of lenses that are designed for more specific purposes. This chapter examines types of specialized lenses (apart from telescope and microscope objectives, which are dealt with in their own chapters) in terms of these specific purposes.

Process Lenses

These are designed to operate at 1:1 magnification, and as they need to be totally free from distortion, they are invariably of symmetrical construction. They also need to be free from lateral chromatic aberration in order to produce color separations that register exactly. They operate at small apertures (*f*/11–*f*/22) and may be equipped with optional elliptical or rectangular Waterhouse stops for specialized halftone work. A typical process lens is shown in Figure 5.1.

A Waterhouse stop is a fixed stop that is inserted in a slot in the lens barrel.

Macro Lenses

These, like process lenses, are designed to operate at around 1:1 magnification, but are not necessarily symmetrical. For large magnifications they may be constructed with a reversible mounting. Macrography is discussed in Chapter 14.

Catadioptric (Mirror/Lens) Systems

The optics of these systems is a hybrid of Cassegrain and Schmidt telescope objectives. They do have all the Seidel aberrations, but these are taken care of by aspheric glass elements called "corrector plates" and field-flattening lenses (Figure 5.2). The

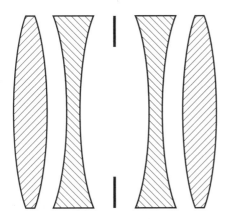

Figure 5.1 A typical symmetrical process lens (Celor type).

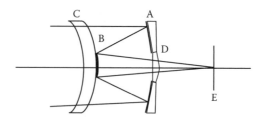

Figure 5.2 A simplified catadioptric system. A and B are the primary and secondary mirrors, corresponding to the positive and negative lens elements in a telephoto lens. C is a correcting element for aberrations, and D is a field-flattening lens. E is the focal plane.

subject of telescope objectives is dealt with further in Chapter 17. A catadioptric system has effectively the same optics as a telephoto lens, but the part of the main lens components is played by mirrors. As the optical path is folded, such objectives are physically short, but have a very long focal length. Another advantage of using mirrors is that they do not suffer from chromatic aberration, and the glass components, being thin, are easy to correct.

"Catadioptric" is a conflation of *catoptric* (mirror system) and *dioptric* (lens system).

Telecentric Lens Systems

A telecentric lens is a lens designed so that its exit pupil is at infinity. The aperture stop is in the principal focal plane, i.e., it is a field stop. Thus all the principal rays (i.e., rays that pass through the optical center of the lens) have to have entered the lens parallel, or very nearly parallel, to the optic axis (Figure 5.3). The image magnification is thus independent of the object distance. This does not mean that there is infinite depth of field—in fact, the depth of field is severely restricted. What it does mean is that the field of view is limited to the diameter of the entrance pupil of the lens system. The main use of these lenses is in machine vision applications, e.g., inspection of manufactured components for quality control. Another form of telecentric lens system consists of two such lens systems positioned back to back so that the rear focal plane of the first coincides with the front focal plane of the second. The stop is placed at their common focal plane. Again, the magnification is independent of the subject depth. The object is located at the front focus of the first lens, and the image is formed at the rear focus of the second lens (a *confocal* system). The angle of field is again very small, and the chief use of this system is in scanning micrography (see Chapter 18).

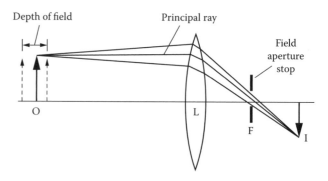

Figure 5.3 Telecentric principle. The aperture stop is set at the principal focus, so that the only rays that can form an image are those close to the principal ray, i.e., the rays parallel to the optic axis. The depth of field is very small.

Ultrawide-Angle Lenses

Any lens that takes in an angle of view of more than about 70° comes under the heading of *wide-angle lenses*. They are subclassified into semiwide-angle (70°–95°), wide-angle (95°–120°), and ultrawide-angle (more than 120°). A special type of this last genre, covering a full 180°, is called a *fisheye lens*, and is dealt with below.

One of the earliest ultrawide-angle lenses was the Goerz Hypergon, which had a field of 130° at *f*/22 with correct drawing. It consisted of a pair of uncorrected hemispherical meniscus glasses, in which the otherwise appalling fall-off in illuminance towards the image edges was more or less taken care of by a little star-shaped windmill swung in front of the lens for half the exposure time, and blown round by an air bulb (or, more often, by the photographer's breath) (Figure 5.4a). Another bizarre manifestation of panoramic lens design was a completely spherical bowl filled with water, and a central fixed stop with butterfly baffles (not shown) to reduce the fall-off (Figure 5.4b).

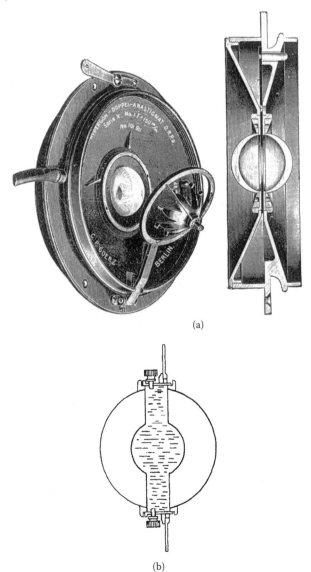

(a)

(b)

Figure 5.4 (a) The 130° Goerz Hypergon lens of 1850. (b) The Sutton water-filled panoramic lens.

Most modern wide-angle lenses for conventional photography are *rectilinear*, i.e., free from distortion. The maximum angle of view for rectilinear wide-angle lenses is about 140°, but at such angles there is severe illumination fall-off, even with the compensations described earlier.

Fisheye Lenses

If distortion is less important than covering the widest possible field, the angle of view can be increased to 180°—and beyond. This is the realm of the *fisheye lens*.

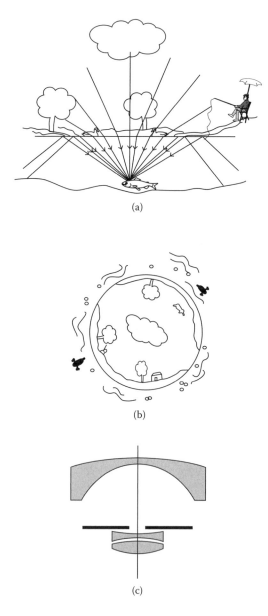

(a)

(b)

(c)

Figure 5.5 What a fish sees. (a) The geometry; (b) the view. The angle of view (about 106° for fresh water) is known as "Snell's window." (c) The Robin Hill Sky lens of 1924, the first true fisheye lens.

> ## WHY "FISHEYE"?
>
> This takes us back to basic optics. A fish at the bottom of a stream has an effective angle of view of 180°, though its visual field may be much less. The critical angle of refraction for water is $\tan^{-1} 1.33$, which is approximately 53°. So within a visual field angle of 106° it can see everything that is above the surface, albeit with much distortion towards the edges. As a bonus, everything beyond the critical angle undergoes total internal reflection, so outside the disc of view it will see a reflection of the bottom of the stream, as well as the underside of any swimming bird (Figure 5.5).

The first true fisheye lens was designed in 1924 by Robin Hill for the Meteorological Office, for cloud studies. It was basically a standard camera lens with an enormous negative meniscus lens in front, producing an extreme retrofocus configuration. As it was to be used with a deep red filter it was not color corrected and was operated at a range of small apertures from $f/16$ to $f/32$. Modern fisheye lenses are designed on a similar principle. The inevitable distortion follows one of three regimes: *equidistant* (radial distance $r = f \times \theta$), *orthographic* ($r = 2f \sin \frac{1}{2}\theta$), or *rectilinear* ($r = f \tan \theta$). These types of distortion are shown schematically in Figure 5.6. The rectilinear geometry has $r = \infty$ for $\theta = 90°$ and is effectively free from distortion but cannot manage 180°, so is not a true fisheye, though it is often so described. Most true fisheye lenses provide orthographic projection.

Panoramic Lenses

The original "panorama" was a cylindrical chamber painted with a 360° landscape (Figure 5.7). As you stood in the center and turned round you could take in the whole scene, in correct perspective.

Panoramas with a Conventional Camera Lens

If you have a camera with a shift lens, you can take a pair of photographs (with a tripod) without moving the camera, but by moving the lens shift between exposures to its two sideways extremes. You will then have a pair of images covering 100° or more, and you can butt-join the prints with a perfect match. The perspective will still show the characteristic stretching of near objects at the edge of the field, and this will appear (to a small extent) in the middle of the composite image as well, but it shouldn't be apparent in landscapes lacking near objects. Alternatively, you can use an ordinary lens and swing the camera between exposures. This time you may find some difficulty in matching the edge detail, especially in the foreground if a straight line (such as a house roof) appears in two adjacent images. You can minimize this by giving about 50 percent overlap, and cropping each image down the halfway overlap point. But you can't get rid of it entirely unless you are prepared to feed the images into your computer, modify them, and stitch them together using Photoshop or similar software.

True Panoramic Systems

The term "panoramic" seems to have become stretched in general parlance to mean any format with an aspect ratio greater than about 1.5:1. True panoramic optical systems cover any angle up to and including 360°, and fall into two categories,

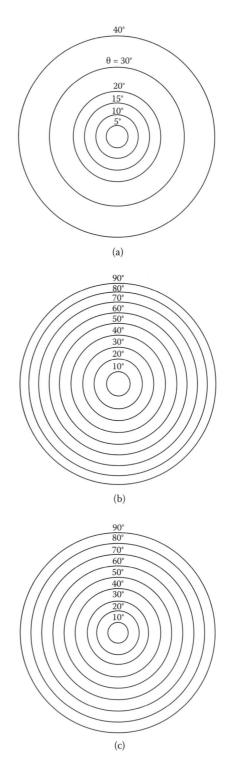

Figure 5.6 Fisheye lens geometry: (a) Rectilinear ($r = f \tan \theta$); (b) orthographic ($r = 2f \sin \frac{1}{2}\theta$); (c) equidistant ($r = f\theta$).

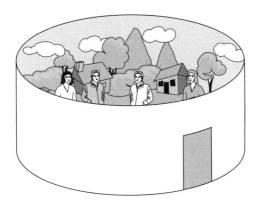

Figure 5.7 A panorama was originally a room painted with a 360° scene.

both employing the slit principle. In the first the film is drawn round the inside of a cylinder centered on the rear nodal point of the lens. The lens itself is pivoted about its rear nodal point, and a narrow slit close to the focal plane is linked to the lens mounting. During the exposure the lens system rotates through its maximum swing angle, about 150°. In the second system the entire camera rotates about the rear nodal point of the lens. The slit is fixed, and the film is advanced in the same direction and at the same speed as the image movement. These two systems are shown schematically in Figure 5.8.

In a panoramic photograph all parallel horizontals record as hyperbolas. This is why it is difficult to fit parts of a house roof together at the edges of a panoramic montage of a town taken from high ground using conventional optics. In theory you would have to take an infinite number of overlapping shots to get them to fit perfectly—and you would still finish up with hyperbolas. Incidentally, you make things worse if you don't keep the horizon centered in the frame. True panoramic views, distortion and all, have recently become popular for display purposes (Figure 5.9).

360° Lenses

A number of lenses that cover a 360° panorama in one shot have been devised for special purposes. The optics system may be a two-stage affair involving a toroidal glass optic and giving multiple contiguous images, or a combination of a more or less orthodox lens with a convex paraboloidal mirror, the axis being vertical (Figure 5.10). A lens system of the latter type is available commercially.

Lenses for Aerial and Satellite Photography

Aerial Reconnaissance Photography

At one time high-level vertical aerial photography was probably the most indispensable component of military intelligence. Even after the emergence of satellite photography in the 1960s it remained an important element for mapping purposes. Reconnaissance aircraft carried fans of four or even six long-focus cameras, covering a swathe some 50°–60° wide, with a lateral overlap of 20 percent and a longitudinal overlap of 60 percent. An additional wide-angle camera took overall covering images, acting like the box lid of a jigsaw puzzle, to enable the identification of the relative positions of the individual images which, of course, were all on

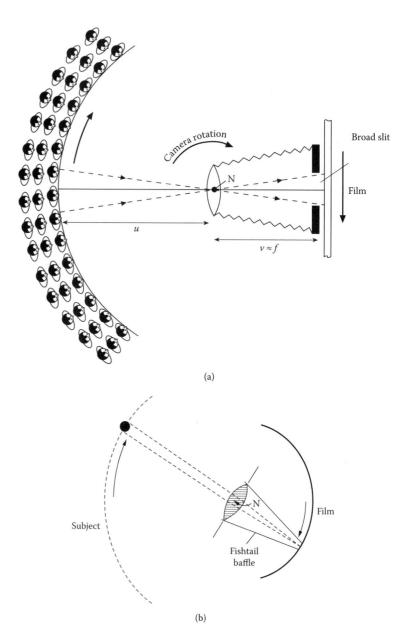

(a)

(b)

Figure 5.8 (a) Swivelling panoramic camera. N is the rear nodal point of the lens. (b) Swing lens camera. The curved film is stationary and the lens rotates about its rear nodal point.

Rectifying printers were scarce in those days. Mostly we just made contact prints from the negatives, soaked them in hot water, and then stretched them until they fitted.

A narrow-angle lens can have the near-axis aberrations (spherical and chromatic aberration and coma) very fully corrected, as the more peripheral aberrations (astigmatism, curvature of field, and distortion) are less important. Emulsion graininess is also less important with a large format. The late George Brock, doyen of Second World War British aerial camera design, used to assert that there was *no* substitute for focal length.

separate films. In the interpretation phase the prints were examined in pairs using stereoscopic viewing systems. If a laid-down mosaic was required, owing to the tilt of the outer cameras it was necessary to rectify the keystoning effects.

Aerial reconnaissance camera lenses for high-level photography typically have a focal length of a meter or so, with a format of up to 250 mm square. The reason for flying reconnaissance sorties at high altitudes was not so much to keep out of range of anti-aircraft missiles, as to make use of what became known as the "ground-glass effect." If you are flying just above the haze level (about 800 m) the haze fills the whole intervening space, and the effect is similar to trying to read a newspaper through a piece of ground glass. The scattered light reduces the contrast to virtually nil. But at a height of several kilometers, nearly all the intervening space is clear,

Figure 5.9 Distortion in a swiveling panoramic camera. The pupils are in a circle centered on the camera, and the horizontal lines of the building behind have become hyperbolas. (The author—aged 12—is in the back row, eleventh from the right.)

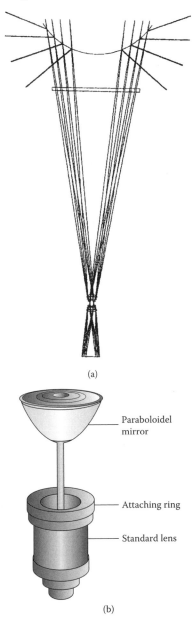

(a)

Paraboloidel mirror

Attaching ring

Standard lens

(b)

Figure 5.10 Lens attachment giving a 360° horizon via a paraboloidal mirror: (a) schematic optics; (b) complete lens system.

corresponding to positioning the ground glass directly on top of the newsprint: the scattered light near the ground largely misses the camera lens. A deep yellow (minus-blue) filter helps to eliminate what is left of the bluish haze.

High-level aerial reconnaissance photography has been superseded by satellite photography, but low-level tactical photography is still carried out, the very short exposures demanding wide-aperture lenses with image motion compensation at the film plane.

Aerial Survey Lenses

The most important attribute of a survey lens is that it should give absolutely correct drawing, and survey cameras are fitted with a register glass bearing a grid specifically calibrated for the lens used in that camera. A graduated filter, densest at the center, may be added to the register glass or to the front of the lens to compensate for illuminance fall-off. There are a number of lenses specifically designed for survey work, mostly based on the quasi-symmetrical Metrogon design (Figure 5.11), with an angle of view (corner to corner) of some 90°. Less formal surveys are often carried out from light aircraft using standard commercial mid-format cameras with lenses selected for freedom from distortion.

Satellite Photography

During the Cold War period high-level reconnaissance photography reached its ultimate stage, with aircraft flying at heights of 20 kilometers and more, but already

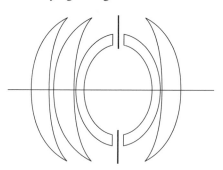

Figure 5.11 Metrogon quasi-symmetrical aerial survey lens.

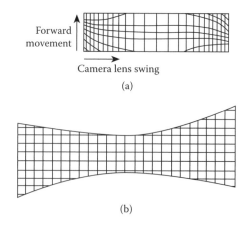

Figure 5.12 Bow-tie distortion combined with forward movement compensation in a swing-lens satellite camera. (a) Distorted grid; (b) shape of corrected area.

satellites were taking over the job. The photography itself was conventional. The satellites operated at heights of 100–200 km, with very long-focus lenses and ultra-fine-grain film, which was ejected after exposure and retrieved by parachute. The authorities have been understandably cagey about the performance of these cameras, but assuming that the lenses were good enough to be aperture-limited at about *f*/8, with an emulsion resolving power of about 200 lp/mm, it would be theoretically possible to pick out objects only about 15 cm across. With these very long-focus lenses (up to 3 meters focal length) an area of only a few square kilometers is imaged at each shot, and more general views are taken using a panoramic slit camera. This produces the usual hyperbolic distortion, and the image is also twisted owing to the forward motion of the satellite (Figure 5.12a). This is corrected by special printing optics matched to the camera system, giving a bow-tie shape to the print (Figure 5.12b).

Much important satellite survey work is now done by means of *multispectral imaging*. Subjects can be identified in the images and analyzed by their spectral signature. The photography can be as simple as the one-shot color camera (see Chapter 7), or as complex as a synchronized array of nine cameras equipped with narrow-band filters covering ranges between 400 and 900 nm. Satellite photography is now invariably carried out using digital sensors and transmission by radio.

Afocal Lens Systems

There are three categories of afocal systems, used as attachments to standard camera lenses: telephoto, wide-angle, and anamorphic. The telephoto attachments increase the focal length of an existing lens system, usually by ×1.5 or ×2, and are simple Galilean telescopes. Wide-angle attachments are in the form of a reversed Galilean telescope, and decrease the focal length similarly. Figure 5.13a and b show the way the optics operate in each case when used in front of the camera lens. In many modern systems they are inserted behind the lens, i.e., in the image space, and in such cases the order of the components is reversed. Anamorphic systems squeeze the image in a single direction, and are used to produce images that can be expanded using a similar system, for projection in wide-screen format. One

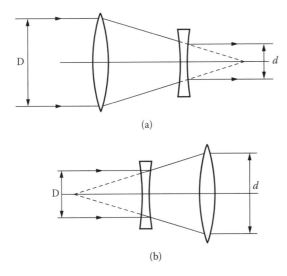

Figure 5.13 Afocal lens attachments: (a) telephoto; (b) retrofocus. These configurations apply to front fitting. D/d is the magnification. When fitted behind the main lens they are reversed.

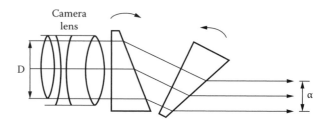

Figure 5.14 Afocal anamorphic attachment using prisms. Rotating the prisms alters the squeeze ratio D/d.

form is similar to the wide-angle attachment except that it has cylindrical instead of spherical components. The other is more sophisticated, and consists of a pair of achromatic prism combinations (Figure 5.14). Altering the angle of these changes the squeeze ratio for different wide-screen formats.

Lens Systems for Underwater Photography

We have become so accustomed to watching superb examples of underwater photography in television documentaries that we tend to forget its technical difficulties (not to mention the hostility of the medium and of some of its residents). The photography itself is hampered by the turbidity of the water and the poor lighting conditions. The main optical problem is that we are no longer working in air, and the refractive index of the water has to be taken into account in the design of the optical system.

If you are in the habit of swimming without goggles or a face mask, you will be aware of the impossibility of seeing underwater objects sharply without an airspace in front of your eyes. Seawater has roughly the same refractive index as your corneas, so that there is no refraction at the front surface; the result is that you become impossibly long-sighted (Figure 5.15a). The flat surfaces of swimming goggles put an air gap in front of your corneas, and restore the nodal points of your eyes' optical systems to their usual position: you can now see clearly. But there is still a difference from viewing in air (Figure 5.15b). There is a kind of reverse fisheye effect, the rays emerging from the flat port being deviated by refraction in such a way that objects appear nearer, by a factor roughly equal to the refractive index of the water (about 1.34). They also appear correspondingly larger.

Optics is not the only problem. At depths as little as 1 meter, roughly 50 percent of red and orange light is lost, and from 10 meters down there is little else but blue-green light. Horizontal visibility is often as little as 1 meter. And although snorkeling around Greek islands with a waterproof disposable camera at only a meter or two depth is great fun (I do it myself) and can produce some striking pictorial effects, it is a very limited area of the technology.

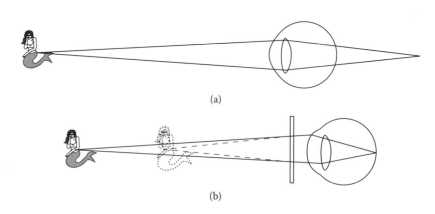

(a)

(b)

Figure 5.15 Seeing underwater. (a) An underwater swimmer without goggles is excessively longsighted. (b) Flat-fronted goggles restore normal focusing, but the flat plate introduces a reverse fisheye effect that makes the object appear nearer.

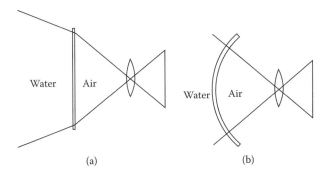

Figure 5.16 (a) A flat camera port gives the same exaggeration as swimming goggles, and the reverse fisheye effect causes pincushion distortion and limits the angle of view. (b) A domed port eliminates the distortion but behaves as a negative lens, reducing the apparent object distance.

Some underwater housings for cameras have optically flat ports, and these not only produce this exaggeration of the image, increasing the effective focal length of the lens by one third, but also limit the field to about 40° for a lens of normal focal length. Because of the reverse fisheye configuration, there is pincushion distortion as well (Figure 5.16). Lateral chromatic aberration is also an important factor limiting peripheral resolution.

A *spherical* or *dome port* is a partial solution. Centered on the front nodal point of the lens, it eliminates the reverse fisheye distortion as well as chromatic aberration. But the curved dome also acts as a concave lens with a negative focal length about three times its radius, and forms a somewhat close virtual image of the subject, on which the camera lens has to focus. The final image has severe curvature of field. Including a positive supplementary lens can alleviate these problems. It is, of course, possible to compensate by substituting a large positive meniscus lens for the dome. A more elegant solution is to build the dome into the front of the camera lens itself, and this is now standard practice in modern specialized underwater cameras.

GRIN Lenses

These are a kind of longitudinal lens, a rod with *G*raded *R*efractive *IN*dex. The refractive index is graded so that it is higher at the periphery than at the center of the rod, and light passing into the rod is turned inwards towards the center and converged to a focus, which may be at or beyond the end of the rod. Long GRIN rods may have many foci along their length, and can be used as image relay systems in optical probes (Figure 5.17). It is even possible to make conventionally shaped lens elements with graded index. Such lenses can correct spherical aberration without having to be aspherical.

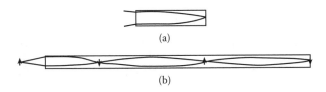

Figure 5.17 (a) GRIN (graded refractive index) rod lens focusing at one end; (b) long GRIN rod acting as a series of relay lenses.

Diffractive Optical Elements (DOEs)

As the name implies, these devices operate by diffraction, not refraction. They can be manufactured holographically, or by computer-controlled engraving. Many optical elements can occupy the same space, so that beams can be combined or split as well as focused. DOEs have considerable potential in optical computing systems, but owing to their high dispersion they are, in general, suitable only for monochromatic (laser) light. However, as the dispersion is in the opposite sense to that of optical glass, this property can be used to counteract chromatic aberration in a conventional optical system. DOEs are discussed further in Chapter 18.

Pinhole Photography

I have included pinhole optics here because a pinhole does indeed have many of the characteristics of a lens. If its inclusion here seems to you to be inappropriate, just think of it as a lens with a very small aperture and a very long focal length.

Pinhole photographs are undoubtedly the cheapest form of photography, needing no more than a cardboard box, a milk bottle top, some photographic film or paper, and double-sided adhesive tape to hold the sensitive material in place. Pinhole cameras can earn their keep, though. They used to be standard equipment for imaging cosmic X-ray sources; they are used for photographic records of aircraft cockpit layouts, and for demonstrating images of solar eclipses. In the sixteenth century a pinhole image was the catalyst that resulted in the adoption of the Gregorian calendar.

In the Tower of Winds in the Vatican, a pinhole in the South Roof casts a solar image that crosses a meridian mark at noon every day. By 1582 this pinhole image showed that the calendar date of the spring equinox had become more than 10 days early. Pope Gregory decreed that 5 October that year would become 15 October, and that a new properly corrected calendar would be inaugurated.

In the field of fine art, the pinhole camera assisted Brunelleschi and his contemporaries in the rediscovery of the laws of perspective. Some modern photographers have become intrigued by the odd and interesting characteristics of the pinhole photograph, and have exploited its properties pictorially. In 1996 the Royal Photographic Society, in its annual print exhibition, made one of its major awards to a pinhole photograph (Figure 5.18, color plate). There is at least one society dedicated to pinhole photography (see Digging Deeper).

Figure 5.18 (See color insert following page 154.) Pinhole photograph by Justin Quinnell: the Royal Crescent in Bath, with fingers. The depth of field extends from near infinity to about 3 cm.

So what does a pinhole camera have that a conventional camera doesn't have?

- It has infinite depth of field.
- It has an extremely wide angle of view.
- The overall slight unsharpness of the pinhole image is aesthetically more acceptable than the unsharpness seen in out-of-focus areas in a lens image.

A pinhole camera, simple as it is, has some optical details that are worth examining.

Optimum Pinhole Size

In conventional geometry, a point object at infinity should produce a uniform disc of confusion the same size as the pinhole. This implies that the smallest pinhole produces the sharpest image. But in practice, as the pinhole becomes smaller the effects of diffraction take over. Now, you will remember that the formula for the diameter d of the central disc of the diffraction image for a circular aperture (the Airy disc) is given by

$$d = 2.44\lambda N$$

where λ is the wavelength of the light and N is the f/no (the f/no of a pinhole is the distance between the pinhole and the film plane divided by the pinhole diameter, distances being expressed in the same units). The two functions (i.e., pinhole function and Airy diffraction pattern) have to be convolved, that is, the Airy disc has to be dealt out to every point on the pinhole function. This isn't a simple task, but as an approximation we can simply make the diameter of the Airy disc equal to the pinhole diameter c. Then if f is the distance from the pinhole to the film (all measurements in meters),

$$c = 2.44\lambda \times f/c$$

or

$$c^2 = 2.44\lambda f$$

i.e.,

$$c = \sqrt{(2.44 \times 0.55f)} \times 10^{-3} \text{ m}$$

$$= \sqrt{(1.342f)} \text{ mm (Remember that } f \text{ is in meters.)}$$

Figure 5.19 shows the variation of optimum pinhole diameter with image distance.

Perhaps surprisingly, pinholes have aberrations the way lenses do:

1. *Chromatic aberration.* As red light is diffracted more than blue light, there will be both longitudinal and lateral chromatic aberration. Small points of light in the central area will show reddish outer edges, and in outer parts of the field they will show a red fringe on the outer side and a blue fringe on the inner side, the effect being proportionally greater for smaller pinholes and greater field angles.
2. *Astigmatism.* The oblique view of the pinhole from the edges of the film is an ellipse with its short axis radial from the optical center. There is therefore more diffraction radial from the optical center than tangential to it. As with lateral chromatic aberration, this effect is proportional to the angle of field.
3. *Cos⁴ fall-off.* You can't compensate for this effect, which follows directly from the obliqueness of the pinhole and the increased distance to the edges of the film. It is aggravated if the pinhole edges are thick, as you then have the equivalent of lens barrel vignetting.
4. *Coma, curvature of field, spherical aberration and distortion.* These are all, mercifully, absent.

The formula recommended by Rayleigh in 1891 was $d = 2\sqrt{(f\lambda)}$, which gives a pinhole with roughly 1.3 times the diameter of mine. Take your choice!

79

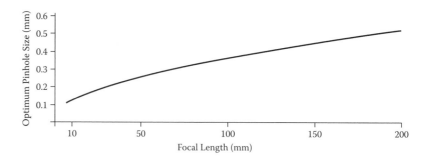

Figure 5.19 Optimum pinhole sizes.

Figure 5.20 A pinhole camera made from a drink can. (Photograph by Justin Quinnell.)

To keep the exposure down you need to use film (or a digital sensor) of at least ISO 400. With a pinhole of effective aperture f/250, the exposure in bright daylight should be around six seconds, but if you are using the full 160° angle of view that a well-made pinhole is capable of, you will need at least four times this exposure. The best camera for pinhole photography is a technical camera with a pinhole cemented to a spare lens board. However, the true pinhole enthusiast uses whatever comes to hand, from cereal packets to soft drink cans (good for semicircular panoramas) (Figure 5.20).

MAKING A PINHOLE

The best material is shim brass, the thinnest you can find, but the aluminium sheet from milk tops or takeaway meal trays is almost as good. Having worked out your optimum pinhole size, find a needle of the right diameter (measure the shank with a micrometer), and push the eye end into a cork. Place the shim on a piece of glass, position the needle on it and give the cork a sharp tap with your fingers. Turn the shim over and polish the bump with an emery board. Now, using a magnifying glass or watchmaker's loupe, center the needle on the bump and repeat the tap. Turn the shim over again and polish it with the emery board. Continue this procedure until you have made a tiny hole. Now enlarge the hole cautiously, twirling the needle and polishing down any burrs, until the needle goes through cleanly to its shank.

To check the quality of the pinhole, shine a laser pointer through it in a darkened room onto a card held some distance away. You will see a small Airy pattern. It should be perfectly circular, with no irregularities. (If you haven't a laser, a powerful focusing torch will do, but the patch will show color fringes). Once the pinhole is to your satisfaction, blacken it by holding it very briefly in a candle flame.

Digging Deeper

In spite of the popularity of panoramic photography, there is little literature dealing with it, though there are plenty of coffee-table books of photographs of this type. Many articles about panoramic photography have appeared in journals, notably in the *British Journal of Photography*. Joseph Meehan's *Panoramic Photography* (Amphoto, 1990) recently went out of print. Sidney Ray has a comprehensive chapter on the subject in *Scientific Photography and Applied Imaging* (Focal Press, 1999). Pinhole photography has had a good innings in the past, but most of the contemporary literature is unhelpful. *Pinhole Photography*, by Eric Renner (2nd edition, Focal Press, 1999), is a notable exception. Although a little short on theory, it is full of imaginative ideas. There is an annual Worldwide Pinhole Day in the last week of April (details at www.pinholeday.org/gallery/index.php). High-level aerial and satellite photography is an awkward area, as it is easy to pick up a promising book only to find that it contains a collection of photographs from inner space with few, if any, clues as to the equipment or even the exposure used. An old book well worth hunting for is N. Jensen's *Optical and Photographic Reconnaissance Systems* (Wiley, 1968), written in the heyday of high-level aerial reconnaissance. A more recent book along similar lines is C. Etachi's *Introduction to the Physics and Technology of Reconnaissance and Survey* (Wiley, 1987). The most recent is S. Drury's *Images of the Earth: A Guide to Remote Sensing* (Oxford University Press, 1998). G. Thomson wrote a regular report on satellite reconnaissance for the *Imaging Science Journal* (formerly the *Journal of Photographic Science*) for a number of years between 1992 and 1999. Do-it-yourself aerial photography is catered for by Kodak's Publication O–27, *The Kodak Guide to Aerial Photography*, by Barrie Rockeach (Silver Pixel Press, 1996). You will find a more thorough treatment in W. S. Warner, R. W. Graham, and R. L. Read's *Small Format Aerial Photography* (Whittles Publishing, 1996). For a comprehensive review of the development of photographic lenses from the beginning of photography, you can do no better than the late Rudolf Kingslake's authoritative (if expensive) *A History of the Photographic Lens* (Academic Press, 1987).

Chapter 6 Resolution in
Optical Systems

In Chapter 4 I asked, "How sharp is 'sharp'?" and suggested that for practical purposes "sharpness" was self-defining: if an image looked sharp, it was, *ipso facto*, sharp. If you felt that this was an unsatisfactory answer, you were right. But even if the detail of your image is sharp beyond this subjective criterion, if you keep on enlarging it you will eventually reach a point where further enlargement won't show any more detail, but will simply enlarge the blur. Microscopists call this condition "empty magnification." The point where this begins to happen is called the *limit of resolution* of the optical system.

You can obtain a rough estimate of the limit of resolution of a camera system by photographing a grid of parallel lines that reduce in spacing with distance across the image, such as the set of park railings seen obliquely in Figure 6.1. The spacing at which you can no longer distinguish the separate railings even under a magnifying glass defines the limit of resolution. Park railings are not a very scientific kind of object, of course. A standard test object of a similar type, called a Sayce test chart, is shown in Figure 6.4a. The resolution of an optical system is usually expressed in terms of its reciprocal, *resolving power*, traditionally stated in *line pairs per millimeter (lp/mm)*, a line pair being a black and a white bar together.

In the real world, that is. I once saw a courtroom-drama film where the crucial piece of evidence was a snapshot of a street scene, blown up to the point where the date on a newspaper held by a man some 50 meters away was clearly readable. In Hollywood anything is possible.

Testing for Resolving Power

The Sayce chart is not well suited to practical applications. Real subject matter seldom includes long parallel bars of high contrast. In practice, objects are usually chunky and fairly low in contrast. This is especially true in aerial and satellite photography, where resolution is very important. The standard test object for optical

Figure 6.1 Park railings seen in perspective represent a one-dimensional object with spatial frequency that increases across the negative.

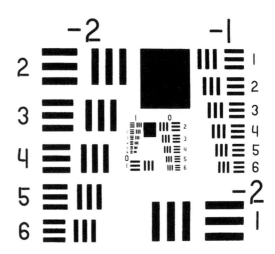

Figure 6.2 U.S. Air Force test target.

resolution was developed by the U.S. Air Force during the Second World War. Each unit consists of three parallel bars with an aspect ratio (i.e., the ratio of width to height) of 1:5, equally spaced, in pairs at right angles to one another (Figure 6.2). A set contains 20 or more of these units in regularly decreasing size, and an array of these is set up so that it fills the field of view of the optics under test. The test object may be of high or low contrast as required, and the limit of resolution is taken as the line pair figure for the smallest block in which the direction of the bars can be distinguished.

Diffraction Limitation

For large lens apertures the limit of resolution of a lens is governed by the residual aberrations, but at small apertures the resolution of a well-corrected lens is limited by diffraction. If the aperture is circular, the diffraction pattern produced by a point object is an Airy pattern (see Figure 1.12c). More than 80 percent of the light intensity is in the central patch, the *Airy disc*.

A camera lens is designed, as a rule, to let as much image-forming light through as possible, so at its maximum aperture there will be some residual aberrations present. With high-quality prime lenses having a maximum aperture of *f*/2.5 or so, the aberrations will have all but disappeared by around two stops down from maximum aperture, by which point the Airy disc will be of a size that matches the diameter of the blur disc resulting from the residual aberrations. This is the point at which the lens is said to be *aperture limited*. At smaller apertures the Airy disc takes over as the main limitation to resolution. Very large aperture lenses (*f*/1.2, say) may become aperture limited only at three or even four stops down. Process lenses are designed to give their best performance at around *f*/22 (this is to allow the halftone screen optics to operate correctly). The older types of wide-angle lenses for large-format cameras also need to be used at a small stop in order to give sufficient coverage, but this kind of performance is unacceptable in modern short-focus lenses for 35-mm and the equivalent digital cameras. High-quality long-focus lenses usually give their best performance at maximum aperture. The intensity profile of the image of a point is called the *point spread function (PSF)*, and its shape changes as the lens is stopped down, from the aberrated image at full aperture to an Airy pattern at small apertures (Figure 6.3).

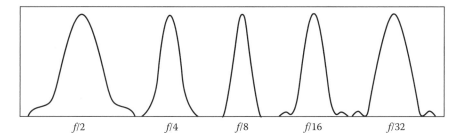

$f/2$ $f/4$ $f/8$ $f/16$ $f/32$

Figure 6.3 Profile of a point spread function with decreasing aperture in a camera lens.

The Rayleigh Criterion

When observing double stars through a telescope, Rayleigh suggested that two stars would just be distinguished as visually separable if the center of the Airy pattern formed by the image of one star fell on the first zero of the Airy pattern of the other. You will remember from Chapter 1 that the radius of the Airy pattern to its first zero is $1.22\lambda N$, where λ is the wavelength of the light and N is the f/no of the optical system. This is the *Rayleigh criterion* for the resolution of a diffraction-limited optical system.

In the early days of photogrammetry it became apparent that the figures found in practice for optical resolution didn't always fit the perceived sharpness of an image, particularly at the edges of details. The concept of *acutance* evolved with reference to the sharpness of image edges. This could (with some difficulty) be quantified, but was generally treated simply as high, medium, or low.

The Inadequacy of Resolving Power Measurements

Aerial reconnaissance played a vital part in the conduct of the Second World War. A group of specialists, known as photo-interpreters, examined the details of reconnaissance photographs under high magnification, and as stereoscopic pairs when possible. It quickly became clear that the resolving power criterion for an optical system was only one factor in the successful interpretation of detail. Indeed, some lenses produced imagery that was more readily interpretable than that produced by others of higher measured resolving power. Some lens types produced actual distortions in the fine detail.

Lens designers laboriously tracing the paths of light rays, and making allowance for the effects of diffraction as well as estimating engineering tolerances, were able to predict the shape and size of the image of a point of light in various parts of the field, and thereby to foretell something of the behavior of the optical system under realistic conditions. The PSF, which at that time was called a "star image," was, however, difficult to quantify mathematically.

The Modern Approach to Image Quality

The solution appeared in the early 1950s, following pioneering work by E. W. H. Selwyn in England and, independently, P. M. Duffieux in France. Both had noted that in the image of a Sayce-type chart, as the spatial frequency increased the intensity profile became markedly sinusoidal and the contrast fell progressively to zero at the resolution limit (Figure 6.4a).

John William Strutt, third Baron Rayleigh (1842–1919), was an outstanding figure in both astronomy and physics. He succeeded Maxwell as Cavendish Professor of Experimental Physics at Cambridge University, and was elected President of the Royal Society in 1905. He discovered the element argon, for which he received the Nobel Prize for Physics in 1904, and gave a scientific explanation for the blue color of the sky. He was the first person to point out the inconsistency in the (then) laws of radiation that led to Planck's quantum model. His work on diffraction was only a small part of his contribution to the advancement of science.

Photogrammetry is measurement carried out through photographic imagery. In silver halide photography, film-processing methods play a large part in the acutance of a photographic image, and in digital cameras the acutance can be controlled to some extent (this is more often done at the print stage). The human retina also possesses an edge-enhancing mechanism (see Chapter 3).

A trained photo-interpreter was usually aware of these optical idiosyncrasies, and would compensate for them mentally. The successful identification of the first V-1 flying bomb by Constance Babington-Smith, from a tiny smudge on a high-level reconnaissance photograph, was an outstanding example of the photo-interpreter's craft—and of incalculable value to the country.

An alternative chart known as the *Siemens star* is used for preference nowadays, as it is particularly good at showing up aliasing effects such as moiré and colored artifacts in digital images (Figure 6.4c).

Selwyn's work was not published at the time as his wartime research was cloaked by the Official Secrets Act, so the credit has usually gone to Duffieux, although his book, *L'Intégrale de Fourier et ses applications à l'Optique*, actually acknowledged Rayleigh as the true originator of the Fourier approach. In turn, Rayleigh's published work refers back to Airy, Helmholtz, and earlier. Shoulders of giants, again!

In some cases the image appeared to be again resolved at a still higher spatial frequency, but with the positions of the light and dark bars interchanged. This is typical of lenses that are slightly out of focus, and is called *spurious resolution*. Figure 6.4c and d show the effect with Sayce and Siemens objects. A test chart with bars having a density profile that varied sinusoidally rather than a "square" profile showed the contrast still falling as the spatial frequency increased, but the bars remained sinusoidal and there was no sign of spurious resolution. It was clear that there was something fundamental about a sinusoidal bar pattern.

ANALYSIS OF A SQUARE WAVE

If you connect the output of an audio signal generator set to "square wave" to a loudspeaker, you will hear not the bland tone of a pure sinewave, but a raspy timbre reminiscent of a badly played clarinet. If you listen carefully you may hear a distinct note an octave and a half higher than the basic (fundamental) note, and if you have a keen ear and the fundamental note has a low pitch you may hear further tones at still higher pitches. Unlike the eye, which perceives a combination of several discrete wavelengths as their mean, the ear is an analytical organ that is capable of separating out individual frequencies. A square wave can be assembled from a series of sinewaves of frequencies 3, 5, 7, 9, etc., times the fundamental frequency, with amplitudes that decrease successively in the same proportions and alternately reversed phase (Figure 6.5).

This is known as *Fourier synthesis*. Its converse, the teasing out of the frequency components of a complex wave, is known as *Fourier analysis*. By putting your signal generator output through a variable low-pass filter and suppressing the higher frequency components one by one, you will finally be left with a pure sinewave at the fundamental frequency. Appendix 3 looks further into the principles underlying the Fourier approach to imaging.

This conceptual breakthrough led to a revolution in optical thinking, and a new era in optics technology was born. It embraced two major insights:

1. Any linear optical system (i.e., a lens free from flare and a light receptor with a linear response) will always produce a sinusoidally varying image from a sinusoidally varying object grating.
2. Any scene can be considered as the sum of a spectrum of sinusoidal gratings of varying spatial frequency, amplitude, phase, and orientation.

This means that if you set up an optical system using an object consisting of a sinusoidal grating of constant amplitude (i.e., constant maximum density) and progressively increasing spatial frequency with distance (in other words, a sinusoidal version of a Sayce chart or Siemens star), and record the amplitude and spatial phase (relative positional shift) of the resultant, you will be able to deduce the response of the system to both coarse and fine detail, and will also be able to predict the extent of any image distortion within the fine detail. In practice this information is plotted graphically as a function of the spatial frequency of the image of the test object.

Modulation

Contrast is a difficult concept to put into figures. The most obvious way would seem to be to define it as the ratio of the luminances of the brightest and darkest areas of the subject matter. This is misleading, however, as a contrast of 1000:1

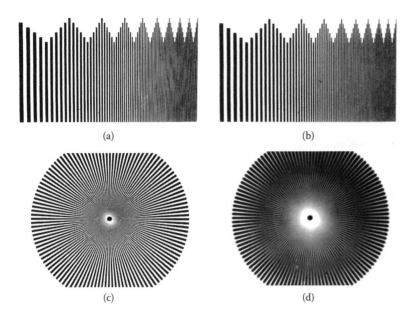

(a) (b)

(c) (d)

Figure 6.4 (a) Sayce chart. (b) The same chart photographed at a very small aperture. Notice the way the density profile of the bars becomes increasingly sinusoidal and the contrast decreases as the spatial frequency increases. (c) Siemens star test object. Notice the moiré effect near the center at the higher spatial frequencies, due to the halftone screen used in the printing. A similar effect occurs with the pixel pattern in a digital camera. (d) The same test object photographed with an out-of-focus lens. Note the spurious resolution beginning at X.

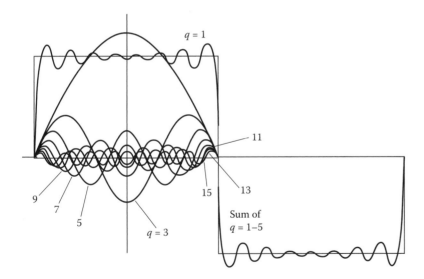

Figure 6.5 Synthesis of a square wave from sinusoidal components. q is the frequency of the component (and the order of the harmonic).

does not look very different from a contrast of 100:1. We obtain a more realistic result by taking the logarithm of the luminance ratio instead. (I discuss this further in Chapter 10.) But this rule doesn't hold accurately once you reach a contrast higher than about 10:1 (a logarithmic contrast of 1). A closer fit to perceived contrast comes from the use of *modulation* as a quantifier for contrast. The modulation *M* of a subject is given by the expression

TABLE 6.1
Numerical Scales of Contrast Compared

Arithmetic	1:1	1.5:1	2:1	3:1	4:1	6:1	10:1	100:1	1000:1	∞
Logarithmic	0.00	0.18	0.30	0.48	0.60	0.78	1.00	2.00	3.00	∞
Modulation	0.00	0.20	0.33	0.50	0.60	0.71	0.82	0.98	0.998	1.00

$$M = (L_{max} - L_{min}) \div (L_{max} + L_{min})$$

where L_{max} and L_{min} are the respective luminances of the brightest and darkest areas of the subject matter. For a transparency, transmittance, and for a print, reflectance, replace luminance.

At low values (up to about 0.7) the numerical scale for modulation doesn't differ by much from that for logarithmic contrast, but above this the two values begin to diverge, as the logarithmic value can in theory go right to infinity, whereas the modulation cannot go higher than unity (Table 6.1).

Modulation does provide a closer approximation than logarithmic contrast to the way we perceive contrast. It is a comparatively recent concept in imaging.

If you have been involved in electronic engineering you will probably already be familiar with the concepts of both modulation and Fourier analysis.

The Optical Transfer Function

The *optical transfer function (OTF)* is a combination of the *modulation transfer function (MTF)* and the *phase transfer function (PTF)*. The MTF is a graph of modulation transfer (image modulation ÷ object modulation) plotted against image spatial frequency in cycles per millimeter (c/mm). The PTF is a graph of the displacement of the image compared with its geometrically correct position, in radians (rad) or milliradians (mrad), plotted against the image spatial frequency (c/mm). A displacement of one whole cycle is $\pm 2\pi$ radians. Figure 6.6 shows a typical OTF for a camera lens. The MTF shows us how the system performs with reference not only to the finest detail (i.e., the limit of resolution), but also to the coarser detail and (indirectly) to the edges of objects. The PTF is less useful for general purposes, but it does predict any distortion of fine detail, in particular whether there is likely

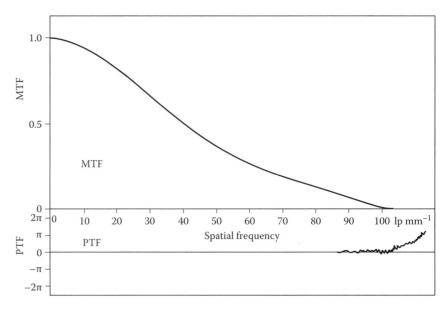

Figure 6.6 OTF for a typical camera lens.

to be a problem with spurious resolution. As a grating of parallel bars is effectively a one-dimensional object and a photographic image is two-dimensional, the practical measurement of the MTF for a camera lens involves two sets of data made with orthogonal set-ups. The directions chosen have the bars of the test object respectively radial and tangential to the optical center of the image field. To evaluate the MTF over the whole image field you have to make measurements at a sufficient number of points between the optical center and the periphery of the field.

The MTF of an "Ideal" Lens

The MTF for an "ideal" or totally nonaberrated lens can be calculated from the diffraction equations, as diffraction is then the only limitation on image quality. It looks like Figure 6.7. The MTF for any real lens must lie below this curve at all points.

OTF AND PSF RELATED

As with the OTF, the PSF is usually measured in radial and tangential directions separately, as *line spread functions (LSFs)*, using a line instead of a point as test object. The tangential LSF will always be symmetrical (provided the lens hasn't somehow become decentered), whereas the radial LSF will be symmetrical only if the lens is completely free from distortion and coma. Quantitative analysis of the LSF is not at all easy, but it doesn't need to be measured in practice as its shape can be readily derived mathematically from the OTF. Nevertheless, when the concept of the OTF first appeared, OTFs had to be computed from the shape of the LSF before equipment for measuring the OTF directly was in existence.

There is a further function, namely the *edge spread function (ESF)*. This is a plot of luminance across the image of a sharp edge. This needs to be measured for both inner and outer edges for the tangential figures. The ESF is related mathematically to the LSF: it is the first derivative of the LSF. Its shape gives us information about acutance, and it is therefore more generally useful in relation to the performance of the light-sensitive material. In practice none of the spread functions is used much, as the OTF is more directly informative, but the PSF often appears (much enlarged) in ray-tracing computer programs.

The OTF is the Fourier transform of the LSF (and vice versa). It may have been the discovery of this that led indirectly to the adoption of the Fourier model for optical imaging, though they are not otherwise connected.

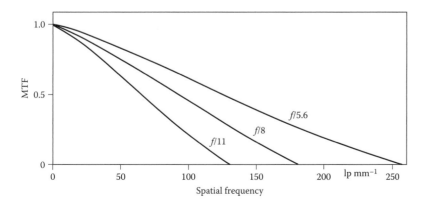

Figure 6.7 The MTF of an "ideal" (diffraction-limited) lens.

Cascading of Transfer Functions

One of the major advantages of the OTF over the various spread functions is that the OTFs of a complex system (such as camera lens, sensor, printer optics, and print material characteristics) can be readily combined—something that is difficult with spread functions.

The final OTF is found by "cascading" the function: multiplying the MTFs together, and adding the PTFs. Some imaging systems (such as the one instanced above) contain a number of components, each with its own OTF. In practice there can be many of these. For the complicated and exacting imagery of aerial or satellite survey you would need to include OTFs for haze, lens flare, camera vibration, recording medium, viewing system, and even the perceptual processes of the human interpreter! Some of these factors may add to the complexity by having a response that is not completely linear, but that takes us beyond the scope of this text.

Spread functions are combined by convolution, i.e., every point on one of the functions is "dealt out" to every point on the other. This is not an easy task.

Granularity and Pixel Size

In a developed silver halide film, the photographic image is made up of millions of tiny grains of opaque silver. Under a jeweler's loupe these look like small pieces of coke, and under a microscope they look like spaghetti bolognese. Although bulk silver metal is shiny, the texture of the grains causes the light to be reflected back and forth within them until it is totally absorbed, so the grains look black. On a microscopic scale, therefore, a photographic image is a binary device: light is either transmitted or blocked. However, as the grains are too small to see, the image appears to the unaided eye to be in varying shades of grey, depending on the number of grains in a given area—in fact, somewhat similar to the halftone image in a newspaper, which is made up of black dots, and is also binary. The silver grains vary in size, and we obtain the figure for *granularity* by statistical methods, using a plot made by a traveling microdensitometer, a light meter measuring density through a tiny pinhole. As the spatial period of the image of the test object (the inverse of the spatial frequency) approaches the mean grain size, the MTF falls rapidly to zero; for the optimum resolving power for the system, this should occur at the point where the lens MTF also falls to zero. But in practice this doesn't necessarily result in optimum final image quality, as the film MTF may be a completely different shape from the lens MTF (Figure 6.8).

This also applies to halftone reproductions, images from digital cameras, and any image (such as that from a liquid crystal display) that is made up of pixels. Because the dots form a regular pattern and are not random as in the silver grains of a photographic image, the cutoff point is very sharply defined. In a pixel image, by its nature, the MTF can remain at unity right up to the cutoff point.

In an ideal lens, using light of a single wavelength (such as laser light) the MTF doesn't fall off at all until the cutoff point (which is determined by the wavelength and lens aperture) is reached.

This indicates one problem with digital imaging that does not occur with silver images. At certain spatial frequencies the regular pixel structure causes artifacts in the fine detail. These are part of the aliasing problem, and sometimes need to be eradicated or minimized by strategies such as "dithering" (very tiny movements of the image during exposure or at the printing stage). If you go back to Figure 6.4 and examine 6.4a and 6.4c carefully, you will see little moiré patterns in the fine detail. In this case these are a consequence of the printing process, which is also a digital technology.

Although quasi-realistic test objects such as the USAF test target may give a good idea of the performance of an existing optical system, it is the OTF that gives us the greatest insight into the nature of the information that can be extracted from

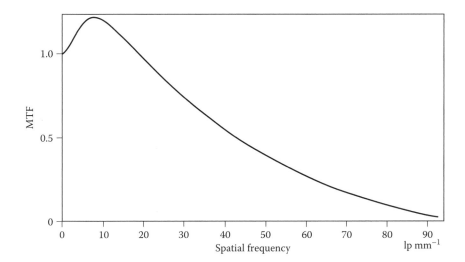

Figure 6.8 The MTF of a typical general-purpose black-and-white film. The peak above 1.0 is the result of adjacency effect, an artifact of the development process.

the image, and knowing its required characteristics in advance can help the optical system designer enormously. The concept of the OTF brought about a revolution in optics. It is sad that although the principle was initially developed in Britain it was left to other nations to take full advantage of it, most notably Japan. As a result, Japan has dominated the camera lens market for more than 50 years.

Digging Deeper

It hasn't always been easy to find good readable textbooks on image quality and optical resolution. Most of the accounts are buried in more general texts, and may vary from sketchy to downright misleading. There is an excellent introduction to the subject in Rudolf Kingslake's *Optics in Photography* (SPIE Press, 1992). There is a good account of practical field-testing of aerial survey cameras in G. C. Brock's *Physical Aspects of Air Photography* (Dover, 1967), which can be applied to other types of cameras; you should be able to borrow a copy through your local library. It is still easy to get hold of copies of *Evidence in Camera*, by Constance Babington Smith (no less), which is an absorbing account of wartime photographic reconnaissance. It sounds dull, but it isn't.

There is an excellent account of the effects of haze in the late Ron Graham and Roger Read's *Manual of Aerial Photography* (Focal Press, 1986). Sidney Ray's *Applied Photographic Optics* (2nd edition, Focal Press, 1988) contains a chapter on the physical optics of lens systems including a vigorous, if somewhat terse explanation of the Fourier approach to the OTF concept. Both granularity and OTFs are covered in some detail in *The Manual of Photography* (10th edition, Focal Press, 2010), and a chapter on Fourier optics is thrown in for good measure. Hecht's *Optics* (again!) has a lucid account. But if you really want to know everything there is to know about OTFs in theory and practice, there is only one choice: Tom Williams's *The Optical Transfer Function of Imaging Systems* (Institute of Physics Publishing, 1999, now CRC Press).

Chapter 7 Images in Color

Early Attempts

When photography first appeared early in the nineteenth century its images were in monochrome; but it was quickly accepted as a promising newcomer to the visual arts. As far as we can tell from contemporary writings, nobody seemed concerned about the absence of color from the image, at least in landscape and architectural photography.

It is perhaps surprising that this lack of color went unremarked, as very few finished paintings seem ever to have been made in a single color (Picasso's "Guernica" springs to mind as a notable exception). It worried many of the pioneer photographers, though. There have been stories of daguerreotypes that showed color when viewed from a particular angle, but these are probably apocryphal: certainly no example from those days now appears to exist. In contrast, today almost all amateur and commercial photographs are in color. Monochrome keeps a tenuous foothold in fine-art photography, and (for its impact) in photo-reportage.

Family portraiture was another matter: Victorian portraits were often hand-colored.

I use the term "monochrome" rather than the more usual "black-and-white," because a majority of exhibition photographs of this type are not a neutral black, but are either toned or made on warm-tone print material.

Lippmann Photography

If not the earliest attempt to produce a photographic image in color, Gabriel Lippmann's was certainly the boldest. Lippmann was a physicist who thought of color in terms of physics, and his method used the interference of light waves to produce colors in a similar manner to the formation of colors in soap bubbles or oil films. In the 1890s, using some of the very first panchromatic materials (i.e., materials sensitive to the whole visible spectrum), Lippmann succeeded in producing color photographs of such high quality that in 1908 he was awarded the Nobel Prize for Physics. The problem Lippmann faced was that if you want an image of (say) a rose to look red and its leaves to look green when illuminated with white light, you

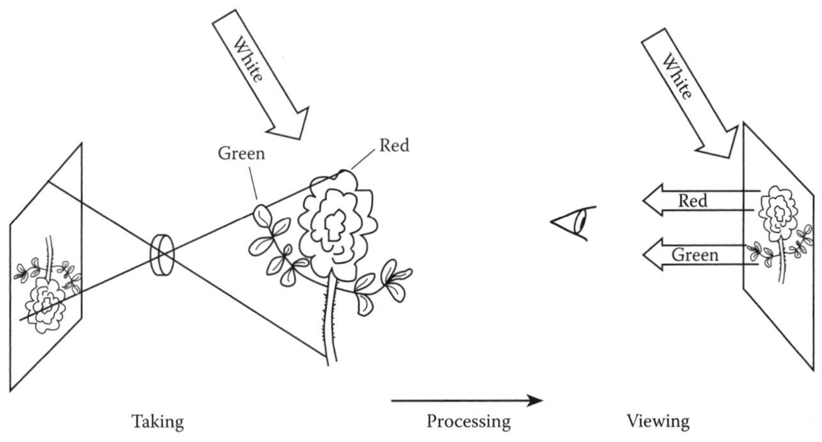

Figure 7.1 In a color photograph, the image must reflect only the colors in the subject.

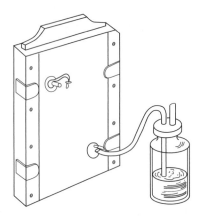

Figure 7.2 Lippmann's special mercury-containing plateholder made by the Penrose Co.

have to arrange for the image of the flower to reflect only red light and for the image of the leaves to reflect only green (Figure 7.1).

Lippmann found an ingenious way to achieve this. He mounted the photographic plate in his camera with the glass side towards the lens and the emulsion in contact with a reflective bath of mercury. Red light forming the image of the flower, and green light forming the image of the petals, would then be reflected back from the mirror through the emulsion (Figure 7.2), interfering with the incoming waves and producing a stationary interference pattern, which on development would produce sheets of semitransparent silver parallel to the surface of the emulsion, spaced exactly one half-wavelength apart. Such a system would reflect strongly only light of the appropriate wavelength, by constructive interference.

But what, you may ask, about desaturated colors: pinks, browns, greys? It would seem at first that only saturated colors with a narrow bandwidth could be reproduced. One would expect that any wider spread of wavelengths would wipe out the delicate structure of interference planes within the thickness of the emulsion. But the pinks, browns, and greys *are* all there, along with startlingly accurate flesh tones. Lippmann's scientific papers do not explain this, although historical research suggests that he did know precisely what was going on. For nearly a century various theories were proposed, but only recently has the correct explanation resurfaced (see Box).

Bragg reflection. The type of interference described here is called *Bragg reflection* (or, more correctly, *diffraction*), after the physicists who successfully investigated the principle (Figure 7.3). William Henry Bragg (1862–1942) and his son Lawrence Bragg (1890–1971) pioneered the techniques of X-ray diffraction that led to the establishment of the atomic structure of crystals, and were jointly awarded the Nobel Prize for Physics in 1915 for this work. Both men were knighted, and both became in turn Director of the Royal Institution. Sir William Bragg is remembered particularly for inaugurating the famous series of Christmas lectures at the Institution. There is more about Bragg diffraction in Chapter 18.

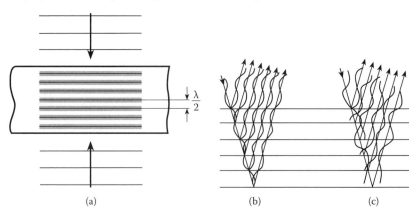

(a)   (b)   (c)

Figure 7.3 Principle of Bragg reflection, (a) where the reflective layers are separated by one half-wavelength: (b) light of the original wavelength is reflected from adjacent layers in phase and is reinforced; (c) light of other wavelengths interferes destructively with adjacent reflections and is suppressed.

LIPPMANN'S DESATURATED COLORS

Light of a single frequency is said to be *temporally coherent*. Such light forms stationary interference patterns when it recrosses its own path. In Lippmann's photographs the image-forming light was reflected directly back through the emulsion, forming interference planes one half-wavelength apart. In this situation, the distance over which the interference planes can occur (i.e., the number of planes that can be formed) depends on the *coherence length*, which is the distance over which the waves remain in phase). The coherence length is inversely proportional to the frequency bandwidth (Figure 7.4).

In laser beams the coherence length can be many meters. Most light sources, though, have a very low coherence length. Even sodium light has a coherence length of less than a millimeter. Light reflected from natural objects has a coherence length of only a few wavelengths, and in the Lippmann process this light would form at most about half a dozen interference planes, so that when the image was illuminated with white light the wavelength selection would be less precise, giving rise to the desaturated colors. In the extreme—white light—only one reflecting plane would be formed, and there would be no wavelength selection: all the light would be reflected.

A number of Lippmann's contemporaries, including the Lumière brothers Louis and Auguste (the inventors of cine photography) and Frederic Ives (an American pioneer of color photography), followed up his work, but the process was so exacting that few photographers were prepared to attempt it, and the technique languished for nearly a century. Figure 7.5 (color plate) is a reproduction of a modern Lippmann photograph.

There are plenty of examples of original Lippmann photographs around, mostly in French museums. Some of these are very beautiful, and most are in nearly perfect condition, owing to the method of mounting (behind a thin glass wedge, to make it possible to view the photograph at the correct angle without one's head getting in the way of the viewing light). A few are in the Science Museum, London. With the renewed availability of ultrafine-grain panchromatic emulsions with a mean crystal diameter of less than 20 nm, interest in the process has begun to revive (see Digging Deeper).

A more rigorous explanation invokes the Fourier model. Lippmann was familiar with the work of Joseph Fourier, and was almost certainly aware of its application to his own work. It is interesting to speculate that had he seen fit to publish this aspect, it might have advanced the "revolution in optics" by more than half a century.

A certain Dr. Spirna of Norway was not put off, though. He took literally hundreds of Lippmann photographs of a stuffed parrot, many of which (unlike the parrot) are still extant.

According to the late Jack Coote, the finest collection of Lippmann photographs is in the Preus Fotomuseum at Horten in Norway (with several parrots).

The Young-Helmholtz Theory of Visual Perception

At the beginning of the nineteenth century Thomas Young showed that almost every possible color could be matched by superimposing on a white screen just

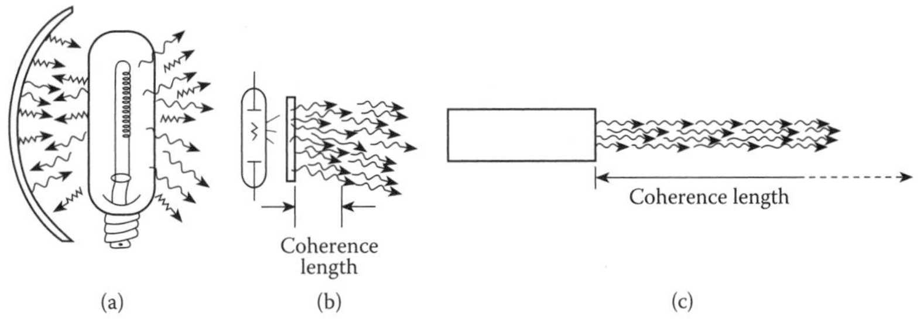

Figure 7.4 Coherence length: (a) incoherent radiation, bandwidth ~60 THz or 1 μm; (b) partly coherent, bandwidth ~300 GHz or 1 mm; (c) highly coherent, bandwidth ~2 MHz or 150 m.

Figure 7.5 (See color insert following page 154.) Self-portrait of Dr. Hans Bjelkhagen made with Lippmann's process. In this printed reproduction it is unfortunately not possible to do justice to the subtle colors of the original.

three lights each of a single color (specifically red, green, and blue) in various intensities. You can easily demonstrate this with three slide projectors loaded with plain slides bearing, respectively, red, green, and blue filters. If you haven't seen a demonstration of this, you may be surprised (most people are) to see that where the red and green patches overlap the color is a brilliant primrose yellow (this was discussed in Chapter 3). Less surprising is the combination of blue and green to form the color we call cyan, a sort of kingfisher blue, and of blue and red to form magenta. The superposition of all three beams gives white (Figure 7.6a, color plate). By varying the relative intensities of the red and green lights you can obtain the gamut of colors, from pure red through orange, yellow, and lime green to pure green. By doing the same with green and blue you will get a similar shift from pure green through emerald green, cyan, "petrol" blue, and finally pure blue. Mixing blue with red similarly takes you through violet, purple, and magenta to crimson and pure red. All the colors from red to blue approximate to pure spectral colors. In colorimetric terms they are 100 percent saturated, as is the magenta range, which

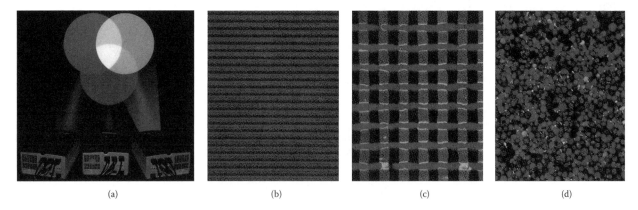

(a)   (b)   (c)   (d)

Figure 7.6 (See color insert following page 154.) (a) Additive color patches, using projected light: (b) enlargement of a section of Lumiére transparency showing colored starch grains; (c) enlargement of a section of Polaroid transparency film, showing color raster; and (d) an enlargement of a Dufaycolor réseau on the same scale.

is not part of the spectrum. To obtain desaturated (pastel) colors such as pinks and blue-greys you need to add the third patch of light and adjust the relative intensities of all three.

In order to explain these results, Young suggested that there were three types of color receptor in the eye sensitive, respectively, to red, green, and blue, and that simultaneous stimulation of pairs of these gave the sensation of the three intermediate colors cyan, magenta, and yellow. Stimulation of all three types of color receptor in varying degrees would result in the perception of desaturated colors—pink, brown, grey, etc. Nearly half a century later, Helmholtz was to take up Young's hypothesis and add a sound physiological basis.

Additive Color Synthesis

The Young-Helmholtz model for color perception was investigated in the 1850s by James Clerk Maxwell, who further extended it in terms of physiology. (Like Helmholtz and Young, he was a qualified medical practitioner.) He constructed various models to demonstrate additive effects, including a spinning disc with adjustable red, green, and blue sectors. In 1861 Maxwell took three separate photographs of a bow of colored ribbon through a red, a green, and a blue filter, respectively. He made positive monochrome slides from his negatives and projected these in register through red, green, and blue filters. The result was an image that approached natural color—at least within the limits of his material.

It was fully 50 years before patents began to appear for color processes based on the additive principle. In 1907 Louis Auguste Lumière patented a method of coating an emulsion with starch grains dyed red, green, and blue. The film was reversal developed to a positive, with the randomly distributed filter grains *in situ* (see Figure 7.6a).

Others such as Paget and Dufay used ruled color mosaics in a similar way (Figure 7.6c). Dufaycolor materials continued to be available up until around 1950, and gave excellent color rendering, particularly of flesh tones, but the material could not be printed successfully, and the transparencies were not very bright. And by this date much brighter materials employing the subtractive principle had long been around.

The principle of additive color imaging underwent a rebirth with the advent of color television. Color TV screens carry a mosaic of red, green, and blue phosphor spots (in some cases short vertical bars), and in the traditional TV tube the red, green, and blue elements of the scene are delivered by three electron guns. A precisely aligned perforated screen allows the beam from each gun to "see" only its appropriate color phosphor spots. (I will be discussing color television in more detail in Chapter 13.)

Quantifying Color: The CIE Chromaticity Diagram

To an artist it may seem vaguely indecent to reduce the mysteries of color to mere physiology; to a physicist it may feel frustratingly difficult to attempt to eliminate the subjective element from the physics of color. The mechanism of color perception became clear in the early years of the twentieth century, and the Young-Helmholtz theory was eventually confirmed by the identification of red, green, and blue sensitive cone cells in the retina. In 1931 the *Commission Internationale de l'Éclairage* (CIE) began to establish color criteria using extensive tests carried out on human subjects,

Hermann von Helmholtz (1821–1894) was a polymath who dominated German science during the mid-nineteenth century. He was a qualified surgeon, and invented the ophthalmoscope. He made the first precise formulation of the principle of conservation of energy. He also developed thermodynamics in physical chemistry, and, apart from his extension of Thomas Young's work to cover after-images and color blindness, made contributions to auditory perception, acoustics (the Helmholtz resonator), hydrodynamics, and electromagnetism (the Helmholtz coil).

Maxwell was lucky to have had any success at all. His plates were completely insensitive to red. Fortunately, as it happened, his red filter was transparent to ultraviolet, and his red ribbon reflected UV strongly. His green filter needed to be very pale for anything to record at all. Hence his separations were not red, green, and blue, but in fact UV, blue, and violet. Maxwell was a theoretical physicist of genius, but he knew very little about photography. Perhaps this explains the apathy that greeted his findings when he presented them to the Royal Society.

Lumière color transparencies have stood the test of time. There are some excellent examples in the archives of the Royal Photographic Society.

The reason for the high density of additive mosaic transparencies is that the maximum transmittance of any tricolor filter cannot exceed 33 percent. In the 1980s additive color photography surfaced again, with an instant-slide material from Polaroid. While still somewhat dark, the resolution was better than with the older films, as the pitch of the color filter mosaic, a raster of plain horizontal rulings, was much finer, and invisible even in a large projected image viewed from a normal distance (Figure 7.6).

It is sometimes difficult to believe that when you are watching color TV there really *are* only three colors—red, green, and blue—present, especially when you are looking at something that appears brilliant yellow or purple. But go up close, and you will see that the yellow consists of only red and green dots, and purple is just red and blue. And if you are wondering how you get brown, look closely again. You will see that it is just red with some green, but less bright, in fact, a dark yellow or orange—as I pointed out in Chapter 3.

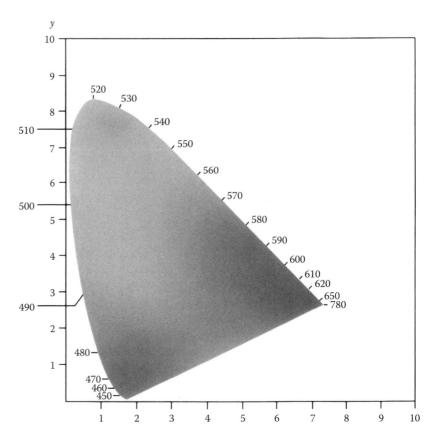

Figure 7.7 (See color insert following page 154.) (a) 1947 CIE chromaticity diagram, showing approximate wavelengths in nm and corresponding colors.

MEASUREMENT SYSTEMS FOR COLOR

In the 1947 CIE chromaticity diagram the axes represented percentages of red (horizontal) and green (vertical). The blue content could be deduced from the missing percentage (e.g., the coordinates 30R, 50G implied 20B). All possible hues and saturations were to be found within the closed curve. The reason the curve does not completely fill the triangle is that the sensitivity ranges of the cone cells of the retina overlap, and a pure spectral green and the deepest blue both provide some stimulation to the ρ (red) cones. In 1978 it was agreed to modify the curve by a transformation, which rotated it anticlockwise through about 40° and then shrank it by about 30 percent horizontally, replacing the axes x and y by u' and v', respectively. The exact relationship is

$$u' = 4x/(-2x + 12y + 3), v' = 9y/(-2x + 12y + 3)$$

Equal distances on this later diagram represent equal perceived changes in hue and/or saturation (lightness does not appear). Complementary hues are at the ends of straight lines drawn through the neutral-color point. Saturation decreases with distance away from the boundary of the curve. In practice the 1947 version is still used for illustration, as the axes correspond to true red and green, making it easier to interpret for some purposes.

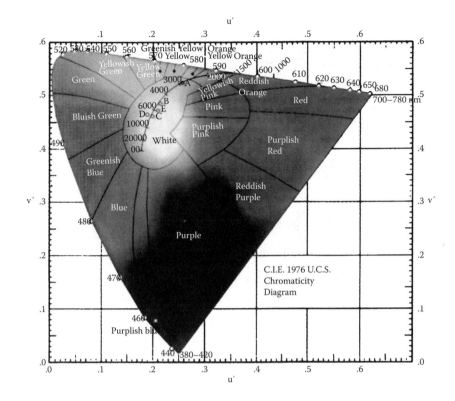

Figure 7.7 (See color insert following page 154.) (b) 1978 modification, representing equal perceived color shifts in hue and saturation by equal distances, and showing the position of various color temperatures.

in order to obtain statistically sound quantitative figures for the perception of color. In 1947 Robert Hunt and his team published the CIE chromaticity diagram (Figure 7.7a, color plate). In its more recent (1978) form (Figure 7.7b) it represents equal perceived changes in hue and saturation by equal distances within the area of the curve.

Other Scales of Color Measurement

There have been a number of other color scales, some decidedly dubious. Those that were most flawed were eventually dropped, but six or seven are still in use. Of these, the most important are probably the Munsell and DIN systems.

The Munsell System

This was originated by the artist A. H. Munsell in 1905, and has been continuously refined ever since. It is the system most used by artists. There are five principal hues—red, yellow, green, blue, and purple—and these hues are spaced equally in terms of perceived hue shift. There are finer divisions of hue between these, designated by numbers and letters. For each hue there is a chart with color patches, which have a constant lightness (called *value* in this system) but decreasing saturation (called *chroma*) from right to left. The vertical scale of patches runs from lightest at the top to darkest at the bottom. All adjacent patches have the same perceived difference. An example is shown in Figure 7.8 (color plate). The *Munsell Book* is an atlas of these patches, with a progressively different hue for each page.

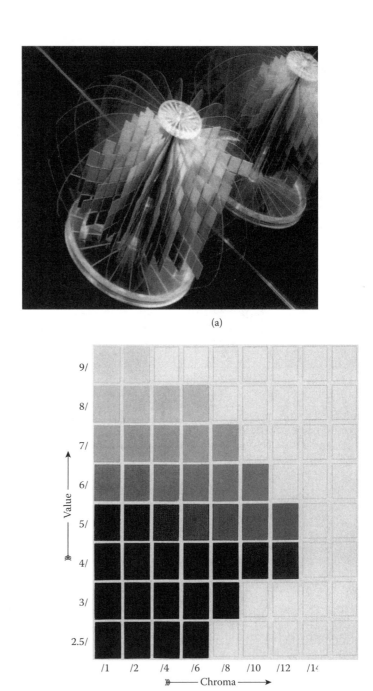

(a)

(b)

Figure 7.8 (See color insert following page 154.) (a) The full Munsell color globe; (b) a page from the Munsell color atlas.

The DIN System

The DIN system (DIN stands for *Deutsches Institut für Normung*) was developed in 1986. It replaces "lightness" by "darkness," and is based on a sphere with white at the upper pole and black at the lower. Hues are represented by successive radial sections and saturation by angular distance out from the grey-scale axis. As with the Munsell system, the DIN scales are obtainable as an atlas, the *DIN Color Chart*.

Subtractive Color Synthesis

Additive color synthesis is fine as long as we are dealing with colored light, as in additive projection systems such as Microsoft®PowerPoint and color TV. But additive transparencies using réseau or mosaic filtering need a great deal of projector light, and additive reflection prints are out of the question. If we are to obtain transparencies and prints where the whites really are white, i.e., transmit or reflect all the light falling on them, we need to use a radically different technique. The way ahead was pointed out by du Hauron, who, in a paper published in 1869, *Les Couleurs en Photographie*, formulated the principles of both additive and subtractive color synthesis.

When painters mix colors they call red, yellow, and blue "primaries," whereas when photographers (and colorimetrists) talk of "primaries" they are thinking of red, green, and blue. There is a terminological confusion here. The photographer's "blue" is a royal blue, like the painter's ultramarine, whereas the painter's "blue" is Prussian blue, nearer to the photographer's cyan. The photographer's "red" is the painter's vermilion, but the painter's "red" is crimson lake, a purplish red akin to the photographer's magenta. So the painter's primaries are the photographer's secondaries (or *complementaries*) cyan, yellow, and magenta.

Now, in the additive process we obtained yellow by adding red light to green light. This is the same thing as starting with white light and removing blue. For this reason photographers often call a deep yellow filter "minus-blue." In a photograph made on panchromatic monochrome material a minus-blue filter will darken a blue sky without affecting red flowers or green grass. Similarly, magenta, which is red light added to blue light, is white light minus green, and cyan (green plus blue) is the same as white minus red. You can test this by putting a cyan filter on top of a red filter, or a magenta on a green, or a yellow on a blue. In each case the result is black—no light. If you take the secondaries and superimpose them in pairs, you get the three (photographer's) primaries. Thus magenta and yellow give red (white minus green and blue); yellow and cyan give green (white minus blue and red); and cyan and magenta give blue (white minus red and green). Again, all three together give black (white minus red, green, and blue). This is illustrated in Figure 7.9 (color plate).

Color Separation Negatives

Until the end of the 1940s the only way to make color prints was to superpose three images in register, one recording all the red component of the image, one the green, and one the blue. As in Maxwell's experiment, the first step was the making of separation negatives, one each through a red, a green, and a blue filter, respectively. These had to be correctly exposed, and developed to the same contrast—a difficult task, as with most emulsions both speed and contrast vary with wavelength. It was usual, therefore, to include a neutral grey scale card at the edge of the scene for control, as well as a set of color patches in red, green, and blue for identification of the negatives.

Louis DuCos du Hauron was born in 1837. His novel ideas, which he never exploited commercially, were recorded in a vast number of scientific papers. In the words of D. A. Spencer (see Digging Deeper): "There is scarcely a principle or process in color photography that his genius did not foreshadow. [...] Rarely has any inventor shown such imaginative foresight, or received so little encouragement. He died, greatly honoured, in abject poverty in 1920."

I agree that these correspondences aren't exact, but the point is that the painter is mixing pigments or dyes, not lights, and is therefore working in subtractive, not additive, color. Many years ago I actually owned a (rather expensive) watercolor paint box that contained only crimson lake, chrome yellow, and Prussian blue. You were expected to make up all the other colors from just these three.

Grey scale cards remain useful for ensuring color accuracy, and are still obtainable commercially. Color patches in red, green, blue, cyan, magenta, and yellow are also obtainable. The preferred card for a more detailed analysis of color performance is the Macbeth chart, which contains a further number of colors known to be difficult to reproduce, including Caucasian and Caribbean flesh tones. Figure 7.10 (color plate) shows an example of each of these.

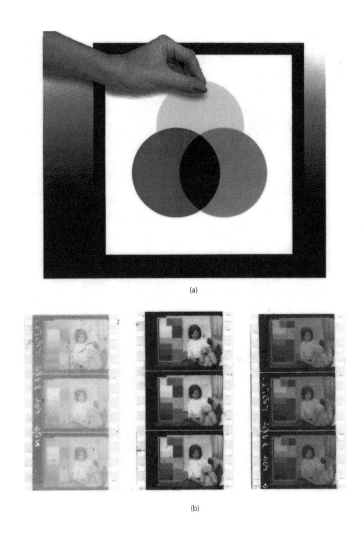

Figure 7.9 (See color insert following page 154.) (a) Subtractive color filters, using transmitted light; (b) Technicolor separations.

To avoid the labor and delay in changing plates and filters, it was possible to obtain repeating backs, which were triple plateholders in which the tricolor separation filters were installed directly in front of the plates. These backs were coupled to the shutter release mechanism so that when you operated the shutter each plate was brought into position in turn and exposed, making the three exposures in rapid succession. The differing filter factors were adjusted by the incorporation of neutral density filters. A better, though much more expensive, solution, was a specialized "one-shot" camera, which exposed all three plates simultaneously, using two dichroic beamsplitters (Figure 7.11).

Color Prints from Separation Negatives

In the subtractive process as visualized by du Hauron, red would be produced by removing green and blue from white light. The red content of a print could thus be controlled by a positive cyan image on white paper. (It would have no effect on the green or blue content of the reflected light.) Similarly, the green content would be controlled by a magenta image, and the blue content by a yellow image. When all

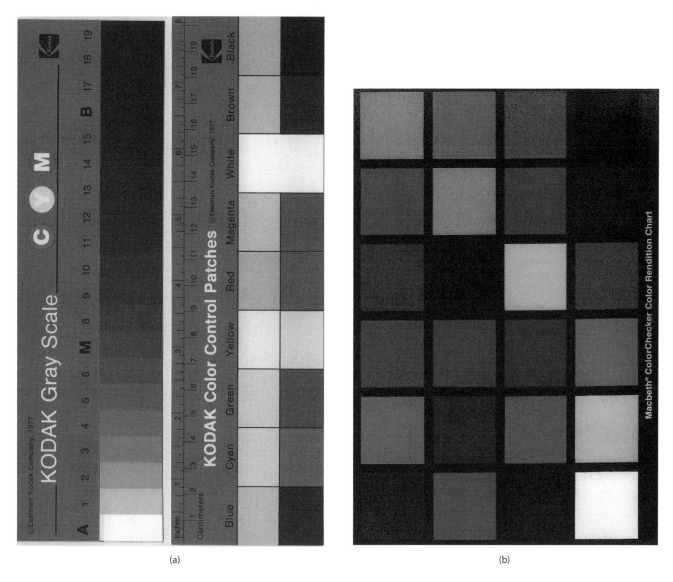

(a)

(b)

Figure 7.10 (See color insert following page 154.) (a) The Kodak color patches and grey scale; (b) Macbeth color chart.

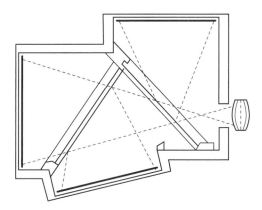

Figure 7.11 Schematic layout of a one-shot color camera using dichroic beamsplitters.

the images were superposed in register, the whole color content—red, green, and blue—would be controlled, and the image would be in natural colors. Two methods emerged for achieving this: "carbro" and "dye transfer." Both were complicated, messy, unreliable, and expensive, but capable of producing outstandingly beautiful (and absolutely permanent) results. Because of the light these techniques throw on modern printing techniques, I have included a simplified description of both below.

Tricolor Carbro

The name is a conflation of "carbon" and "bromide" as it involves pigmented gelatin papers ("carbon tissues") and silver halide prints ("bromide prints"). The process was introduced by Thomas Manly in 1899 (hence the old names). When a silver halide print is bleached in a dichromate solution in contact with a pigmented gelatin paper, the gelatin is hardened to a depth proportional to the image density. When the soft gelatin is washed away with warm water a positive image in pigment remains. You begin by making matched prints from the separation negatives, and make gelatin positives from these using cyan, magenta, and yellow pigmented papers. You strip these gelatin images onto transparent plastic sheet (originally waxed celluloid) and transfer them, one at a time, in register onto a final white paper base. The process was never easy, and only one company offered carbro printing as a service, using a variant of the technique named Vivex. Production of carbro materials ceased in the 1950s.

This is a much-simplified account. To say it was not an easy process is no understatement. Variables such as pH and temperature were critical; material varied from batch to batch; and it took up to 10 hours to produce a finished print. The delicate gelatins often tore during registration and at worst slid off the substrate into the sink and disappeared down the plughole. But when it did work the beautiful results made all the hours of frustration seem worthwhile.

Dye Transfer

This evolved over the first years of the twentieth century from the Woodburytype principle, in which a relief image in gelatin was used in the manufacture of a printing plate. Eastman Kodak introduced the dye-transfer (then called "wash-off relief") process for color prints in 1946. In this process you made relief positives from the separation negatives employing large films coated with very soft yellow-dyed emulsion, using a hardening developer and bleach, and washing away the unwanted gelatin with warm water (later, dissolving it chemically). In order to prevent the hardened image also washing away, you needed to expose the emulsion through the back. These matrices were pin-registered. You then soaked them separately in cyan, magenta, and yellow dye baths, the dye being taken up in proportion to the thickness of the gelatin. Finally, you squeegeed the matrices in turn in register on to mordanted paper, leaving a print in full color. Dye transfer printing was easier and more reliable than carbro, and had the advantage that once you had borne the expense of producing the matrices the prints themselves were very cheap to produce. The pin-registration equipment was expensive, and there were dodges for obtaining exact superposition without it, but modern printing processes capable of image modification by dodging and shading superseded the technique.

In 1994 materials for dye transfer ceased abruptly to be available, but a few dedicated workers managed to stockpile a large quantity of matrix material before it was too late, and continued to produce dye-transfer prints, some using Epson printer dyes once the Kodak supplies ran out.

Technicolor

After its inception in the 1920s, Technicolor motion picture technique went through a number of stages, the most successful being the simultaneous exposure of three separate films in a special camera. A blue/yellow dichroic filter reflected the blue component onto a blue-sensitive film. The remaining light fell on two films running in contact: one green-sensitive (orthochromatic) and the other red-sensitive. These films were then used as masters for the dye-transfer process on a clear film coated with mordanted gelatin (Figure 7.9(b), color plate). After 1950 standard color film was used, the separations being made from this. A grey image was added to strengthen the shadows in the final print. Although

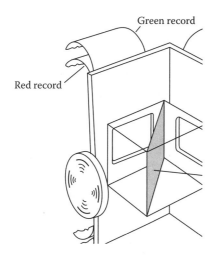

Figure 7.12 Schematic layout of a Technicolor camera (1948).

the early Technicolor prints are now faded beyond hope, the original black-and-white separations remain in good condition, and it has been possible in many cases to resurrect old color films in the kind of quality we expect from present-day cinema. The Technicolor camera is shown schematically in Figure 7.12.

Tripack Color Transparencies

Kodachrome

The first color film based on the subtractive process was Kodachrome. It was invented, surprisingly, by two professional musicians, Leopold Mannes and Leopold Godowsky.

The principle, which had been foreshadowed in one of du Hauron's papers, was to expose three emulsions through red, green, and blue filters, and finish with three superposed positive transparencies in cyan, magenta, and yellow, respectively. In the Kodachrome process, launched in 1935, the three emulsions were coated on the same base; this became known as a *tripack* emulsion. The outermost emulsion layer was blue-sensitive only, and behind it was a yellow (minus blue) filter layer. The middle layer was orthochromatic, i.e., sensitive only to green and blue, and therefore recording only green. The innermost emulsion was sensitive to red and blue only, and thus recorded the red.

The processing of a Kodachrome film was a long and complicated business, at first requiring 28 separate steps, all under precise control. Over the years the processing became somewhat more simplified, though it was still too complicated for home processing. Popular demand, however, kept the process alive. Processing began by developing the film to a negative.

It was then exposed to red light and developed in a developer containing a cyan dye, which coupled only to the developing grains (which were in the red-sensitive emulsion). The film was next exposed to blue light and developed again, this time in a yellow-coupling developer. The remaining emulsion was now chemically fogged and developed in a magenta-coupling developer. The positive color image was now complete, and the final stage was to remove the silver with a bleach-fix solution, leaving the color image in pure dyes. As might be expected, only a few specialized units were licensed to process Kodachrome material.

Mannes and Godowsky were not particularly outstanding musicians compared with their respective fathers: Mannes *père* was a distinguished violinist and conductor, and Godowsky *père* was arguably the most technically accomplished pianist ever. Both the sons had been groomed for a career in music, but they were fascinated by the idea of color photography, and carried out their clandestine experiments in kitchens and bathrooms. They were discovered by the far-sighted Kenneth Mees, then Head of the Eastman Research Institute, and joined his team, abandoning the musical profession—much to the dismay of their distinguished parents. However, both men eventually returned to their musical careers.

Development is explained in detail in Chapter 10.

Kodachrome materials ceased production in 2010. Processing remains available.

105

The run of photographs consisted of two-color photographs, a black-and-white flash shot (a four-inch magnesium-powder "bomb" of about 10^6 cd, dropped with a parachute) and two black-and-white exposures without flash, one short and one longer, to record traces of fires and assist the photo-interpreters to deduce the precise location and path of the aircraft. The camera used for this purpose had no exposing shutter, and the exposures depended on an automatically timed rapid film wind mechanism controlled by a preprogrammed intervalometer.

In professional photography the standard term for this type of processing is *reversal processing*, and the result is called a *transparency*. In other European languages the term is *diapositive* (or a similar variant). At the time of writing, digital imagery has still not ousted color transparency film for many purposes, though it is rapidly becoming universal practice to scan the images in order to project them via PowerPoint, rather than use a traditional slide projector.

Color density is measured with a densitometer through a complementary filter.

Ektachrome

During the Second World War it became necessary to secure color aerial photographs on night bombing missions, to establish from the images of colored flares dropped by "pathfinder" aircraft whether the targets had been found. The sections of color film were to be spliced into the (black-and-white) films that were used to photograph the target itself, using flash. These color inserts would have to be processed separately under ordinary darkroom conditions—quite a tall order at the time.

Eastman Kodak responded with a new idea. Instead of the dyes being in the developers they would be formed in the emulsion itself. Colorless molecules anchored in the emulsion layers would react with the second developer to form the appropriate dyes wherever an image was formed. The first development was to a plain black-and-white negative, after which the film was exposed to white light, and then the remaining silver (the positive image) was developed in the special color-forming developer. Finally, as in Kodachrome, all the silver was removed, leaving the positive image in dyes. This wartime film, called Kodak Aerial Color Reversal Film, was the forerunner of Ektachrome, which first appeared in 1946. The processing procedure has been steadily improved and the processing times shortened, but the principle remains the same today. The reexposure step, which was somewhat unreliable, has been replaced by the inclusion of a fogging agent in the second developer.

Other Color Transparency Films

Other film manufacturers followed Kodak as soon as they had found ways round Kodak's patents: Agfa, Ilford, Ferrania, Gevaert, and Orwo were among the frontrunners. To begin with, each make of film had its own processing chemistry, but eventually all the emulsions were made compatible with the Kodak processing routine. The principles and processing are illustrated in Figure 7.13 (color plate).

Prints from Transparencies

Owing to the convenience of inkjet prints that nowadays can compete in quality with anything that comes out of the darkroom, "wet" printing has all but disappeared. Nevertheless, there is a minority who prefer to make their prints using traditional enlarging methods. If you want to make a print from a transparency using darkroom chemistry there are two routes you can adopt. The one most used is a direct positive print. There are two technologies available to you. *Direct positive* is a reversal material that works in exactly the same way as transparency film, but has a low inherent contrast to compensate for the high contrast of the original. This is also the usual way for making copies of transparencies. An alternative print material uses a *dye-bleach* process. Here, the print material contains latent dyes, which are destroyed progressively by exposure to light. The processing chemistry is completely different from the reversal processing system, and involves a single development. For both types of material you can correct any color imbalance by using CP (color print) filters, which are available in all six primary and secondary hues in gradations of 0.25 density.

Internegative

The second route is to make a color "internegative" from the transparency and a color print from the internegative. This was the standard method for making release

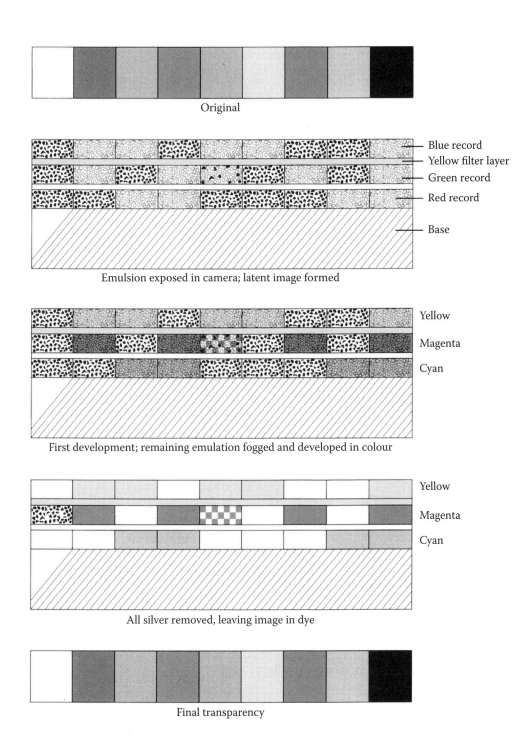

Original

Blue record
Yellow filter layer
Green record
Red record
Base

Emulsion exposed in camera; latent image formed

Yellow
Magenta
Cyan

First development; remaining emulsion fogged and developed in colour

Yellow
Magenta
Cyan

All silver removed, leaving image in dye

Final transparency

Figure 7.13 (See color insert following page 154.) Color reproduction in a reversal color film.

copies of color movies shot originally on tripack reversal material. Although the process involves two stages, it is theoretically possible to obtain an exact replica of the original, whereas in a "positive-positive" system the inherent color errors of the reversal process are doubled. (It is only fair to say, however, that with present-day materials such errors are virtually undetectable except in extreme examples such as magenta cactus flowers.)

Polaroid Color

The Polaroid Corporation was originally formed by Edwin Land to manufacture and market a sheet material that would polarize light passing through it. Land was conducting a campaign to have polarizing antiglare screens fitted to car headlights throughout the United States, and in order to finance this he invented the Polaroid Land Instant Camera, which first reached the public in 1948. The campaign failed, but Polaroid photography went from strength to strength. There were two main markets: the casual photographer, who wanted instant party and beach photographs; and the professional, who needed to check setups, lighting, and composition before the all-important final shot.

Polaroid photography has traditionally used a diffusion transfer process. In the black-and-white version, the action of ejecting the film and paper in an opaque package releases a developer solution. This contains chemicals that cause the undeveloped residue of the negative to diffuse into the paper, where a reducing agent blackens it. In some versions you can wash, dry, and reuse the negative for making prints.

There are two versions of Polaroid color. The first type, used chiefly by professional photographers, also uses a diffusion transfer principle. In the tripack emulsion, the cyan, magenta, and yellow dyes are rendered immobile in the developing negative; in the unexposed regions the dyes migrate freely into the print paper, resulting in a positive color image. The process takes about a minute and allows some manipulation, especially while the gelatin is still soft and easy to lift off and distort.

The Polaroid Corporation encouraged this treatment. At one time it actually made awards to creative Polaroid photographers. At the time of writing Polaroid material is no longer available, having been eclipsed in the amateur (and to a large extent professional) market by digital imaging, but rescue operations have been mounted, and no doubt such a valuable asset to professional photography will not be allowed to disappear altogether.

The second type was aimed at casual photographers, and required a special camera. It worked on a broadly similar principle, but there was no separation of negative and positive after development. Instead, the dyes diffused through an opaque white layer of titanium oxide, which became the substrate for the developing picture. One consequence of there being no stripping procedure was that the optical image needed to be reversed right to left, and this was achieved in the camera by the use of an inverting mirror in the optical path. Again, as noted here in the margin, this aspect of instant photography has atrophied due to the instant-access possibilities of digital imaging, and it has become difficult to obtain either equipment or materials (hence the change to the past tense in this edition).

Color Negative-Positive Systems

Most amateur photographers who prefer to work with film choose color negatives rather than transparencies, as they afford more latitude in exposure, and there is less need to compensate for variations in the color temperature of the lighting. In this system the color film is developed in a single stage to a negative, using a color-forming developer. The silver and unchanged silver halide are then both removed, leaving the negative image in dye. The light parts of the subject record dark and vice versa, just as in a black-and-white negative. But the red-recording layer produces its negative image in cyan (minus-red), the green-recording layer in magenta (minus-green), and the blue-recording layer in yellow (minus-blue). So the colors are negative too. As you might expect, saturation is unaffected: saturated colors remain saturated. The print material is also coated with three emulsions, so that after exposure to the negative and a similar development, the tones and hues are again reversed, and the final print has the correct scales of hue, lightness, and saturation.

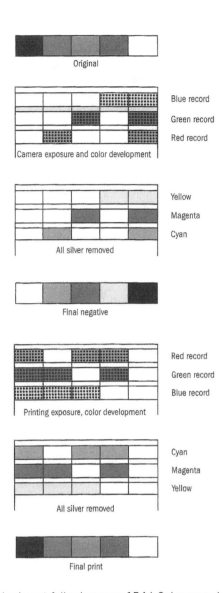

Figure 7.14 (See color insert following page 154.) Color reproduction by the negative-positive color process.

Color print materials are usually balanced to give a print in approximately correct color rendition when used with a normal enlarger filament lamp, but may need some correction by means of CP filters. Unlike positive-positive printing, the filters need to be the same hue as the error (e.g., a magenta cast in the test print demands a magenta CP filter). The filtration has a much stronger effect than with positive-positive prints too, because of the much higher inherent contrast of negative-positive print paper. The principles of negative-positive color are shown in Figure 7.14 (color plate). The clever thing about the negative-positive process is that instead of doubling the errors inherent in color recording as in the positive-positive process, they can be almost totally eliminated. As the color negative is an intermediate and not for display, its physical appearance is immaterial. Its tones and colors can be, as it were, pre-distorted to compensate for subsequent errors in the print material. You will no doubt have noticed that the backgrounds of color negatives have an orange-yellow cast, and that their contrast is low. That is part of the pre-compensation mechanism.

COLOR MASKING

The colors of dyes are dependent on specific atomic vibrations within the dye molecules, and there are limits to what we can do to modify these. In our present state of knowledge we cannot produce a perfect magenta that absorbs all green and transmits all red and all blue light. The best magenta we can manage still absorbs around 20 percent of blue light, and thus behaves as if it had 20 percent yellow dye mixed into it. Thus magenta shows a hue shift towards red. Cyan is an even poorer performer: it absorbs about 20 percent of green and 15 percent of blue. It can thus never demonstrate 100 percent saturation, and behaves as if it had 15 percent red, plus 5 percent magenta, mixed into it. Fortunately, we hardly ever need to record 100 percent magenta or cyan. So in principle all we need to do to get the colors correct is to remove some yellow dye wherever magenta occurs, and to remove some red wherever cyan occurs. This is achieved by incorporating 20 percent yellow dye in the green-record emulsion and 20 percent red dye in the red-record emulsion, both dyes being bleached wherever the image is formed in that emulsion. The result is that the magentas and cyans are adequately corrected, and the whole negative is effectively overlaid by a uniform layer of 20 percent yellow and 20 percent red. This overall orange cast is allowed for in the manufacture of the print paper by extra sensitivity to blue and green. By careful balancing of this integral masking, as it is called, the similar errors in the dyes of the print material can be allowed for in advance. Masking is dealt with more fully in Chapter 12.

The figures given here are approximate, as they vary from one type of film to another. There is usually less than 5 percent error in the yellow, so there is no real need for compensation here.

Cross-Processing

It is possible to process transparency material to a color negative using standard color negative processing techniques. This is called *cross-processing*, and a print made from such a negative shows high contrast and color saturation, usually with some anomalous shifts in hue because of the lack of integral color masking. Commercial and fine art photography sometimes makes use of the technique, and it can be useful in producing diagrams for projection, using the negative itself as a slide. However, these methods are not 100 percent predictable, and you can usually obtain the result you want more easily by digital modification of the image (Chapter 11). Although the photographic darkroom still holds its attractions, and in some circumstances has no substitute, most professional, and almost all amateur, photographers now prefer scanning and printing systems that bypass the whole enlarging and printing stage.

Digging Deeper

In this chapter I have cut a good many corners. In particular I have had to leave out any discussion of dye-coupler chemistry. For a comprehensive account of this and color processes in general, *The Theory of the Photographic Process* (4th edition, ed. T. James, Macmillan, 1977) is the standard work (earlier editions were by Kenneth Mees, then Mees and James). You can keep up to date with *The Imaging Science Journal*, published bimonthly by the Royal Photographic Society. For a simple expansion of the material in this chapter, *The Manual of Photography* (10th edition, Focal Press, 2010) is helpful, but the older 8th edition has more of the chemistry background. If you are interested in the way color photography developed, don't miss the late Jack Coote's *The History of Color Photography* (Fountain Press, 1997), which is a fine exposition of the principles of color photography from a historical viewpoint. A classic to hunt down in second-hand technical bookshops is D. A. Spencer's *Color Photography in Practice* (3rd edition, Pitman, 1937), which is a gem. Spencer was Kodak's senior technologist and managing director of Vivex, the only company to make carbro prints commercially. After his death the book was regularly updated by Focal Press's authors, but the original, with its beautiful illustrations printed separately and individually tipped in, is a book to treasure. Spencer discusses both the carbro and dye-transfer processes in detail, as well as more recent techniques. The marginal note on du Hauron is a quotation from Spencer's dedication of this book to du Hauron's memory.

If you were in any doubt about the validity of colorimetry as a scientific discipline, have a look at Robert Hunt's *The Reproduction of Color* (5th edition, Focal Press, 1998) and his companion volume *The Measurement of Color* (3rd edition, Focal Press, 1998), where there is enough mathematics to satisfy the most masochistic seeker after colorimetric enlightenment. (Robert Hunt was the begetter of the CIE chromaticity diagram.) I haven't said much about subjective color effects. This is a large and controversial subject. Among those books I can safely recommend are *Eye and Brain*, by Richard Gregory (5th edition, OUP, 1998), *Visual Perception*, by Nicholas Wade and Michael Swanson (Routledge, 1991), and *Perception*, by Ian Rock (Scientific American Library, 1984).

The development of reflection holography (Chapter 17) has sparked off a renewed interest in Lippmann color photography, with which it has some common elements. Some of the best recent research papers may be found in the *Proceedings of the 5th Lake Forest Symposium on Display Holography* (SPIE Press, Vol. 3358, 1997). In the United Kingdom, Dr. Hans Bjelkhagen, one of the protagonists of modern Lippmann photography, currently heads a research team at OpTIC Technium, St. Asaph, North Wales.

Chapter 8 Still Cameras

Early Cameras

Although cameras come in a large variety of shapes and sizes, they all consist basically of a lightproof box with a lens at one end, a light-sensitive recording material at the other, and a shutter to control the exposures. A further necessity is a sighting device or viewfinder. The earliest cameras of Louis Daguerre in France and W. H. F. Talbot in England were indeed little more then wooden boxes with a lens cap for a shutter. Talbot's wife called his cameras "Henry's mousetraps."

Talbot's full name was William Henry Fox Talbot. "Fox" was a forename, not a surname, and he never used it.

Shutters

The earliest photographic materials were not very sensitive to light, and often demanded exposure times of many minutes, but with the advent in 1841 of Josef Petzval's wide aperture lens, with some 15 times the transmittance of other current lenses, and of Frederick Scott Archer's wet collodion process 10 years later, exposure durations of a second or less became common, and it became necessary to add shutters to cameras. At first these were crude roller blinds fitted to the front or rear of the lens. George Eastman's Kodak camera, introduced in 1888, had a rotating sector shutter driven by a spring. Rotating sector shutters are still used in modern cine cameras.

Today there are two different types of shutters used in cameras for still photography. The first is the *leaf shutter*. This is usually situated within the lens close to the iris diaphragm (the *intralens* position), though in studio cameras it may be behind the lens. It consists of a number of metal leaves (usually two to five) mounted on pivots on a ring which, when rotated, flicks the leaves rapidly outwards (Figure 8.1) and returns them at the end of the exposure time. On nonautomatic cameras with this type of shutter the series of exposure settings is a geometrical progression of halving: 1 s, then ½, ¼, 1/8, 1/15, 1/30 s, down to 1/500 s (sometimes 1/800 too).

The other main type of shutter consists of a blind situated immediately in front of the film; for this reason it is known as a *focal-plane (FP) shutter* (Figure 8.2).

The figures are rounded off, by international agreement, though it is hard to see why, as "256" takes up no more room on a dial then "250." It would have been more logical to have agreed to rate exposure durations in milliseconds rather then fractions of a second.

Figure 8.1 Action of a leaf shutter. The outer ring is fixed. Rotation of the inner ring causes the blades to open or close.

In modern cameras FP shutters have two separate blinds (or curtains). The exposure duration is controlled by the length of time between the release of the leading or front blind and the release of the trailing or rear blind. In nonautomatic cameras the series of exposure durations also follows a progression of halves, but the minimum figure can be as low as 1/8000 s. To achieve the shortest timings the rear blind is released before the leading blind has completed its traverse, so that the two shutter blinds form a moving slit.

Shutter Efficiency

With both leaf and FP shutters it takes a finite length of time to uncover the whole lens aperture. The total time for which the lens is partly uncovered is greater than the effective exposure duration, and this gives rise to the concept of shutter efficiency. The effective exposure duration is approximately equal to the time from the shutter "25 percent open" point to the "75 percent closed" point, and dividing this by the time taken from the beginning of opening to the end of closing gives the approximate efficiency figure. More accurately, if you make a plot of the illuminance at the focal plane against time, the shutter efficiency is the area under this curve divided by the area of the rectangle enclosing it. This is illustrated in Figure 8.3.

From the plots in Figure 8.3a and b you can deduce that the opening and closing patterns for the leaf shutter are asymmetrical, and that shutter efficiency is higher at small relative apertures and lower at shorter exposure durations. At the shortest durations the shutter spends its entire time opening and closing. Under these conditions the efficiency for the leaf shutter is only about 40–45 percent, at full aperture. You may have to take this into account when trying to "freeze" the subject in an action shot, as the shutter may be at least partially open for up to twice the time set. FP shutters are generally more efficient than leaf shutters, as the efficiency depends chiefly on how close the shutter blinds are to the film plane, whereas the efficiency of a leaf shutter depends not only on its mechanical properties, but also varies inversely with the diameter of the iris diaphragm.

Timing Mechanisms

Mechanical leaf shutters achieve their longer exposure times by a ratchet delaying mechanism, and their shorter times by varying the spring tension on the movable ring that operates the leaves. Both of these mechanisms can become worn with time, and older cameras need checking occasionally. FP shutters use a delaying mechanism for longer exposure times but obtain the shorter times by

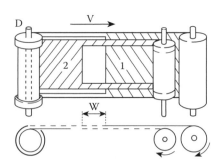

Figure 8.2 Action of a focal-plane shutter. The exposure duration is controlled by the interval between the release of the leading blind (1) and the trailing blind (2). For the shortest exposure durations the two blinds form a slit, width *w* in the diagram.

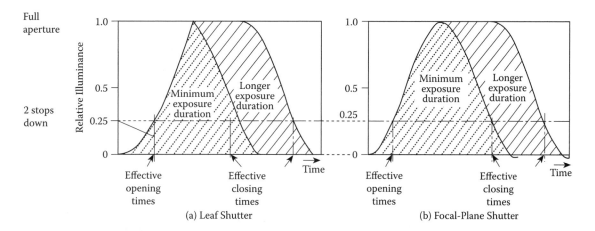

Figure 8.3 Shutter efficiency: At full aperture the shutter spends its entire time opening and closing, and efficiency is a minimum: 42 percent for a (a) leaf shutter and 50 percent for a (b) focal-plane shutter. At longer durations and/or smaller apertures the efficiency is increased; at smaller apertures the effective exposure duration is also increased.

varying the slit width. Checking the timing mechanism of a shutter for accuracy is easy with a film camera: you simply shoot off a series of exposures with the lens pointed at a plain white wall, and make a series of exposures varying the duration but keeping the exposure itself nominally the same, e.g., 1/500 at f/2.8. 1/250 at f/4, 1/125 at f/5.6, and so on. Place the developed film on a sheet of white paper and check the density (darkness) of the negatives. They should all be equally dark. If they are not, you will be able to see which way the exposure error goes, and compensate as necessary by adjusting the f/no. (You can usually take the 1/60 setting as approximately correct.) The most common errors in an older camera are a gradual change from one end of the exposure series to the other, and a jump between 1/30s and 1/60s or between 1/15s and 1/30s, i.e., the point where the internal shutter timing mechanism changes to a spring loading system.

Remember, the f/no series goes in steps of $\sqrt{2}$, each step representing a halving of the light transmission of the lens. These tests only apply to mechanical shutters on older film cameras: most shutters on modern cameras are electronically controlled, and both exposure and aperture may be altered in smaller steps.

Focal-Plane Shutter Distortion

There is another important property of FP shutters, namely that for short exposure durations the slit is narrower than the frame, so the time it takes to traverse the frame is longer than the time it takes to expose a single point on the film. With a fast-moving object the image may have moved some distance in the frame between the start and the finish of the exposure. For example, if you have a shutter blind that moves downwards (most of them do), as the image is inverted the shutter slit will record the bottom of the scene a fraction earlier than its top, and if you are photographing, say, a racing car, it will appear to be leaning forward. With a horizontally moving blind the car will be compressed or elongated, depending on whether the shutter moves in the opposite direction to the (inverted) image of the car, or in the same direction (Figure 8.4).

Other Types of Shutters

There are other shutter types, such as louvre and contrarotating sector shutters, but these are used only in specialized cameras. There are also shutters operating on electro-optical principles, and these do not have any moving parts. The most

115

(a)

(b)

Figure 8.4 Focal-plane shutter distortions. (a) When a focal-plane shutter aperture moves in the same direction as a moving image the result is elongated. (b) When it moves in the opposite direction the result is compressed. (Photograph by Jon Tarrant.)

important of these is the Pockels cell, a device consisting of an optically active crystal mounted between crossed polarizers. When an electric field is switched on, the crystal rotates the polarization through 90°. Such shutters are capable of controlling exposure durations measured in nanoseconds.

Types of Camera

Simple Cameras

The simple film camera has a long history. The first commercial model was marketed by George Eastman in 1888, and this led to the Brownie series, which was the archetypal model for the genre, with its roll film, sector shutter, and single f/14 meniscus lens set to the hyperfocal distance. Some later models had a range of fixed stops and a crude focusing arrangement. Folding models appeared around the turn of the nineteenth century. So many variants of the simple camera have been marketed over the past hundred years that a full historical account of them would demand a thick book.

Today the simple camera is (24 × 36 mm) on perforated 35-mm film, or APS format (16.7 × 30.2 mm) on 24-mm film, the various other formats having fallen by the wayside. It is still a box with a fixed-focus lens ("focus-free" in marketing-speak), though it may have some extras such as a built-in flash.

Single-Use Cameras

These cost little more than the price of the film they contain. Manufacturers reclaim and reload them for resale (a throwback to George Eastman's original concept).

Any volunteers?

Don't scorn the simple camera. Within its limits it can provide results as good as professional cameras costing a small fortune, and if you leave them around they are less likely to be nicked. The first photograph of mine to be hung in an international exhibition was taken with one. One of the best (and cheapest) is the Russian Lomo, which even has its own fan club.

Compact Cameras

These were developed to fill a marketing niche for people who wanted a comparatively simple camera that was more versatile than the basic box. Present-day digital compact cameras have automatic exposure control, automatic focus, and a flash that operates when the light level is low. Many have zoom lenses. The most sophisticated compacts rival professional cameras. In fact, professionals often keep them as a second camera to use in the field. Although some cameras are equipped with optical viewfinders, most rely on LED displays. Film cameras are invariably 35-mm or APS format, though the latter are beginning to disappear from the market.

Rangefinder Cameras

Since the widespread adoption of automatic focusing systems, the amateur rangefinder camera has all but disappeared, though there are a number of professional cameras of this type. They stem from the old Leica and Contax designs and their offshoots, in which an optical rangefinder is coupled to the focusing mechanism. (I discuss mechanisms in more detail later in this chapter.) Rangefinder cameras are particularly useful in poor lighting conditions where finding focus on a focusing screen is difficult, and they are a boon for a photographer cursed with poor eyesight, or in circumstances where automatic focus control is inappropriate. Medium-format rangefinder cameras using 120-size film are versatile and light compared with their single-lens reflex equivalents.

The forerunner of the medium-format rangefinder camera was built by Fairchild for the U.S. Forces. It took 70-mm perforated film in reloadable cassettes. Because of its large size it became known as the "Texas Leica."

Twin-Lens Reflex (TLR) Cameras

These have a viewfinder that is virtually a second camera, with a ground glass viewing and focusing screen and a lens coupled to the taking lens. The term "reflex" refers to the fact that the image in the finder is formed via a 45° front-surface mirror so that the viewing screen is horizontal. Most finders incorporate a Fresnel lens to improve the image brightness at the corners, and a foldaway loupe to aid focusing. Because of the mechanical and optical excellence of the better models, they were usually the choice of photojournalists and sports photographers during the 1940s and 1950s, but they had several serious drawbacks. The viewfinder image was laterally reversed; there were parallax problems in close-ups; they had to be operated from waist level; and (with one notable exception) the lenses were not interchangeable. They are now little used except by a small die-hard coterie.

Waist-level (as opposed to eye-level) operation did have the advantage that photographs of celebrities at parties came out with their figures in proportion, rather than with a huge head and tiny flat feet.

Single-Lens Reflex (SLR) Cameras

The earliest SLR cameras appeared in the 1890s, and were used chiefly for portraiture. The reflex mirror for the viewing and focusing screen was situated in the camera body, and the focusing lens was the actual camera lens, which was interchangeable. Before taking the photograph it was necessary to close down the lens aperture and raise the mirror, which blacked out the viewfinder, making the use of a tripod essential. (The TLR camera was designed to overcome these difficulties.) The first SLR camera to feature instant mirror return appeared on the market soon after the end of the Second World War, and this innovation was quickly followed by the pentaprism viewfinder (which gives an upright unreversed image at eye level), and by a stop system that allowed focusing at full aperture, the iris diaphragm closing down for the exposure itself.

SLR cameras have almost completely taken over from other cameras for photoreportage, action photography, and most social events. The more advanced models contain computer programs that can be switched in to provide for all types of photographic situations, and complex systems for computing correct exposure and focusing. Nearly all such cameras are now digital, and the standard format for these has until recently been 18 × 25 mm. Some manufacturers prefer a slightly larger ("four-thirds") format, and there is a general move towards full 35-mm format (24 × 36 mm), with its higher pixel count. There is a growing professional market for medium-format SLR cameras; these are often based on designs originally intended for film, and some have interchangeable film or digital backs.

Field and Studio Cameras

These are of two main types: *baseboard* and *monorail*. The baseboard type, usually called a *technical camera*, has a base plate with rails on which the lens assembly slides for focusing. The lens panel is connected to the back by bellows, and has sideways and vertical movements as well as tilt and swing (rotation about horizontal and vertical axes); the back rotates to give either portrait or landscape format, and may have adjustments for tilt and swing. You can use a technical camera in the open for landscape and architectural work on a tripod, and can also operate it handheld. Fine focusing is by a ground glass viewing screen, but for handheld work most technical cameras have a coupled rangefinder adjustable for several focal lengths. The cameras can be folded up for transport. Technical cameras, as you have probably guessed, are the descendants of the old wooden stand camera with a black cloth and of the big press camera with a baseboard and a huge flashgun. They are usually 4 × 5 inches, though there are smaller formats. They can be used with interchangeable backs taking sheet film, roll film, Polaroid material, plates, or a digital sensor.

The monorail camera is not so much a single camera as a system. It is based on the optical bench principle, and all the components are interchangeable. The lens panel is mounted on lockable gimbals, usually (though unfortunately not invariably) pivoted so that the rear nodal point remains stationary, and so the image remains stationary when you operate the swing and tilt movements. The swing and tilt axes of the camera back pass (or should pass) through the optical center of the format. The bellows set is also interchangeable, with long bellows for close-up work and bag bellows for wide-angle photography with a short-focus lens. Tilt and swing angles are calibrated in degrees, and shift movements in millimeters. The format is usually either 4 × 5 in or 8 × 10 in. Again, there is a choice of backs.

Self-Processing Cameras

Polaroid and other *in situ* processing cameras differ from other cameras chiefly in their film transport system, which releases the processing chemicals and either delivers the film ready to separate from the print after an appropriate interval, or (in more recent models) simply delivers the developing print. The film-stripping cameras have normal camera optics, but the newer models have a folded imaging beam path including a mirror, so that the image emerges the correct way round.

As I pointed out in Chapter 7, the future of this particular technology is uncertain.

Specialized Cameras

There are many different types of specialized cameras, for example, the fundus camera for photographing the inside of the eyeball, and the periphery camera for making a continuous record of the outside or inside of a pipe (or any other complete circumference). The more important specialized cameras turn up in the appropriate chapters of this book. Some of them that you may come across in a working environment are:

Underwater cameras. These are sealed in watertight containers, or are themselves watertight. They have special controls that are easy to operate under water. As a rule they have fixed lenses of short focal length, to compensate for the magnification due to the refractive index of water. Some single-use cameras can be used under water, too.

Aerial cameras. Aerial survey cameras are carried on fixed antivibration mounts, and are operated by an *intervalometer*, which computes the time interval between exposures for correct overlap. Survey lenses are wide-angle (90°) and are designed to have less than 0.1 percent distortion, each film register glass being individually calibrated. Cameras for aerial reconnaissance are no longer the long-focus monsters of a few decades ago (satellites have taken over their job), but small cameras using 70-mm film or charge-coupled device (CCD) chips, with rates of exposure of 12 pps or more. Handheld aerial cameras are rugged box cameras with fixed-focus lenses and handgrip operation.

Wide-angle and panoramic cameras. These usually have a noninterchangeable lens with a rectilinear field of around 110°. The rotating type of panoramic camera has a normal-focus lens that rotates about its rear nodal point, coupled to a moving slit field aperture. Alternatively, the whole camera may rotate. In this case the slit is fixed and the film moves in synchronism with the image.

Rostrum cameras. The rostrum camera has joined the scanner in replacing the enormous processing cameras that used to grace press darkrooms. A rostrum camera is a sophisticated and versatile copying camera, set up vertically like a photographic enlarger, which can be used for copying drawings and transparencies and for animations.

High-speed cameras. These are important research tools, and are dealt with in Chapter 9.

Figure 8.5 shows examples of the major types of cameras.

Viewfinders

Focusing Screens

Studio cameras are not equipped with viewfinders, because you compose and focus the image on a ground glass screen. Such screens have always presented a problem: if they are coarsely textured the image is bright overall, but focusing is imprecise; if they are finely textured (acid-etched), focusing can be more precise, but the image becomes very dim towards the corners of the frame. The situation is improved by positioning a Fresnel lens in contact with the ground glass to redirect the light into the viewer's eye. The lands of the Fresnel lens need to be closely packed so as not to be obtrusive, and the central area is usually left plain. New screens have been developed using microlithographic and holographic techniques that redirect the light by diffraction.

Frame and Optical Finders

Some technical cameras, particularly the larger ones formerly used as press cameras, have a wire frame finder; such finders are easy to use in bad light, and move

The Science of Imaging, Second Edition: An Introduction

with the lens panel when using the rising or cross front to keep the picture correctly framed. There are several patterns of optical finder, mostly based on the optics of a reversed Galilean telescope. Some, designed for cameras with interchangeable lenses, have variable framing to suit the different focal lengths. The *Albada* finder (Figure 8.6) has a front component surrounded by a partially silvered rectangle that is projected into the view to indicate the edges of the field. This is useful in sports photography where the subject matter may be continually moving in and out of the frame. More recent versions have the framing rectangle generated within the camera and projected into the viewfinder field via a beam combiner. However, as mentioned above, most amateur and small professional cameras incorporate an LED screen, usually with a somewhat magnified image.

Twin-Lens Reflex Finder

The twin-lens reflex (TLR) camera has a viewfinder that is optically a replica of the main camera, except that it has a 45° front-surface mirror to erect the image on a horizontal focusing screen. The two lenses are mounted on a single panel, so that focusing the finder also focuses the taking lens. One advantage over optical finders is that the image is full size, but there are several disadvantages, as we

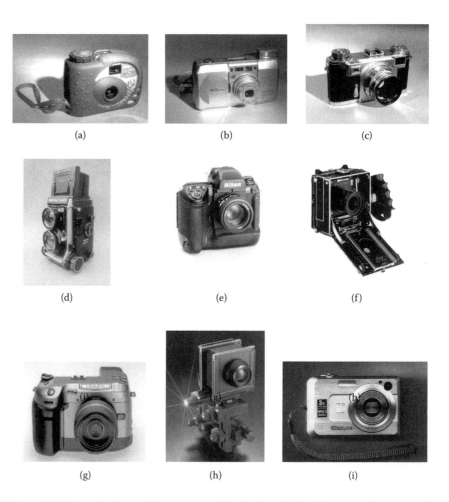

(a)   (b)   (c)

(d)   (e)   (f)

(g)   (h)   (i)

Figure 8.5 The main types of camera are: (a) simple, single-use camera (waterproof); (b) compact camera; (c) classic rangefinder camera; (d) twin-lens reflex camera; (e) single lens reflex camera; (f) technical camera; (g) medium-format digital camera; (h) monorail camera; and (i) compact digital camera with movie facility.

120

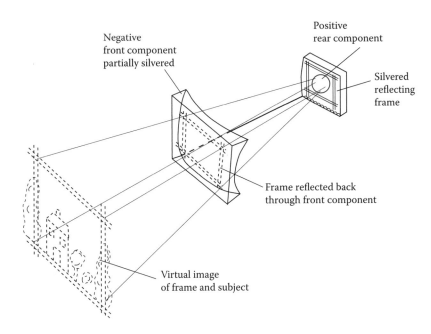

Figure 8.6 Optical arrangement of an Albada finder.

have seen. The fact that the viewfinder lens is displaced from the taking lens by up to 7 cm means that the actual frame is much lower than the viewfinder frame, and in close-up the top of the composition may be cut off. In some TLRs a moving mask on the finder screen compensates for this, and supplementary close-up lenses come with an optical wedge for the finder lens to further compensate for parallax, but the viewpoints of finder and taking lens still differ, and there may even be an obstruction your finder doesn't see, for example, if you are photographing through a grille.

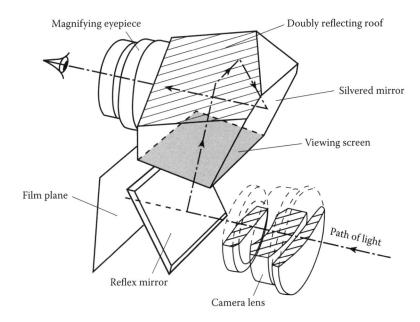

Figure 8.7 Pentaprism viewfinder for single-lens reflex camera.

Single-Lens Reflex Finder

In early large-format SLR cameras the viewfinder image was reversed right to left as in TLR viewfinders. Some smaller SLR cameras with additional sports finders were made during the 1930s, but the breakthrough came in the 1950s with the appearance of the instant-return mirror and the pentaprism finder. A pentaprism is a kind of roof prism with an extra reflecting surface (Figure 8.7). Early SLR cameras were difficult to focus at small lens apertures, and this led to the development of full-aperture focusing, with the iris diaphragm remaining wide open except during the actual exposure.

Rangefinders and Focus Finders

The coupled rangefinder appeared at the same time as the 35-mm camera. The optical principle was an old one, used in the First World War by artillery regiments and naval vessels. Two 45° mirrors a meter or more apart produced images in a telescope sight via a beam combiner (Figure 8.8). One of the mirrors was rotated until the two images coincided, and the range was read from a scale attached to it.

The rangefinder in a camera is basically a smaller version of this. In the rangefinders of technical cameras a system of levers connects the mirror movement to the focusing system. In 35-mm and medium-format rangefinder cameras both mirrors are usually fixed, with a combination of moving wedges in one of the two optical paths; this raises the mechanical advantage and reduces possible backlash (Figure 8.9).

TLR and SLR cameras have a ground glass focusing finder, and since the image is small, it can be difficult to find an exact focus with the unaided eye. In the studio, of course, you can use a loupe, and most TLR cameras have one hinged to the hood. SLR film cameras have an optical focus finder at the center of the field. It consists of two small wedges, called (rather misleadingly) a "biprism," one above the other,

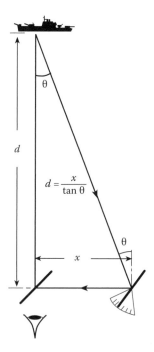

$$d = \frac{x}{\tan \theta}$$

Figure 8.8 Principle of the rangefinder.

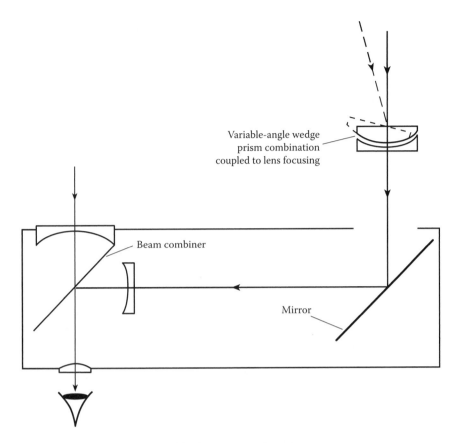

Variable-angle wedge
prism combination
coupled to lens focusing

Beam combiner

Mirror

Figure 8.9 Rangefinder as used in the post-war Contax camera.

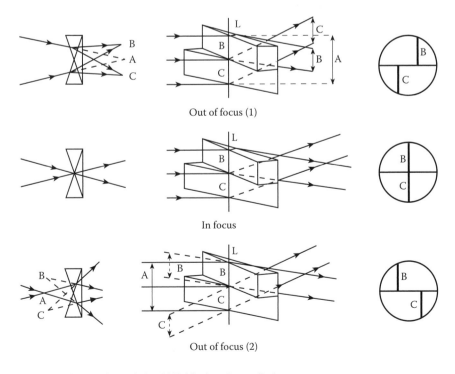

Out of focus (1)

In focus

Out of focus (2)

Figure 8.10 Principle of the SLR biprism focus finder.

with equal and opposite slopes. The crossing point is in the plane of the focusing screen (Figure 8.10). Light entering the biprism is deflected in opposite directions to give a horizontally split image. The two halves coincide when the image is in the plane of the focusing screen.

Instead of a single biprism an array of "microprisms" can be used. This gives a shimmering mosaic pattern that disappears when the image is in correct focus. Both systems depend on the use of a fairly large lens aperture. At apertures smaller than about $f/11$, as well as with very long-focus lenses, you simply see a black disc or half-disc at the center of the field unless your eye is very carefully aligned.

Automatic Focus Control

There are three techniques for automatic focus control in use at present: ranging, contrast measurement, and phase detection.

Ranging

Little used now, the sonar autofocusing principle was the earliest successful method of ranging. In this system the default setting has the lens at its closest focusing distance. In operation, a piezoelectric transducer emits a pulse of ultrasound, and at the same time the lens begins to move towards its infinity position; the returning echo is used to stop it. As the elapsed time is proportional to the subject distance, the rate of movement of the lens needs to be suitably matched to the speed of sound. This system will operate in the dark, of course, but it is thrown by the presence of an intervening glass window. Other ranging systems include a scanning system, where the output of a swinging mirror synchronized with the focusing movement is correlated with the output of a fixed mirror (Figure 8.11); when the correlation is a maximum the subject is in focus. This principle is similar to that of the optical rangefinder, and works best in bright light.

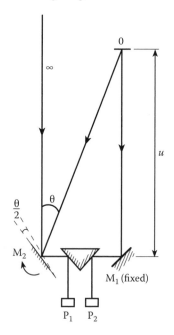

Figure 8.11 Ranging autofocus (scanning system). The swinging mirror M_2 scans the scene until the outputs of the photoreceptors P_1 and P_2 are correlated. From the mirror deflection $\theta/2$ the distance of the object O can now be computed and set.

Image Contrast

When you focus an image on a focusing screen your main clue to correct focus is not so much the fine detail as the contrast of edges, which is a maximum when the image is in focus. Most autofocus compact cameras use this principle. A row of charge-coupled devices (CCDs) is set at the equivalent focus (that is, at an optical path distance equal to the focal length, using either a beamsplitter or a separate optical system) and its intensity profile analyzed by a signal-processing unit for sharp/unsharp characteristics. A variant is to use two or even three sets of CCDs at different focusing distances (Figure 8.12). The signal-processing unit then selects from the outputs (equal output from two rows, sharpest output from three). This system is capable of very rapid response. The multiple-set CCD systems show unambiguously the direction in which the focusing is to move. As with the ranging system, the image contrast system does not work very effectively in poor light.

Phase Detection Focusing

This is at present the preferred system for the better quality SLR cameras. The CCD arrays are mounted behind the equivalent focal plane, so that the pencils of rays have come to a focus and are diverging again. They are refocused onto the CCD array by small lenses (Figure 8.13); the distance apart of the images depends on the amount of divergence of the rays, being closer together for too far a focus and farther apart for too near a focus. This works very rapidly, and in poor light can be boosted by a dedicated flash head programmed to give a short burst of weak red light.

Automatic Exposure Control

Exposure meters measure illuminance: light from the subject falls on a light-sensitive device, which may be photovoltaic (selenium), photoconductive (cadmium sulphide),

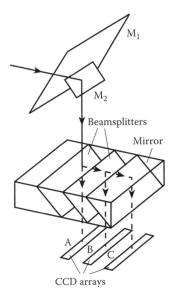

Figure 8.12 Image contrast autofocus (multi-array system). A small mirror M_2 on the back of the main reflex mirror M_1 directs part of the image-forming light onto three arrays of CCDs, the middle one of which (B) is in the equivalent film plane. The lens is driven until B shows the sharpest image edge characteristics.

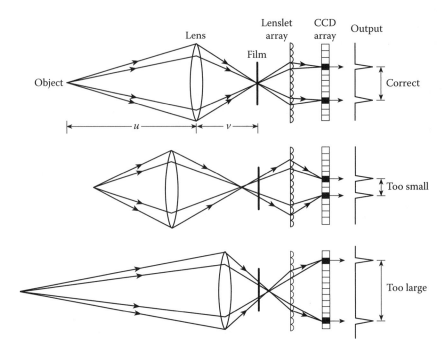

Figure 8.13 Phase detection autofocus. A CCD array is behind the film plane, and the diverging rays are focused on it by a row of lenslets. The separation of the images depends on the divergence of the rays, being too large for too near a focus and too small for too far a focus.

or a photodiode or phototransistor (silicon). These last react many times faster than the first two types; they can even measure the luminous output of an electronic flash. With a separate handheld meter you can take a measurement at the subject with the meter pointed at the camera: this is called "incident light measurement." You will then get the same reading regardless of the brightness of the subject, and this is the way it should be, certainly for transparencies where you want a black object to look black and a white object to look white. A meter that is built into the camera can only point forward at the subject matter, and it will give very different readings for a black object and a white one. If you believe the meter, you will finish with two images the same shade of grey. This is the reason incident light meters are de rigueur in professional studios.

Out of doors, built-in meters come into their own. They are calibrated on the basis that the average subject has an overall reflectance of 18 percent, which is, visually, a mid-grey. However, if your subject is a little unusual, such as a glade with dark shadows, or a snow scene where almost everything is white, you have a problem if you are working with transparencies. With a separate meter you can always go up close and measure a part of the subject that *is* a mid-grey (or take a reading on a standard 18 percent grey card), but when, as is usual with modern amateur cameras, the meter actually controls the exposure, your glade will be overexposed and your snow scene will be grey. The better amateur cameras cope with this, after a fashion, by weighting the center of the picture more heavily than the outer part (Figure 8.14a), though you still need to make allowances for nonstandard subject matter (usually by adjusting the ISO setting). More sophisticated systems can be switched to measure only a very small spot, which you choose yourself, holding the reading while you make the exposure (Figure 8.14b). The most advanced systems take simultaneous readings over five or so more areas of the field and ignore anomalous areas according to an inbuilt program that you can set according to the nature of the subject matter (Figure 8.14c).

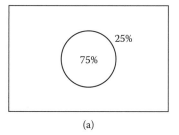
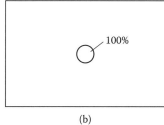
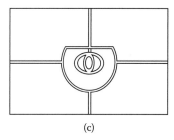

(a)   (b)   (c)

Figure 8.14 Through-the-lens metering options: (a) center weighted; (b) spot metering; (c) matrix metering.

How does the metering system control the exposure? With the older, cruder semi-automatic systems you had to match the meter needle, turning the iris diaphragm or shutter-setting ring until the needle coincided with a marker. With fully automatic operation, the signal from the meter is amplified and fed to servos that control the aperture setting and exposure duration. In through-the-lens metering in SLR cameras, a part of the area of the reflex mirror is made sufficiently transparent to allow a small amount of light through, and this is reflected onto the light-sensitive device(s). The simplest form of automatic exposure control is a program that starts off at the smallest aperture and shortest exposure duration for the brightest scene, then for dimmer scenes progressively opens to full aperture, thereafter lengthening the exposure duration as necessary. Another program shares the exposure increase between the aperture and the exposure duration. Professional and semiprofessional cameras offer a choice of *aperture priority*, where you set the aperture and the meter controls the exposure duration, or *shutter priority*, where you set the shutter and the meter looks after the aperture. These cameras also have a control that lets you uprate or downrate an exposure (bracketing) for unusual scenes. Top-grade amateur and semiprofessional cameras also contain specific exposing programs for such projects as sports shots, close-ups, silhouettes, and so on. In these cameras there may be as many as 60 sampling points across the format.

Flash Synchronization

Until fairly recently, you could still come across cameras that employed expendable flashbulbs to illuminate dimly lit scenes or to fill in dark shadows. The flash was produced by the electrical ignition of a bundle of fine ribbon of zirconium or magnesium metal, the bulb being filled with a matched quantity of oxygen. As the full power of the flash developed only after a delay of 5 to 40 milliseconds, depending on the type of bulb, it was necessary to delay the full opening of the shutter by a matching time interval, or at least to ensure the shutter was held open until the flash had passed its peak power. Flashbulbs are seldom seen today. Even the giant flash bombs once used by the RAF for nighttime reconnaissance have given way to electronic flash.

Modern electronic flash heads have a flash duration of no more than one or two milliseconds, often much less, and synchronization now is simply a matter of ensuring that the shutter is wide open when the flash fires (Figure 8.15). The flash-tube contains a mixture of gases, mainly xenon at low pressure, which, when ionized by a brief high-voltage pulse (the "trigger"), conducts electricity freely, allowing a large current to flow briefly and generate a burst of light energy that is spectrally a close match to daylight. The power comes from a bank of capacitors that have been precharged. The procedure is quite safe as it is only the very small trigger current that passes through the camera circuitry.

The photometry of flash was dealt with in Chapter 2.

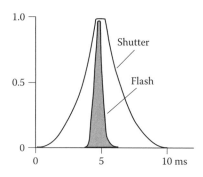

Figure 8.15 Synchronization of electronic flash (leaf shutter).

Most compact cameras have integral flash heads, which can be set to operate automatically when the light is inadequate for an exposure by ambient light, and can also be set manually for fill-in purposes in daylight. In more advanced models the flash duration may be controlled in real time by the measured subject illumination. Flash heads for professional and semiprofessional cameras are made to match specific cameras, and are said to be *dedicated*. They operate in a much more sophisticated way than general-purpose flashguns. Most of them measure the light actually falling on the film during the exposure, and quench the flash when sufficient exposure has been given. This is called *through-the-lens (TTL)* metering. It is not normally possible to use electronic flash with an FP shutter at very short exposure times, as the shutter is then only a narrow slit, though the more advanced flash heads can be programmed to break up the flash into a series of rapidly repeated short flashes so that the illumination is effectively constant for the whole of the shutter transit.

Flash heads are usually mounted close to the lens axis, which is not the best place for the main light. However, most heads can be angled to reflect from a side wall or ceiling, to produce a more general illumination. A specialized type of head called a *ring flash* fits round the camera lens and thus produces shadowless frontal lighting. It was originally intended for medical photography, but is often used in fashion work for special effects. Flash systems can also be used with extension leads, though more often slave flashes are operated remotely, their operation being triggered by the master flash on the camera. No such facility exists for the cheaper compact cameras, and social photographs taken using older models show the dreaded "red-eye" effect caused by reflection of the flash from the subjects' retinas. Some compact cameras,

Figure 8.16 Use of rear curtain synchronization (focal-plane shutter). The blur is behind the subject, giving a natural impression of movement. (From Nikon F-90X handbook.)

and all professional equipment, now include the facility for a carefully timed pre-flash, which closes the subjects' pupils down sufficiently to avoid the effect.

In a night flash shot containing moving figures, any bright or self-luminous part of the subject will show a trail added to the sharp image. If the flash is synchronized to the opening of the front blind, and the shutter remains open for a short time after the flash, the trail will be in front of the moving subject, and the result may look odd. Newer dedicated flash heads have a "rear curtain" option that fires the flash at the end of the exposure instead. Any trails now appear behind the moving subject, and look more natural (Figure 8.16)

GUIDE NUMBERS

A guide number is given to a flash head (not the dedicated kind) to relate the ISO speed, the *f*/no, and the flash-to-subject distance. The relationship is

Guide no ÷ Distance = *f*/no

Guide numbers are usually given for distances measured in meters, and for various ISO speeds. For studio equipment they are usually also listed for half and quarter power.

Camera Shake and Stabilization Mechanisms

In a static camera system with a stationary subject, the only kind of unsharpness you are likely to encounter will come from residual lens aberrations, incorrect focus, and light scatter within the thickness of the film. But most amateur and many professional photographs are shot handheld, and here the major cause of unsharpness is what is known by the unlovely title of *camera shake*. It is such an important item in the photographer's tally of gremlins that it is worth examining in some detail. First, it isn't shaking, of course, just movement. A handheld camera has six degrees of freedom, three of which are translational (movement laterally, vertically, and longitudinally) and three rotational (rotation about the camera axis, horizontal rotation, and vertical rotation) (Figure 8.17).

Translational movements have very little effect on the image sharpness (except in close-ups): you would need to move some distance to get any unsharpness. But rotation is another matter. Rotation about the camera axis can be discounted, as it is very unlikely unless you do it deliberately. Most camera shake comes down to a combination of sideways and vertical rotation at the instant you press the shutter release. Of course, the effect is reduced by your giving the shortest exposure time you can (and not being afraid to use full aperture when you need it). But even that may not be enough. Suppose your camera swings at the rate of just one radian per second (a radian is about 53°). This is fairly slow: try it with your arm. With an exposure of 1/1000 s (1 millisecond) you will have swung the camera through 1 milliradian. And that, remember, corresponds to the resolution of an average human eye. Now, as I pointed out in Chapter 3, half the population has better visual acuity than this, so you can see that it doesn't take much movement to make your picture visibly unsharp. It is a nasty kind of unsharpness, too. Every point in the image is turned into a short straight line, usually diagonal, and there may be spurious double edges to details.

The use of long focus lenses increases the risk of camera shake for handheld exposures. The rule of thumb is to use an exposure duration that is not more than the reciprocal of the focal length in millimeters, e.g., 1/200 s for a 200-mm lens—and that is for a steady pair of hands. However, technology has come to

When using your handheld camera you can minimize camera shake by leaning against something solid and holding the camera to your face. Heavy cameras are easier to steady, too. Small, lightweight digital cameras that you have to hold at arm's length to see the image in the finder are at the most risk. Unfortunately, they are also the most common.

An analysis of the OTF for uniform camera rotation reveals a phase switch of 180° at high spatial frequencies, which accounts for this effect.

129

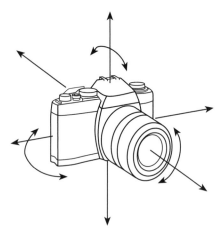

Figure 8.17 The six possible movements of a camera during a handheld exposure.

the rescue here: no less than a servo system that detects movement and immediately compensates by adjusting the position of a floating component in the lens. This system was originally developed for use in high-power binoculars, but is now becoming standard in top-quality long-focus lenses, and even in prime lenses. Any small movement of the lens is detected by a piezo-operated accelerometer that triggers the compensating component movement within microseconds.

Moving subjects are another aspect of unsharpness. Here, the simple answer is to follow the subject by "panning" the camera. This is common practice in sports photography, where it doesn't matter if the spectators are blurred. Panning does not affect stabilization systems, which are programmed to be sensitive only to rapid small movements, and to ignore large-scale swings. But camera shake does in any case seem to be less of a problem in panned shots.

Image Motion Compensation

A special case of image movement occurs in aerial photography with a vertical camera, where the highest definition is essential. As the aircraft flies forward the image moves in the same direction, at a speed given by the relation

$V \times f/h$

where V is the forward speed of the aircraft, h is its height, and f is the focal length of the camera lens, all in the same units (Figure 8.18).

During an exposure the image may move a sufficient distance for the result to be unacceptably blurred. The light-sensitive surface therefore needs to be moved during the exposure at a matching speed. This technique is known as *forward motion compensation*, or *FMC*. The role of high-level reconnaissance aircraft has now been taken over by satellites, but the problem exists for those too, and with present-day tactical photographic sorties being made by ever faster aircraft from ever lower altitudes, even small side-oblique cameras with their very short exposures need FMC.

An oddity among cameras—but an important one—is the photo finish camera. It has no shutter as such (apart from a capping shutter), but instead has a narrow vertical slit in the focal plane, behind which the film can be moved at a controlled

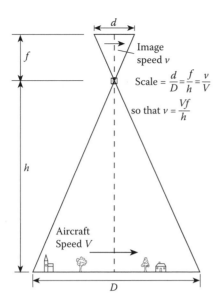

Figure 8.18 Forward motion compensation in aerial photography.

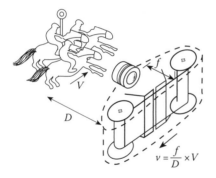

Figure 8.19 Principle of photo finish camera.

speed. The camera is aimed at the finishing line, and the film speed adjusted to match the speed of the horses, athletes, etc. (Figure 8.19).

Slit panoramic cameras, mentioned earlier, share a somewhat similar arrangement. A *periphery camera* is a slit camera that either moves round the object in a 360° circle, or remains stationary while the subject is rotated.

Digging Deeper

Much of this chapter seems to be more technology than science. But it is difficult, and unrealistic, to try to draw a hard and fast line between what is science and what is technology. You can't have one without the other. The modern camera is a blend of roughly equal parts of science and technology, with craftsmanship thrown in too. I have kept to generalities about construction and mechanical operation, partly because for any given camera the machinery itself quickly becomes last year's model, but mostly because this aspect of photography has been covered, and continues to be covered, much better by specialists writing in dedicated periodicals. Most photographic magazines devote regular articles to both new and classic photographic equipment. Of them the *British Journal of Photography* is perhaps the best. Among the textbooks, the *Manual of Photography* (10th edition, Focal Press, 2009) is again a strong recommendation. Sidney Ray's monumental work *Scientific Photography and Applied Imaging* (Focal Press, 1999) is a mine of information about specialized cameras. One of the best books about flash photography is *Images Below*, by Chris Howes (Wild Places, 1997). Ostensibly about cave photography, it contains more hints and tips about flash than the average overground photographer might ever need, as well as some magnificent photography. A more general book, aimed specifically at digital camera users, is *Mastering Digital Flash Photography*, by Chris George (2008). *Mastering Flash Photography*, by Susan McCartney (1998), is a more comprehensive guide including studio work, though a little short on technical details.

Chapter 9 Motion and High-Speed Photography

Persistence of Vision

A photographic flashbulb has a total flash duration of about 1/25 second, or 40 milliseconds (ms). An electronic flashtube produces a flash lasting about 1 ms. A pulse laser's flash lasts a mere 25 nanoseconds (ns). Their durations are very different, though the total energy they emit is similar.

To the eye the durations of the flashes, too, look similar. This is because of a lag in our visual perception processes. We don't perceive the start of the flash until a fraction of a second after it has begun, and we continue to perceive it for a similar time after it has ceased. This is known as *persistence of vision*. And without it cinema and TV simply wouldn't work. They would seem to flicker, just as a slowed-down cine picture does.

In fact, they do flicker, 50 times a second (60 for TV in the United States). Neon lights do, too, as do the LEDs on your hi-fi. They seem steady when you look at them, but as you look away the visual image breaks up into a line of short dashes. If you deliberately look quickly away from a TV picture the result is more interesting: flick your gaze to the right and the chain of images appear as rhomboids tilted to the right; flick your gaze to the left and you will produce rhomboids tilted to the left. Flicking upwards stretches the image vertically, and flicking down compresses it.

The reason for these effects is that the picture is built up line by line successively. You can see flicker in the cinema, too, but as the whole image is displayed at once there isn't any distortion as you look away. A cine camera takes 25 pictures per second (pps) (24 on older models), but as 25 pps is only just above the rate at which the average person can see flicker in a bright image, the projector shutter runs at twice this speed, and shows each image twice. Home movies on 8- or 16-mm film are shot at 16 pps (as were the old silent films) and for these the projector shows each image three times.

The Phi Phenomenon

When there is action during the filming, each successive image is a little different from its predecessor, and these successive changes give the illusion of movement. Even so, the more rapid movements would seem to be jerky except for a quirk in visual perception, called the *phi phenomenon*.

If you watch a row of lights that are lighting up in succession, the impression that the light itself is moving is very strong. This effect is exploited in the "moving" alphanumeric displays you see in public places. The phi phenomenon is so powerful that it will operate with as few as three lights. This effect, added to persistence of vision, and the slight blurring of images of fast-moving subjects, goes to reinforce the impression of genuine movement in the image.

A nanosecond (ns) is one thousand millionth (10^{-9}) of a second. It is a very short time indeed: in 1 ns a beam of light travels only about 30 cm. So in 25 ns a light beam would just about cross the average optics lab.

If you flick your eyes down rapidly you can squeeze the images to single bars, and if you do this fast enough (flick your head and neck, too) you may actually succeed in inverting the picture. (But make sure nobody is watching you.)

In order to show an old silent film at its original filmed speed on a modern projector, every other frame has to be shown twice. Unfortunately this is not always done.

Animated cartoons made before the computer age do seem jerky. This is because most of the drawn images were shown two or more times, thus saving a good deal of work in the draftsmen's studio. Modern computer techniques can calculate how often a frame needs changing to avoid this jerkiness, and can also blur edges of rapidly moving images.

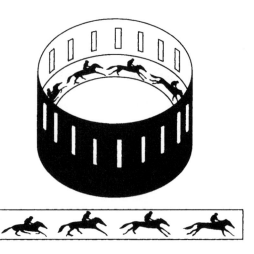

Figure 9.1 The zoetrope.

Early Experiments

Hand-drawn moving pictures have been around since at least the eighteenth century. In Victorian times the *zoetrope* was a familiar toy (Figure 9.1). When it was rotated you could see the animated object moving as you looked through the slits at the inside of the cylinder.

Figure 9.2 (See color insert following page 154.) This is a simple form of zoetrope. To operate it, make a photocopy, mount it on card, and cut round the outside, including the slots. Push a pencil through the center. Now hold the disc in front of a mirror so that you can see an image through one of the slots, and rotate the disc clockwise.

A simple version of a zoetrope display is shown in Figure 9.2 (color plate). If you make a photocopy of this (copyright has been waived for this purpose), mount it on card, cutting out the slots; push a pencil or skewer through the center; then hold it in front of a mirror while you rotate it, looking through the slots with one eye—you will see the moving image.

The first moving pictures made by a photographic process were shot by Eadweard Muybridge (1830–1904), who began work on his system in 1872.

He had been commissioned to take photographs to settle a bet over whether a trotting horse had all four feet off the ground at some point. In order to do this he set up a line of 24 cameras, the shutters operated by threads stretched across the path of the horse, which broke them as it passed. The experiment was a success (the horse did have all four feet off the ground), and the bet was won. Muybridge subsequently made many more series, often of nude athletes (and himself), for the purpose of physiological research. These have several times been made into zoetrope images. The principle has recently been revived for making animated stereograms (see Chapter 16), using a bank of cameras fired under computer control.

His real name was Edward Muggridge. He spent the first part of his life in a large house in Kingston-on-Thames (which bears a blue plaque in his honor), but emigrated to California, where he became famous for his photographs of Yosemite Valley, shot with an enormous camera taking 20 × 24 inch glass plates. He continued his motion recording experiments into the 1890s, before retiring and returning to England.

The Modern Cine Camera

For many years the standard format for motion pictures has been 18 × 24 mm on 35-mm perforated film moving downwards through the projector gate. In order to fill a wider screen the optical image may be squeezed horizontally by an afocal cylindrical lens or by a prism combination (see Chapter 5), the optical system being reversed for projection. The schematic layout of a typical cine camera and projector are shown in Figure 9.3.

Smaller film widths include 16 mm, 8 mm, and the obsolete 9.5 mm. The normal framing rate for these smaller formats is 16 pps, but some cameras have optional speeds of 32 and 64 pps for slow-motion shots. Professional 16-mm cameras, however, operate at 24 or 25 pps, with optical soundtracks. These are recorded on the side of the frame, and are usually generated by a light that illuminates a slit, which follows the sound waves by fluctuating in width; this is focused on the film. In the projector this track modulates a light beam passing through a fixed slit on to a photoreceptor. Eight-millimeter film can have a magnetic stripe coated outside the perforation, on which you can record commentary.

In professional cinematography, the trend towards larger screens has led to the employment of larger formats on horizontal 35-mm film, and even on 70-mm film and with stereoscopic viewing for the giant IMAX screens. The soundtrack has expanded, too, with many modern films equipped with seven-track optical sound.

Amateur cine work using film has almost completely disappeared, owing to considerable improvements in digital movie recording. Indeed, some amateur compact digital still cameras can now shoot several minutes of movie footage on a single memory card, with optional sound recording thrown in.

There is so much slow-motion and time-lapse imaging on television wildlife programs that some viewers must think large birds and animals really do move like that, and, at the opposite extreme, that toadstools and cumulonimbus clouds grow and tropical flowers open within the space of a few seconds. It would be a good idea if TV film companies were to add an "S" in the corner of slow-motion sequences, and a "T" for time-lapse.

Slow Motion and Time Lapse

To make a slow-motion sequence you simply shoot at a higher speed than the projection speed (usually at twice or four times the normal framing rate), and then project at the normal rate. In TV and video recording you can replay a normally recorded sequence at any speed you like, but the result may be jerky (like those irritating slow-motion clips you see behind some TV news items). Time-lapse photography needs more careful preparation, and some calculation. For example, if a flower takes six hours to open fully, to show this in 12 seconds at a framing rate of 25 pps, you need a picture frequency of $(60 \times 60 \times 6) \div (25 \times 12) = 1/72$ pps or

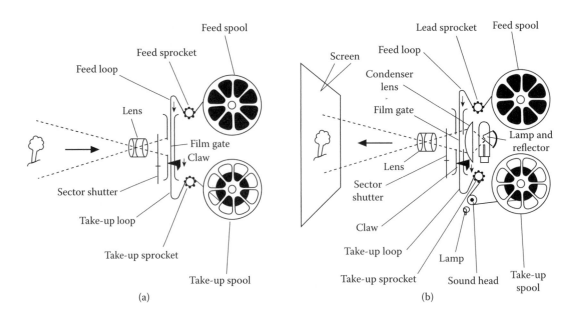

Figure 9.3 Schematic layout of a conventional cine camera (a) and projector (b).

1 picture every 72 seconds. With a time interval as long as this, you usually need to rely on artificial lighting, preferably flash. Most high-quality cine cameras, and some camcorders, have facilities for making single exposures at timed intervals.

High-Speed Cine

Up to about 500 pps it is possible to use a conventional cine camera system, usually with a strengthened shutter and a modified claw movement with two claws, and possibly a beater in the bottom loop to aid film pull-down. However, there is a physical limit to a stop-start action, usually dictated by the maximum allowable stress on the film, and above about 500 pps the film needs to be moved continuously. This means that the image must also be moved in synchronism with the film. The usual way to achieve this is to replace the normal shutter with a rotating parallel-sided glass block (a square prism). The rotation of this prism (Figure 9.4) moves the image in synchronism with the film movement, and at the same time acts as a shutter. This system allows the maximum exposure duration at the speed. Some models use a many-sided prism with a synchronized sector shutter, permitting framing rates up to 10,000 pps.

These cameras provide sufficient speed to capture most natural phenomena such as insect flight, a chameleon's tongue, or an exploding seedcase, and to show them in ultraslow motion with excellent resolution. At the highest framing rates the format

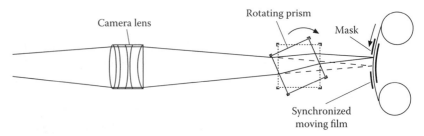

Figure 9.4 High-speed framing camera with continuous film movement.

height is usually halved. Using such speeds it becomes difficult to obtain adequate exposure using a steady light source, as sufficiently powerful lamps can fry the subject matter, so it is usual to employ a synchronized stroboscopic flash. This may be a xenon discharge lamp, or, increasingly commonly, a solid-state laser. Such a laser can produce pulses lasting less than a nanosecond, and can "freeze" the fastest motion.

Mirror and Drum Photography

The highest framing rates put a severe strain on the capabilities of film. There is a mechanical limit to the speed at which you can run a film through a gate. At over 10,000 pps the responsibility for imaging is usually handed over to digital recording systems, which are more sensitive than film and can be refreshed rapidly. However, there are still important uses for film, if a method can be found to make all the exposures with the film stationary. One solution is to fix the film to the inside of a drum, and to rotate the image using a mirror system; the other is to rotate the drum itself. In both cases, of course, it is still necessary to stabilize the image with respect to the drum for each exposure interval.

Mirror Cameras

In a mirror camera the drum is stationary and the beam of image-forming light rotates (Figure 9.5). The separate images are formed by a ring of lenses trimmed to a rectangular shape to save space. The optical principle underlying the technique is called the Miller principle, after its inventor David Miller. The image is formed by the main camera lens onto a field lens that is itself focused on an entrance stop. The light passing through this stop is focused by a relay lens on a further field lens just in front of a rapidly rotating mirror. The image of the entrance stop fills each of a ring of lenses in turn, and these in turn focus the final image on the stationary film. By careful adjustment of the distance between the second image and the rotating mirror, the final image can be made to remain stationary as the image of the stop sweeps across each lens. As the sensors are stationary, this second method lends itself more readily to digital recording.

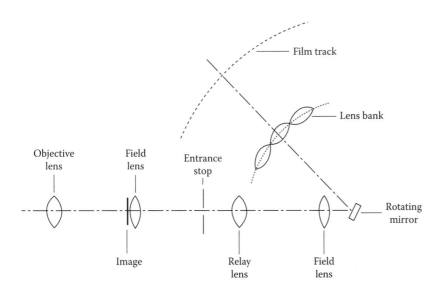

Figure 9.5 Optical configuration of a rotating mirror camera using the Miller principle.

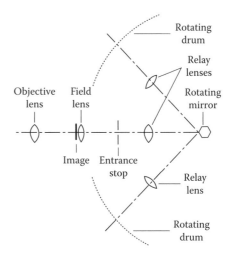

Figure 9.6 Optical configuration of a rotating drum camera.

Drum Cameras

In a drum camera the film is fixed to the drum as before, but the drum itself rotates. The system again uses the Miller principle, but omits the ring of lenses. Instead, the image is focused finally by a single lens. As the drum is moving, the image has to move, too. If the second image is formed ahead of the rotating mirror it will move in one direction, and if it is formed behind the mirror it will move in the other direction. By adjusting this distance the image movement can be synchronized to match the film movement. There is no second field lens. The camera uses a multifaceted mirror that takes in only the width of the final relay lens in its scan. Thus as one exposed frame moves out of the exposure path the next mirror facet begins to record the next frame. If there are two relay lenses offset horizontally, one above and one below the mirror as in Figure 9.6, two sets of frames can be recorded side by side, and the framing speed effectively doubled. Up to four relay lenses have been used in this manner, giving an effective speed of 200,000 pps.

Smear and Streak Photography

These two methods of imaging are superficially somewhat similar: both differ radically in principle from those of conventional photography. The photofinish camera records motion in one direction, employing a film movement that matches the speed of image movement. Thus there is resolution in one spatial dimension and in time, instead of the usual two spatial dimensions. Smear and streak photography employs a similar principle, but with a subtler use of optics.

Smear Photography

This is the more basic of the two methods. In its fundamental form it is a camera used with an open shutter and film moving through the gate at a fixed speed. Such a system is standard for making oscillograph records, where circumstances rule out pen recorders. The camera records the movement of an oscilloscope spot as a graph. The movement of the spot along the *x*-axis is switched off, so that the spot only moves up and down; the time component is contributed by the movement of the film. There is usually a device to place timing pulses on the bottom of the film (Figure 9.7).

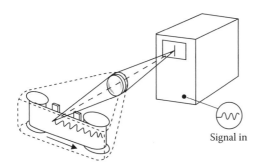

Figure 9.7 Principle of the oscillograph camera.

The second mode of use for the smear camera is analogous to image motion compensation in aerial photography. The camera system produces an image in the normal way but, since the object is moving, the film has to move synchronously with the image. This is fine when you can predict the speed of the object (as, for example, a bullet in flight), but is of little use when you can't. The speed criterion is less critical when you use a very brief flash for the exposure instead of a continuous light. In this case you can also obtain an acceptably sharp image of objects that are not traveling in quite the same direction as the test object (e.g., fragments of smashed glass). You can obtain a succession of images like a cine record if you move the film somewhat faster than the image, and use ultrashort strobe pulses from a pulse laser as your light source.

Streak Photography

This is an adaptation of smear photography that resembles photo finish photography rather more closely. It can operate in two modes, *cross slit* and *parallel slit*. Instead of being in the film plane, the slit is in the focal plane of the main objective lens, the image of the slit being projected on the moving film by a relay lens (Figure 9.8). This slit image is as narrow as is feasible in order to give optimum resolution: ideally, the width of the projected slit should match the resolution of the optical system. In cross-slit photography the slit is at right angles to the motion of the subject and the film, and the image, as it were, paints itself on the film during the time of transit, just as the competitors do in a photo finish camera. The result is a sharp image of everything that is moving in roughly the right direction.

For very high-time resolution, the streak camera has a rotating mirror system with the mirror sweeping the slit image round the film mounted inside a stationary drum. Alternatively, the image may be projected on the film mounted in a revolving drum (Figure 9.9). Such systems can achieve a time resolution of just a few nanoseconds. Again, the stationary drum technique lends itself readily to digital recording.

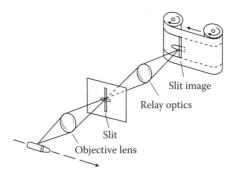

Figure 9.8 Cross-slit streak camera for ballistics photography.

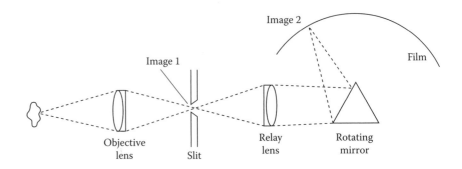

Figure 9.9 Schematic of a streak camera with rotating mirror. Alternatively, the mirror is fixed and the drum rotates.

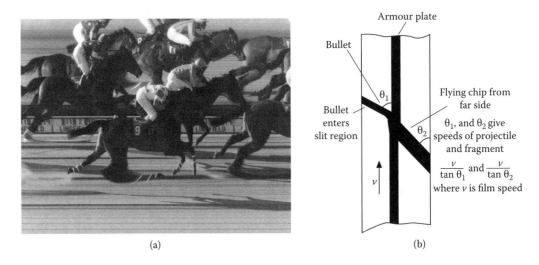

(a)   (b)

Figure 9.10 (a) Photo finish. Note the smear effect on several of the horses' hooves, where the optical image has coincided with the slit as they were on the ground and momentarily stationary. (Photograph courtesy of Racetech Ltd.) (b) Parallel streak (shadowgraph) record of bullet striking armor plate, and a fragment flying off the opposite side of the plate.

In the parallel-slit configuration the subject moves parallel to the slit, whereas the film still moves perpendicular to the slit image. The result is in principle similar to the oscilloscope camera above. The movement of the various parts of the subject is recorded in the manner of a time-distance graph.

You will seldom see any recognizable image on the negative, but if you wind the film past the slit you can replay the movement. Figure 9.10b is an impression of the sort of shadowgraph image you would get on the film from a bullet striking an armor plate, lit from behind.

Of course, with such very high-speed events the film is used up in a fraction of a second, and the moving parts of the camera have to be brought up to speed first, so some kind of capping shutter is essential unless the exposure is made by a very brief flash or the event is self-luminous (as with an electrical discharge). Most events need very precise synchronization, and there is a whole range of techniques for achieving the start of the exposure. To avoid overwriting the film the exposure has to be terminated after one revolution of the mirror or drum, and in some cases this is done by a so-called blast shutter, which fires a small explosive charge to shatter a mirror in the optical path, or to render a plastic window opaque. Synchronization is much less of a problem with electronic recording, which can be triggered in nanoseconds.

You can see this effect occasionally in photo finishes of horse races where a horse's hoof comes to the ground exactly in line with the camera slit (see Figure 9.10a). There are diagonal smears at the beginning and end of the movement, as the hoof goes down and up, and a longer horizontal smear indicating the time it is on the ground.

Lighting for High-Speed Photography

For medium- to high-speed cine photography, continuous light sources such as metal halide or xenon arc lamps are the rule, but for very high speeds it is more usual to employ a flash source. For running times longer than about 40 milliseconds (the effective duration of a flashbulb), a set of flashbulbs can be arranged to fire successfully ("ripple firing"). For much shorter exposures the earliest, and one of the simplest, is an electric spark from a capacitor charged to a high voltage, which ionizes the air and gives a very brief, intense white flash (no doubt you will have noticed the brilliant sparks produced by the contact shoes of London Underground trains going over points). The most commonly used source today for exposures of around 10 μs to 1 ms is electronic flash, which can be very accurately timed. You can also use electronic flashtubes in stroboscopic mode (see below).

The most dramatic flash source is the *argon bomb*. It is used mainly in ballistics photography, as it is so powerful that it can swamp even the light from an explosion. It consists of a cardboard or plastic tube filled with argon gas, with a clear window at one end. At the other end is a small explosive charge. Detonating this causes a shock wave to travel through the gas, ionizing it and producing a brilliant flash of white light with a duration of about 100 μs. Needless to say, if you are going to use such a device you need a little more protection than just goggles and a white coat.

Henry Talbot made a spark photograph in 1851 of a rapidly rotating disc bearing part of a newspaper. The exposure duration for this has been estimated at less than 1 ms.

Stroboscopy

As I mentioned earlier, electronic flashtubes can be programmed to produce short repeated flashes. If you want a sharp photograph of an object that is in a steady state of vibration or rotation, you can make it appear stationary by synchronizing the flash rate with the cycles of its motion. You can also make it appear to move slowly forward by setting the rate of flash repetition slightly slower than the cycling time, or slightly backward if you set it faster. This technique is known as *stroboscopy*.

This technique is handy for examining the behavior of loudspeaker diaphragms, valves in internal combustion engines, and even human larynxes, by synchronized cine photography or television. Strobe light sources are usually either a continuous source interrupted by an optical chopper (multi-bladed fan shutter) or a repeating electronic flash. A comparative newcomer to this field is the laser, which can be designed to operate up to very high speeds (up to 100 kHz) and very short flash durations (down to 100 femtoseconds, or 10^{-13} s).

Another type of stroboscopic photography puts all the images onto one composite photograph. You may have seen multiple images of falling objects in physics textbooks, or, more interestingly, of golf swings or tennis serves (Figure 9.11). Again, you usually need some kind of trigger system to start the strobe action at the beginning of the operation and switch it off at the end.

Stroboscopic effects sometimes appear gratuitously in movies: for instance, when you see an aircraft propeller starting up, and it seems to turn first one way, then the other. An unfortunate example occurs regularly in Western films, where wagon wheels seem to be turning backwards while the wagon goes forwards.

The term "strobe" is often used loosely in the United States to refer to an ordinary (single) electronic flash. If you use an American textbook, don't be misled by this.

Digging Deeper

There don't seem to be many books around dealing with the principles of cine photography *per se*, though *The Haunted Gallery* (Lynda Nead, Yale University Press, 2007), is an excellent account of its earliest days, treated from a sociological point of view. There are dozens of books on the practicalities of film and video: Focal Press and its associate label George Newnes have a large section of their title list devoted to all aspects of vision and sound recording. Most of today's manuals

Figure 9.11 Stroboscopic composite photograph of a martial arts student in action. (Photograph courtesy of Sidney Ray.)

are concerned with the practicalities of movies and video, and go into the underlying principles only sketchily, if at all. There are two notable exceptions, however, and between them they cover the field completely. *High Speed Photography and Photonics* (ed. Sidney Ray, Focal Press, 1997) is a multi-author work, and as might be expected, there is some overlap of material. But it does no harm to see the same concept through different eyes, and the sections relevant to this chapter are wide ranging and insightful. Sidney Ray's own book *Scientific Photography and Digital Imaging* (Focal Press, 1999) is also invaluable. It contains a long chapter on cine and high-speed photography, with some 140 references at the end of the chapter. So if you are enthusiastic about the subject, there is plenty of ground to dig.

Chapter 10 The Silver Halide Process

In the silver halide process the sensitive material consists of a suspension of tiny silver microcrystals in gelatin, coated on a transparent polyester film base. It is called an *emulsion*.

It records the light intensity at each point on the optical image, in the form of a chemical change that produces a *latent image*, and this is the first stage of the photographic process. "Latent" means "unrevealed," as there is no visible change at this stage, except perhaps under an electron microscope.

The next stage is to render this latent image visible as a *photographic image*, by a process called *development*, which reduces each crystal that bears a latent image to black metallic silver, thus producing a negative image: negative, because the most light energy (the highlights of the subject) results in the most silver, and the least energy (the shadows) results in the least silver. After development, the remaining silver compounds are still light sensitive, so a further stage called *fixation* removes them. It only remains to wash out the leftover chemicals and dry the negative.

You make the print by a similar route. This time the emulsion is coated on white paper onto which you project an image of the negative for an appropriate time, using a photographic enlarger. This is an optical projector using the same principle as a camera. The print is a negative of the negative, i.e., a positive, and the processing is the same as with a negative, chemically speaking.

The Uniqueness of Silver

Silver occupies a unique position among the metallic elements. It forms compounds with other elements more readily than the other "noble" metals such as gold, platinum, and iridium. On the other hand, its compounds are much less stable than those of the "base" metals such as sodium, magnesium, and aluminum. But its uniqueness lies in its being the only element whose compounds are readily broken down by light energy. Silver (symbol Ag) readily forms compounds with the halogen elements chlorine (Cl), bromine (Br), and iodine (I), and the action of light on these compounds is to release the halogen, leaving pure silver behind.

This phenomenon fascinated eighteenth-century chemists, and both Joseph Priestley in England and Karl Wilhelm Scheele in Sweden produced shadowgrams of leaves and paper cutouts on silver chloride precipitated in a beaker. They were unable to make the image permanent.

It is easy to duplicate their experiment: you need about half a teaspoonful each of silver nitrate and common salt, dissolved separately in distilled water and then mixed; a white crust of light-sensitive silver chloride will fall to the bottom of the vessel. You can make shadowgrams on it in the same way as the pioneers, and like theirs your image will disappear as soon as you disturb the surface. Early workers such as Thomas Wedgwood (son of the great Josiah) impregnated white leather and other surfaces with silver chloride and made shadowgrams on them, but again were unable to fix the images. In his early

Incorrectly, as it happens. An emulsion is a stable mixture of two or more liquids (such as oil and water) that don't normally mix: mayonnaise is a familiar example. A photographic "emulsion" is actually a solid suspension.

The latent image actually consists of microscopic specks of silver, which act as nuclei for the development process.

There are two other halogens, fluorine (F) and astatine (At). These cannot be used in photography as astatine is an unstable element with a half-life of barely 8 hours (!), and silver fluoride is soluble in water.

Joseph Priestley (1733–1804) was one of the earliest chemists (as distinct from alchemists). He is best known for his discovery of oxygen. He was an outspoken religious dissenter, and his opinions were so unpopular in England that he was forced to emigrate to America. Karl Wilhelm Scheele (1742–1786) was a Swedish apothecary; he discovered many previously unknown elements (including oxygen, independently), and showed how to synthesize a large number of compounds.

Sir John Herschel (1792–1871) was the son of Sir William Herschel, the famed organist, astronomer, and discoverer of the planet Uranus, and an astronomer himself. He spent four years in South Africa, where he mapped the whole of the southern sky. He was well-read, and coined much of the present-day nomenclature of photography such as "positive," "negative," and "snapshot," as well as many of the "-scopes," "-graphs," and "-ologies" of contemporary science.

Joseph Nicéphore Niépce (1765–1833) made the first ever photograph in 1822, using a form of light-sensitive bitumen; his exposure was more than eight hours. Louis Jacque Monde Daguerre (1789–1851) invented the daguerreotype in 1839.

Collodion is a solution of nitrocellulose in a mixture of ether and ethanol. It is porous as long as it remains tacky. It is also dangerously flammable (nitrocellulose is also known as "guncotton."

Richard Leach Maddox (1816–1902) was an inventor who, like Du Hauron, published his ideas for the public good instead of reaping a fortune from patents. Like Du Hauron, he died a pauper.

Latitude means, roughly, what you can get away with in terms of over- or underexposure and still get a good-quality print. Photographers refer to both undeveloped and developed crystals as "grains." This can be confusing, and I am using the term "grain" only when referring to the developed image.

experiments Henry Talbot found that a strong solution of common salt would do the trick. Soon afterwards, John Herschel discovered that sodium thiosulphate, $Na_2S_2O_3$, then called hyposulphite or "hypo," would dissolve silver halides (or rather, turn them into a soluble compound) more efficiently, so that the remaining light-sensitive material could be washed away, leaving a "fixed" photographic image.

There was still a need for a good binding medium to hold the silver halide. Talbot used sized paper; Louis Daguerre prepared his "Daguerreotypes" by forming silver iodide directly on polished silver plates by the action of iodine vapor. By 1848, Nièpce de Saint-Victor, a relative of Nicéphore Nièpce, had discovered the merits of albumen (egg white).

In 1851 Scott Archer introduced collodion as a substrate, and this proved an excellent medium, provided it was not allowed to dry out before processing.

The real breakthrough came in 1871 when Richard Maddox introduced the gelatin dry plate. Gelatin is a form of collagen, a substance found in animal skin, bone, and sinew, and is manufactured by boiling these inedible parts of cattle and pigs and refining the resulting goo (sorry, but I had to tell you). The presence of gelatin prevents the precipitation of silver halide from the solution. Instead, it forms a suspension of silver halide in the form of microscopic crystals.

The manufacture of an emulsion is fairly complicated, and usually involves several "ripening" stages, during which the size and sensitivity of the crystals increase. Most black-and-white films are coated with two layers of emulsion, one "slow" (low sensitivity) and one "fast" (high sensitivity), to increase the exposure latitude.

Color Sensitivity of Emulsions

As I explained in Chapter 1, the energy of a photon is proportional to its frequency. The lattice strength of a silver chloride crystal is fairly high, and needs an energetic photon to initiate a latent image. Thus a plain silver halide emulsion is sensitive to only blue and UV radiation. In order to sensitize an emulsion to longer wavelengths it is treated with special types of dye. The earlier emulsions were made sensitive to blue and green but not to red. Such emulsions were termed *orthochromatic*. They can be handled in a red light (called a "safelight") without becoming fogged, and are still used in some photomechanical processes. Around 1890 a family of dyes that would sensitize emulsions to red (and later to infrared) was found, and these emulsions were dubbed *panchromatic*. Figure 10.1 shows the spectral sensitivities of the various types compared.

Development

Talbot's first negatives were printed out, that is to say, the silver halide was exposed until the light had actually decomposed it into a visible image, a process that could take several hours. He discovered that he could strengthen his weakest images with a solution of gallic acid. This was the first historical example of the use of a chemical developer. Development is the first stage of the photographic process. As chemical reactions go, it is very slow. Just as well, perhaps, as if it were to be allowed to go to completion the entire emulsion would turn black. In practice what happens is that the largest crystals, those with the most latent image specks, decompose into silver grains first, then the medium-sized ones follow, and finally the smallest ones. Thus the contrast is built up, after which the whole image gradually becomes darker, and

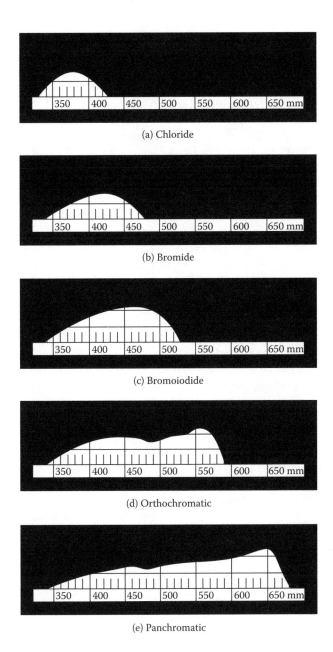

(a) Chloride

(b) Bromide

(c) Bromoiodide

(d) Orthochromatic

(e) Panchromatic

Figure 10.1 Spectral sensitivities for the various types of emulsion.

eventually even the unexposed crystals begin to be attacked. Plainly, a developer should turn the silver halide crystals into grains of metallic silver if, and only if, they bear a latent image. There are obviously some restrictions on the type of substance we can use for this. First, it has to work sufficiently slowly for us to be able to stop the process before it goes too far. Second, we have to choose a substance that will attack the crystals that already bear a latent image, but not the ones that do not have one. The process of removing the halide ions and leaving silver atoms behind is called *reduction*, and the active constituent of the developer is called the *reducing agent*.

In conventional chemistry, a reducing agent is any substance that can give away electrons, i.e., is an *electron donor*. Its opposite number is an *oxidizing agent* or *electron acceptor*. For our purpose, a reducing agent is something that will separate silver ions from their halogen partners and make good their missing electrons.

Gallic acid is a close relative of pyrogallol, a classic developing agent. In dietetics, reducing agents are called "antioxidants," and are alleged to be good for your long-term health. However, this doesn't apply to all reducing agents. You won't extend your lifespan by taking a swig of photographic developer, even if some developers do contain vitamin C.

145

OXIDATION AND REDUCTION

When an atom of silver loses an electron to an atom of (say) bromine, it has lost a unit of negative charge and become a positive ion, Ag^+. The atom of bromine has become a negative ion, Br^-. In an *ionic crystal* such as silver bromide the bonds are the electrical attraction between the oppositely charged ions of silver and bromine. A photographic developer can break such a bond and supply electrons to the silver ions, but only in the presence of a speck of metallic silver (i.e., a latent image). The silver nucleus is in effect a catalyst. By donating electrons the developing agent turns the Ag^+ ions into Ag atoms, which collect on the latent image. Eventually the whole crystal is reduced to an opaque tangle of filaments of silver.

A catalyst is a substance that promotes a chemical reaction without itself being changed.

Metol is 4-methylaminophenol hemisulphate, hydroquinone is 1,4-dihydroxybenzene, and phenidone is 1-phenylpyrazolid-3-one. You don't have to remember this.

In order to function efficiently, a developing solution needs more than just a reducing agent. Apart from the solvent (water) there are usually four main constituents in a developer: a *developing agent, alkali, preservative,* and *restrainer*. There are many developing agents (Henney and Dudley list more than 200!) but only a few are in general use, in particular metol, hydroquinone, and phenidone.

As developing agents in general will not work except in alkaline solution, an alkali, usually sodium carbonate, is added. The preservative, usually sodium sulphite, is added to prevent the developing agent from combining with the oxygen in the air, which would otherwise make it useless, and a restrainer such as potassium bromide is added in small quantities to ensure the developing agent does not attack unexposed silver halide crystals.

Fixing, Washing, and Drying

After development the emulsion contains unchanged silver halide that is still sensitive to light, so it needs to be removed. Sodium and ammonium thiosulphates ($Na_2S_2O_3$ and $[NH_4]_2S_2O_3$) react with silver halides to form soluble compounds that can be washed out of the emulsion with water. You can see the process of fixation in the clearing of the milky appearance of the negative. Washing is a diffusion process that takes upwards of 10 minutes. Washing thoroughly is important, as the compounds of silver with thiosulphate tend to decompose after a few weeks to form brown stains of silver sulphide. It is common practice to add a few drops of wetting agent to the final wash water to promote even drying.

Printing

Print paper works in the same way as black-and-white negative film, so when you project the image of the negative onto it with an enlarger and develop the result it produces a positive. Although often called "bromide prints," the emulsion is usually mainly silver chloride. As it is insensitive to red and green light, you can handle it under a yellow safelight. Printing emulsions come in up to seven grades of contrast to match the contrast of the print to that of the negative, but many (probably most) photographers prefer to use variable-contrast paper, as this saves having to keep several boxes of paper of different grades. Variable-contrast papers have a second, green-sensitive layer of low-contrast emulsion. Consequently, you can control the contrast of the print by filtering the enlarger light with magenta for higher contrast and yellow for lower contrast. (Most enlarger heads are fitted with filter drawers; the more sophisticated ones have built-in variable density filters in cyan, magenta,

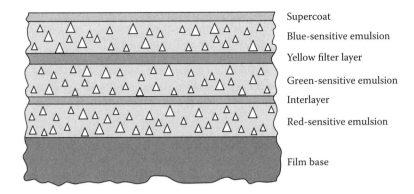

Supercoat
Blue-sensitive emulsion
Yellow filter layer
Green-sensitive emulsion
Interlayer
Red-sensitive emulsion
Film base

Figure 10.2 Construction of a color emulsion (simplified).

and yellow for color printing. For black-and-white variable-contrast paper the cyan filter is set to zero.)

Most print papers for general use are resin coated (RC). They have a base of paper impregnated with a polymer, and need only a few minutes' wash. Some papers, though, more especially types of paper intended for exhibition prints, are fiber based, and prints made on this base need up to an hour's washing in running water, as the fixation products adhere strongly to the paper fibers, and any sulphur that remains in the paper base will eventually cause the image to fade and discolor.

Color Emulsions

Color film is much more complicated to manufacture than black-and-white film. In principle, a color emulsion contains three layers of emulsion: an inner layer sensitive to red light, a middle layer sensitive to green light, and an outer layer sensitive to blue light. In practice all three layers are sensitive to blue, so blue light has to be blocked off from the inner two layers, by including a layer of yellow dye beneath the blue-recording layer (Figure 10.2).

As in black-and-white emulsions, to extend the response of the film two emulsions of different speeds are coated for each color, and with further layers incorporated to inhibit unwanted migration of dyes between emulsions, an opaque-dyed undercoat to suppress internal reflections, and an abrasion-resistant supercoat, a color emulsion may have as many as 14 layers, in a total dry thickness of less than 50 μm.

Processing of Color Emulsions

The processing of a color negative film is broadly similar to that of a black-and-white film, except that the final stage is not simply fixing but bleach-fixing, that is, the removal of all silver and silver halide, using a solution that oxidizes the negative image to a substance that is soluble in thiosulphate (which may be in either the same bath or a subsequent one). The developer is a special color-forming developer giving reaction products that couple to latent dyes in the emulsion layers wherever a silver image is being formed, so that the red recording layer produces a negative image in cyan, the green an image in magenta, and the blue an image in yellow. When the silver image is removed along with the unchanged silver halide the dyes remain, and in the final negative the hues of the original are reversed as well as its tones.

147

Kodachrome is not suitable for home processing. The dye-couplers are present in the developers, not in the emulsions, and there are three separate, carefully controlled development stages for cyan, magenta, and yellow, respectively. Sadly, information on Kodachrome now seems to have become history (see earlier note).

The print material is broadly similar, though the layers are in a different order, and there is a gap in light sensitivity between the red and green wavelength bands to allow the use of a yellow sodium safelight.

Transparency material is processed in a different manner. The initial development is with a more or less conventional black-and-white developer, after which the remaining undeveloped silver halide (which is a positive image) is fogged and developed in a color-forming developer. Bleach-fixing then leaves this positive image in dyes.

Printing from a transparency uses a similar method. However, in one commercial process (the *dye-bleach* process) all the dyes are present in the emulsion to begin with, and are selectively bleached by exposure and processing.

Sensitometry

The word *sensitometry* has struck fear into the hearts of many a budding technician, perhaps unjustifiably. It does involve a thorough understanding of the principles of logarithms, and a lot of graph plotting, but with present-day equipment, an assistant in a processing house has little more to do than to feed readings into a computer. However, if you are involved in technical photography, and especially if you are working with film, you need to be able to match its characteristics of the material to the job it is to do, e.g., high speed and wide exposure latitude for social and sports photography; accurate tone reproduction for copies and duplicates; high contrast for line diagrams and halftone work. In short, you need to know the *characteristics* of the film.

Even though you may never work with film, don't skip this section altogether. The characteristic curve is just as important in digital as in chemical photography.

The scientific foundations of sensitometry were laid down by two scientists, Hurter and Driffield. Their research resulted in the establishment of a *characteristic curve* for an emulsion, often called the H&D curve after their initials, and of the first speed rating for emulsions, the *H&D system*, based on the inertia point (see Figure 10.3). Their results were published in 1880.

Ferdinand Hurter (1844–1898) was a Swiss-born chemist who worked in a British chemical factory with Vero Charles Driffield (1848–1915). They wanted to market an exposure meter they had invented, but in order to successfully do so they had to establish a rigorous system for quantifying the speed and contrast of a photographic emulsion. For their pioneering work they were awarded the Royal Photographic Society's Progress Medal, its highest honor, in 1898.

Hurter and Driffield needed to establish meaningful figures for the speed and inherent contrast of an emulsion. As has since become standard practice, they decided to do this on a basis of graphically plotted data. However, plotting the transmittance of the negative image against the luminance of the subject gave a very nonlinear result. After much experimentation they settled on a plot of the logarithm of the reciprocal of the transmittance against the logarithm of the exposure (duration × luminance). With the emulsions in use at that time this gave a straight line with only a short "toe" (Figure 10.3).

It seems odd that they shouldn't have tried plotting logs immediately. These days every science student doing practical experiments knows that if you get a regular curve when what you want is a straight line, you simply plot the logs of one or both of the variables instead. (The reason this works is revealed in Appendix 1.) In addition, had the pair been familiar with the Weber-Fechner law they would have used this strategy at once.

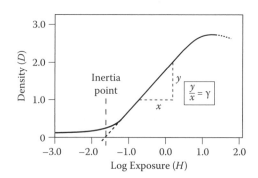

Figure 10.3 Hurter and Driffield's emulsions had a straight-line characteristic, with only a short toe.

Practical Units of Measurement

I mentioned the Weber-Fechner law in Chapter 3. It relates the intensity of a physical stimulus to the intensity of perception. One way of stating this is that when a stimuus increases in equal multiples, e.g., 1, 2, 4, 8, 16, … , what we perceive is equal increments 1, 2, 3, 4, 5, … . This relationship is the same as that between a number and its logarithm (see Appendix 1). As the log of 1 is 0 and the log of 2 is 0.3, the series

1   2   4   8   16   32   64 128 has log values

0 0.3 0.6 0.9 1.2 1.5 1.8 2.1

and so on. The log of 10 is 1 and the log of 100 is 2.

The Characteristic Curve

The log of the reciprocal of the transmittance of the negative is called the *density*, and is measured with a device called a *densitometer*.

To obtain the characteristic curve you simply plot a series of densities on the horizontal axis of a graph and the corresponding values of the logs of the exposures that produced these densities on the vertical axis. The method of obtaining the figures is usually to give the emulsion a single measured exposure under a calibrated density step tablet. This gives a set of log-exposure values with a minimum of arithmetic. The point from which the ISO speed is determined is the intercept on the log-exposure axis corresponding to a density of 0.1 above the fog-plus-base level (Figure 10.4). The ISO speed is calculated from the exposure corresponding to this point according to the formula

ISO Speed Index = $0.8 \, H_M$

where H_M is the exposure at the speed point in lux seconds (lx s).

Strictly, *photographic density*, to distinguish it from density of a substance (kg m^{-3}) and from optical density, which is an alternative term for refractive index.

Inherent Contrast

The inherent contrast of an emulsion is represented by the slope of the straight-line portion of the characteristic curve, and was dubbed *gamma* (the Greek letter γ) by Hurter and Driffield. It represents the contrast of the negative relative to the contrast of the subject. Modern emulsions do not have a straight line but a gentle S-shaped curve, and the average gradient is taken between the speed point and a point corresponding to a value 1.5 log-exposure units along the horizontal axis (Figure 10.5). This gradient is called $\bar{G}$ (pronounced gee-bar), the bar indicating an average value.

The fog-plus-base level is the density corresponding to zero exposure, usually about 0.01 in clear films. It is higher in 35-mm films as these have a grey-dyed base to absorb internal reflections. The standard step tablet is produced by Kodak in several sizes. It contains 20 approximately equal density steps from 0 to 3.

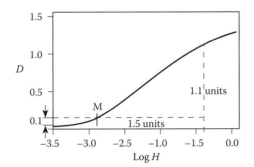

Figure 10.4 Determination of the speed point M (here approximately ISO 64).

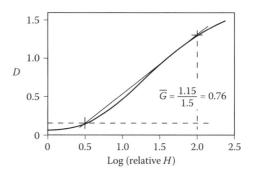

Figure 10.5 Determination of $\bar{G}$ (here approximately 0.76).

What the Characteristic Curve Tells Us

To obtain the ISO value from the characteristic curve you need to know the exact exposure in lux seconds that corresponds to the speed point. For most purposes this absolute value is not needed, and instead we simply use the log of the *relative* exposure, as the shape of the curve, the important item, will not be affected. We can learn a great deal, especially when we compare curves made using different emulsions or different processing techniques. Thus we can examine the effect on speed and contrast of variations in development time and temperature, and of varying the processing technique; we can also assess the exposure latitude of a particular emulsion, and the effect of any after-treatment.

The characteristic curve for a digital sensor can tell us an equal amount, and its shape is particularly helpful when we are playing with contrast within the image itself using Photoshop or a similar program. The default curve will usually be a straight line, but as the image is a positive rather than a negative, it may slope downwards rather than upwards, depending on your setting.

You can skip this bit if you never deal with silver halide film.

Effect of Varying the Development Time

Manufacturers of developers usually give lists of recommended development times for various types of film, sometimes in the form of a graph. So far we have considered only a single characteristic curve derived from a single development time at a standard temperature. If we plot a family of curves for different times we will see that the value of $\bar{G}$ increases at first rapidly, then more slowly, until it reaches maximum and the fog level begins to creep up (Figure 10.6). The recommended value for G is usually between 0.6 and 0.7.

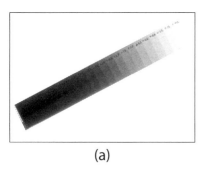

(a)

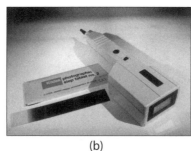

(b)

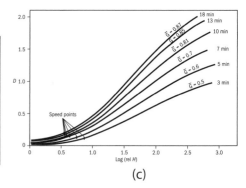

(c)

Figure 10.6 (a) A Kodak density step tablet; (b) a simple densitometer (X-ograph); (c) family of characteristic curves corresponding to variations in development time of the negative.

Reciprocity Failure

The Law of Reciprocity, due to Bunsen and Roscoe, states that photochemical effect of an exposure H is directly proportional to the product of the illuminance, E, and the duration, t, i.e.

$H = Et$

At very low light levels the Law of Reciprocity breaks down, because during long exposure durations some of the interatomic bonds within the silver halide crystals recombine, so that latent image formation is inefficient; emulsion speed is lowered, and contrast is increased. This has always been the bugbear of low-light photography, and in particular of astronomical photography. Fortunately the effect is virtually nonexistent in digital sensors, which are effectively photon counters (see Chapter 11).

Print Materials

There is an important difference between a photographic negative and a print: the negative is only an intermediate stage, so it doesn't matter how it actually looks, but a print is made to be looked at. (This applies equally to the electronic record produced in a digital camera, of course.) In silver chemistry it is particularly true of color negatives, which are made with deliberate color distortions in order to counterbalance those of the print emulsion. The other, less obvious difference is that a silver halide negative can have a very high contrast, up to 400:1, i.e., a density range of 2.6, whereas a print (photographic or digital) is limited to not much more than 32:1, or a density range of 1.5–1.6. We define *reflection density* as the log of the reciprocal of the reflectance, and although in theory this can go as high as the transmission density of a negative, in practice there is scarcely any visible difference between reflection densities of 1.6 and 2.6: they both look black.

Print papers have a characteristic curve too, with reflection density plotted against the log of relative exposure, but there are important differences in the curve. It reaches its maximum value quickly during development, and as development continues its shape and slope stay the same while the whole curve moves progressively towards the left (Figure 10.7). This means in practice that you can obtain a good quality print, even though your exposure in the enlarger was not ideal, simply by shortening or extending the development time.

Wilhelm Bunsen (1811–1899) didn't invent the Bunsen burner, he merely refined it, but he did invent the Bunsen cell, the Bunsen calorimeter, and the Bunsen photometer. He isolated magnesium, pioneered spectroscopy in analysis, and discovered rubidium and cesium. He never married: contemporary opinion seems to have attributed this to an aversion to soap and water, but his unattractive aura more likely stemmed from his lengthy research into the malodorous compounds of arsenic (which nearly killed him). Sir Henry Enfield Roscoe FRS (1833–1915) was professor of chemistry at Owen College, Manchester, where he isolated vanadium. He became President of the British Association for the Advancement of Science and Vice-Chancellor of London University, combining these activities with a career in politics.

Even if you intend to do all your printing via a computer, and don't possess an enlarger, you may miss something useful if you skip this section.

Printing specialists call this "snatching" and "stewing," respectively, and tend to be contemptuous of this amateurish practice. However, with modern high-quality materials it is usually (fortunately) undetectable.

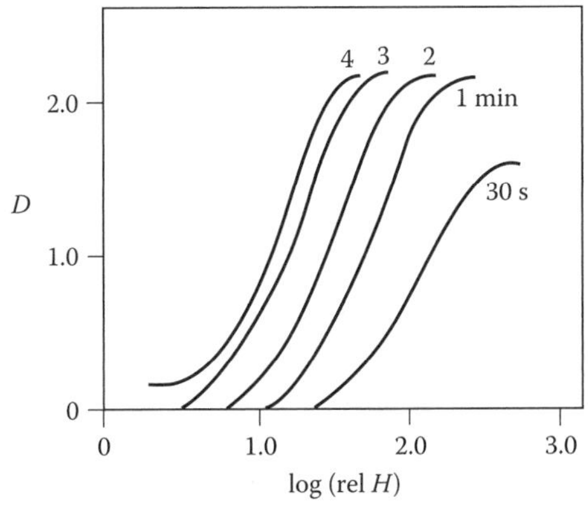

Figure 10.7 Effect of development time on characteristic curve for a print.

This section also applies to inkjet papers.

The Density Range of a Paper

I used the term "contrast" to refer to the density range of a negative (it also applies to a digital record), and it is a valid term for a print, too. In general we try to match the range of the negative to that obtainable in the print, to obtain a full range of tones in the final result without any blocked-up shadows or washed-out highlights. A negative with too high a density range will give a print with either or both.

Black-and-white printers call this "soot-and-whitewash." Of course, it may actually be the effect you want.

Some print processes, for example, bromoil, platinotype, and some special types of display material, have a fixed exposure range that can't be extended or compressed in processing, so the negative density range has to be carefully controlled to match the print material characteristics. This is also broadly true of color print processes. Of course, any good darkroom printer is capable of controlling print densities by dodging and burning in tricky areas on the enlarging easel, but this is not really relevant to the point I am making here.

See below for an explanation of these terms.

Digital photographers have a big advantage here, as they can call up the characteristic curve of their image, and change its slope and modify its shape to obtain the required quality in shadows, mid-tones, and highlights without having to use dodging shapes or wave their hands over the enlarging easel.

Color Print Papers

Papers for Direct Positive Printing

This applies to making prints from positive transparencies. There are two types of material. The reversal-process type employs a process similar to that of a transparency, with first development, color development, and bleach-fix stages. The dye-bleach type employs a developer followed by a dye-bleach bath that destroys the dye already present in the emulsion in proportion to the amount of silver present. The final stage is a bleach-fix that removes the silver. All direct positive papers have a long log-exposure scale designed to be a match for the high contrast of the original transparency.

Unfortunately, these deficiencies are all too obvious in prints made from transparencies of flowers, when you compare them with the original. Seedsmen's catalogues are notorious for their inaccurate color renderings. That is why botanical textbooks still prefer watercolor illustrations to photographs.

However, any color errors in the original will be aggravated. Specifically, as the magenta dye inevitably has a slight yellow cast, the contrast of the yellow image has to be lowered a little to balance it, and this results in the yellows being too light and the greens bluish. For the cyan image with its spurious red content, the red (i.e., yellow + magenta) has to be lowered too, lightening the yellow further and lowering the saturation of blues and greens. This sounds terrible, but in practice the effects are slight, and you have to go through several generations of copies before the effects become obvious.

Papers for Negative-Positive Printing

As color negative material contains integral masks to correct for errors in the magenta and cyan dyes, you are not likely to encounter any serious color errors. The print material has a high inherent contrast, to match the low value of $\bar{G}$ in the negative material. It takes only a small change in color balancing filtration to produce a large swing in the color balance of the print. You can modify the contrast of a print by curtailing or extending the development time up to minus 20 or plus 40 percent, but you will almost certainly need to change the filtration too. A well-corrected negative-positive print can give a much more accurate rendering of the colors of flowers than a print from a positive transparency can, even when the latter is printed digitally.

Image Modification

It used to be said that the camera never lies. Your camera can certainly provide better evidence of some event than your memory can. But at best a photograph tells only part of the truth. We saw in Chapter 5 that the detail in a photograph is limited by the optical transfer characteristics of the recording system. We can tackle some of those deficiencies by modifying the image in various ways. Of course, by "modifying" I don't mean putting anything into the image that was not there in the first place. There are plenty of techniques that can do that. Modern digital methods can produce extravagant visual fabrications that would have gladdened the hearts of the artists who modified news photographs during Stalin's wicked reign (see Digging Deeper).

Even though a photograph may not show the whole truth, much of it may be buried there in the image, if we could only get at it. We can go a long way towards finding out what *is* there from the optical transfer function (OTF) of the system, which enables us to obtain an objective assessment of the way the system depicts an image.

Shading, Dodging, and Burning In

One of the most common problems with a photograph is blocked up shadows, due to the contrast of the subject being too great to reproduce accurately on the limited density range of the print, even though the detail may be present in the negative. In silver halide printing, a good printer will bring out shadow detail by shading, to prevent the print from becoming too dark over the offending area, or, on a smaller scale, by 'dodging' small areas to hold back the exposure, or, at the highlight end of the scale, by 'burning' in areas such as stained glass windows. In black-and-white photography the use of variable-contrast paper introduces the possibility of shading with color filters: magenta to give increased contrast, and yellow for reduced contrast. In color printing it is possible to correct small local errors in hue by dodging with small pieces of CP filter in the appropriate hue. These tasks present fewer difficulties for the digital photographer, who can modify the tones and hues of chosen areas with comparative ease, using a program such as Photoshop.

Professional printers have to do this a lot. It helps if you are good at making rabbit shadows with your hands.

Unsharp Masking

This is an ingenious method of increasing the sharpness of an image.

It may perhaps seem illogical that you can make an image sharper by coupling it with an unsharp one, but it is in fact logical enough. In silver halide photography the method is a bit complicated, though. What you need to do is to make a slightly unsharp positive transparency from your negative with a low contrast (say a $\bar{G}$ of about 0.3). You fix this in register with the negative, and print the combination on paper one grade more contrasty than you would have used for the original negative. As the mask is unsharp, it will have lowered the contrast in the larger areas only, leaving the fine detail unaffected. Hence this fine detail will be printed in enhanced contrast, while the larger features will remain at the original contrast. The sharpness will be noticeably improved (Figure 10.8). The principle is easily appreciated in terms of the modulation transfer function of the combined system. Figure 10.9 shows diagrammatically the way it works.

Sharpness is a subjective concept. It is correlated with acutance and resolving power, and has been quantified in these terms, though without much justification (see Chapter 5). Unsharp masking has no effect on resolving power.

Making an unsharp mask in the darkroom is a tedious business. Various printers have been designed to produce the effect, usually using a flying spot to do the

(a)   (c)

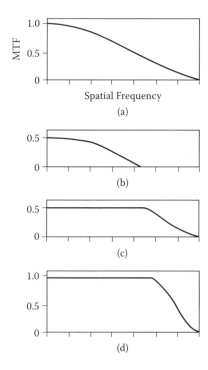

(b)

Figure 10.8 (a) Print without masking; (b) negative and unsharp mask; (c) masked print.

Spatial Frequency

(a)

(b)

(c)

(d)

Figure 10.9 Effect of unsharp masking on the MTF.

(a)

(b)

Figure 1.9 Spectra formed by (a) a prism; (b) a diffraction grating.

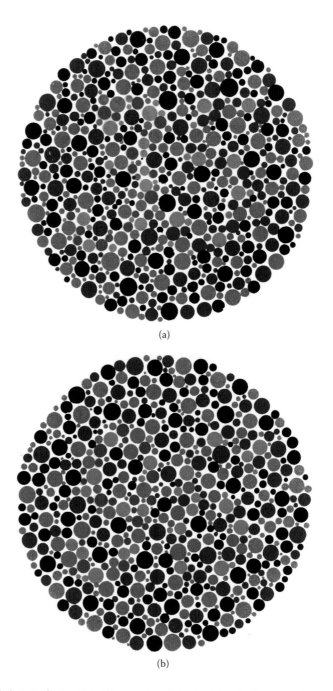

(a)

(b)

Figure 3.7 A page from the Ishihara color vision test book. Persons with normal color vision see the number 35. Persons with protanopia (red cone deficiency) will see only the figure 5 and those with deuteranopia (green cone deficiency) will see only the figure 3.

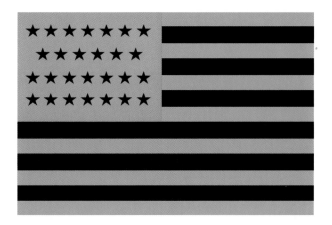

Figure 3.8 If you stare at the center of this flag in bright light for about half a minute and then look away at a white surface, you will see the U.S. flag as an after-image, in its correct colors.

Figure 5.18 Pinhole photograph by Justin Quinnell: the Royal Crescent in Bath, with fingers. The depth of field extends from near infinity to about 3 cm.

Figure 7.5 Self-portrait of Dr. Hans Bjelkhagen made with Lippmann's process. In this printed reproduction it is unfortunately not possible to do justice to the subtle colors of the original.

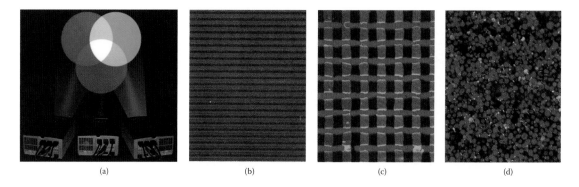

(a)   (b)   (c)   (d)

Figure 7.6 (a) Additive color patches, using projected light: (a) enlargement of a section of Lumiére transparency showing colored starch grains; (b) enlargement of a section of Polaroid transparency film, showing color raster; and (c) an enlargement of a Dufaycolor réseau on the same scale.

(a)

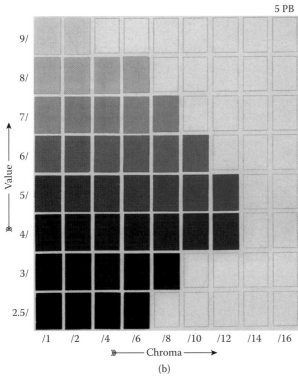

5 PB

(b)

Figure 7.8 (a) The full Munsell color globe; (b) a page from the Munsell color atlas.

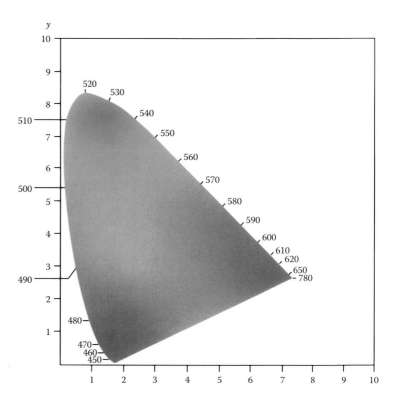

Figure 7.7 (a) 1947 CIE chromaticity diagram, showing approximate wavelengths in nm and corresponding colors;

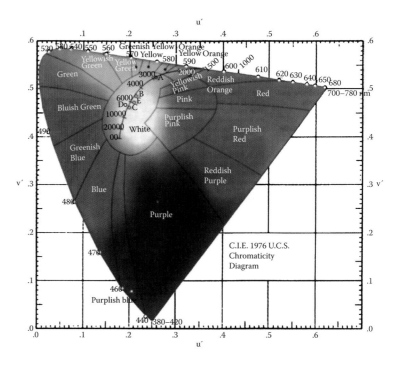

Figure 7.7 (b) 1978 modification, representing equal perceived color shifts in hue and saturation by equal distances, and showing the position of various color temperatures.

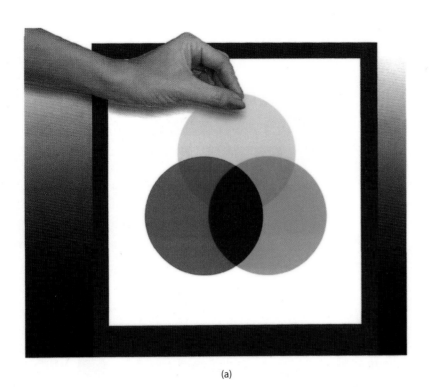

(a)

(b)

Figure 7.9 (a) Subtractive color filters, using transmitted light; (b) Technicolor separations; (c) Color reproduction by the negative - positive color process.

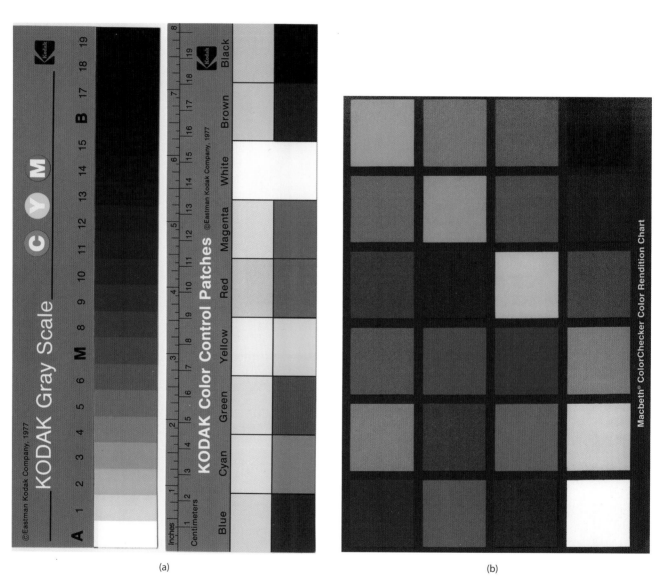

Figure 7.10 (a) The Kodak color patches and grey scale; (b) Macbeth color chart.

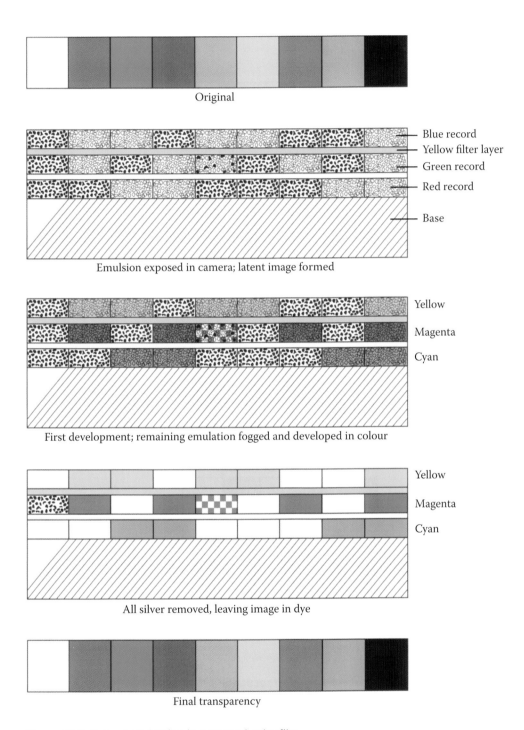

Original

Blue record
Yellow filter layer
Green record
Red record
Base

Emulsion exposed in camera; latent image formed

Yellow
Magenta
Cyan

First development; remaining emulsion fogged and developed in colour

Yellow
Magenta
Cyan

All silver removed, leaving image in dye

Final transparency

Figure 7.13 Color reproduction in a reversal color film.

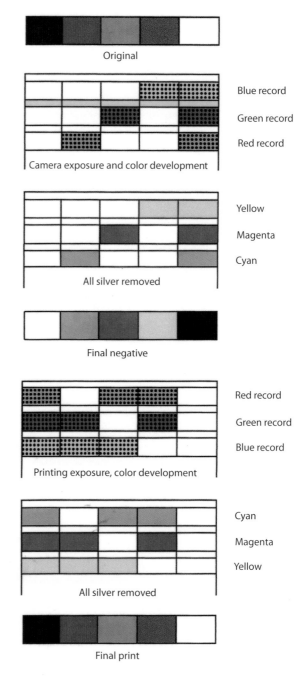

Figure 7.14 Color reproduction by the negative-positive color process.

Figure 9.2 This is a simple form of zoetrope. To operate it, make a photocopy, mount it on card, and cut round the outside, including the slots. Push a pencil through the center. Now hold the disc in front of a mirror so that you can see an image through one of the slots, and rotate the disc clockwise.

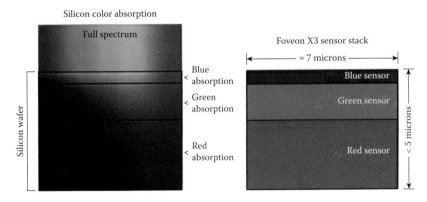

Figure 11.15 (a) Color absorption in silicon; (b) relative thickness of subpixel segments.

Figure 15.1 (a) A famous seventeenth-century *trompe l'œil* painting on the barrel (cylindrical) ceiling in the church of Sant'Ignazio in Rome by Fra Andrea Pozzo.

Figure 15.1 (b) A photograph showing clues 1–4. Notice (1) the relative size of the two pillars, (2) the overlap of objects, (3) the slight haze reducing the contrast of the distant fort, and (4) the modeling and texture resulting from the side light on the nearer pillar.

Figure 15.2 A stereoscopic pair. To view these images stereoscopically without an optical aid, hold the book about 40 cm away from you just below eye level, while you look at a distant object in line with them. Then bring the book up into your line of vision, keeping your eyes relaxed. You will see one blurred image in the middle, flanked by two other (also blurred) images. Now slowly bring the middle image into focus without losing fusion. (This may take some practice.) You may find it easier to deliberately cross your eyes, in which case the stereoscopic depth will be reversed.

Figure 15.13 Monochromatic anaglyph. This is an example of hypostereoscopy. To view this illustration and Figure 15.14 in stereo you need a pair of red/cyan glasses, with the red filter in front of your left eye. (Red/green will do for this figure, but will not give accurate colors on Figure 15.14.)

Figure 15.14 Color anaglyph. The red and blue/green images are in register in the plane of the photograph, so that the tiger's eyes appear to be in this plane.

Figure 19.3 Infrared false color aerial photograph.

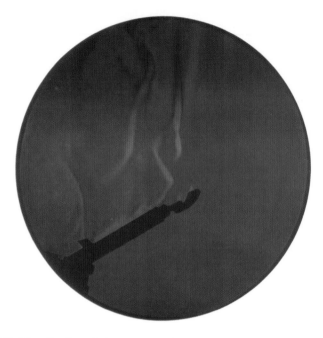

Figure 19.21 Visualization of air currents using color filters instead of a knife-edge.

exposing, but these cumbersome and expensive devices have been overtaken by modern scanners with built-in programs; digital photographers should be grateful for this. The "unsharp mask" routine, which is often built into cameras under the guise of "sharpness control" does exactly the same at the basic image stage, and with a program such as Photoshop you can take it as far as you like, even to the extent of ending up with a line drawing.

Skewed Characteristic Curves

This fault occurs when the individual characteristic curves for red, green, and blue have different shapes. The symptoms are that shadows have a different color cast from highlights, and the fault occurs most commonly in color film that is out of date or has been badly stored; it can also occur when a color negative is underexposed, and the effect may be aggravated by unsuitable lighting. An experienced printer can minimize the fault by shading with pieces of CP filter material, but it can only be cured by scanning the original, digging out its three-color characteristic curves and manipulating them in the computer until they coincide as nearly as possible. This fault is not the same as the odd color balance you will get if you mix tungsten and daylight or fluorescent light. This gives a result you can do little to correct (unless, of course, you actually wanted this effect).

Keystoning and Rectification

At some time or other you will no doubt have photographed a building with the camera tilted upwards, with the result that in the image the verticals (which you know to have been parallel) converge towards the top (see Chapter 4). If the convergence is small, this may not be important (after all, they *do* converge). But our brains have a built-in template that tells our perceptual mechanism that the apparent convergence is an optical artifact, so we perceive the actual building as having parallel sides. Unfortunately, the camera sees it as it is. When faced with this problem the usual darkroom strategy is to tilt the enlarging easel until the verticals in the projected image are parallel or nearly so (and a surprising amount of tilt it needs, too), and close down the lens aperture until the image is sharp overall. However, if the convergence is more than a few degrees the building will appear noticeably elongated. You can see this even in the minor correction employed in Figure 4.24d.

Some enlargers have provision for tilting both the easel and the negative stage so that the Scheimpflug condition is satisfied. This allows the use of a larger lens aperture, but does little to ameliorate the elongation problem.

Converging verticals are a form of geometrical distortion known as *keystoning* (see Chapter 4). Keystoning is a perennial problem in aerial survey photography. It led to the development of the rectifying enlarger, in which a system of interconnected levers adjusts the position of the negative and easel planes to satisfy the Scheimpflug condition, while a built-in computer adjusts the position of the optical center of the negative. Once again, if you are working digitally, there is a program that does this with one click.

Although the equation for the required displacement is a complicated one, you can get a good approximation for your building by moving the negative so that the top of the building moves up by about 10–20 percent.

Digging Deeper

In this area of imaging science you can dig very deeply indeed. *Basic Photographic Materials and Processes*, by Ira Current, John Compton, Leslie Stroebel, and Richard Zakia (2nd edition, Focal Press, 2000) gives a comprehensive description of emulsion making. There is a good analysis of theories of latent image formation in *The Theory of the Photographic Process* (4th edition, ed. T. James, Macmillan, 1977), which is still valid in spite of its age. For a detailed description of processing techniques, go to *The Manual of Photography* (10th edition, Focal Press, 2009). This manual also deals with image measurement more thoroughly than I have been able to do here. For those who are diffident about a graphical approach, the Zone System, a method of approaching exposure and print quality in terms of numbered subject brightnesses or "zones," was adopted wholeheartedly by the great photographer and teacher Ansel Adams, and you could do worse than read his two books, *The Negative* and *The Print* (Little, Brown; republished 1995). If you want the whole works on sensitometry, try Jack Eggleston's *Sensitometry for Photographers* (Focal Press, 1984, out of print, but still easy to find). The principles and practice of unsharp masking, together with other masking techniques, are set out in some depth in Sidney Ray's *Scientific Photography and Applied Imaging* (Focal Press, 1999). The same chapter contains the fundamentals of image manipulation by digital methods. For a fuller account of rectification in aerial photography, look for a copy of Ron Graham and Roger Read's *Manual of Aerial Photography* (Focal Press, 1986), now (inexplicably) out of print, but easy to find via the Web. If you want to know more about the machinations of the Soviet press under the Stalinist regime, David King's *The Commissar Vanishes* (Canongate, 1997) is a revelation.

If you need information about Photoshop or any of its rivals, try the Internet. Amazon lists more than 1300 titles! One I can recommend without reservation is Martin Evening's *Adobe Photoshop CS4 for Photographers* (Focal Press, 2008). But the best way to learn digital image manipulation is to take part in a hands-on workshop. Both the Royal Photographic Society and the British Institute of Professional Photography run such workshops several times a year in various parts of the United Kingdom.

Chapter 11 Digital Recording of Images

The Digital Principle

Picture an ornamental gate topped with spikes, something like Figure 11.1, with its upper outline tracing an elegant curve. But it doesn't, though, does it? It's only a row of spikes. All the same, the designer must have given a template of the curve to the craftsman, who would need to make each spike the correct length. Just as the craftsman would have fitted the spikes to the template, so now, looking at the spikes, you can make a fair guess, from the height of the spikes, as to the shape of the original template.

Figure 11.1 illustrates the fundamental principle of the analog-digital relationship. If you have a quantity that is known to vary smoothly with distance or time, such as light intensity, voltage, or any similarly measurable varying quantity, you can represent its behavior by a series of sample values and still recover the original smooth curve, provided the samples are sufficiently close together. The smooth curve is called the *analog signal*, and the row of samples is called the *digital signal*.

Of course, the sampling principle is much older than its modern applications. You use it every time you plot a graph of a continuous function. You would use it, for example, if you were plotting the characteristic curve of a film from a set of density readings. You have a set of discrete values of density corresponding to a chosen set of equally spaced log-exposure values. If these values are sufficiently close together you can get a reasonable result just by joining the dots with straight lines, like a puzzle in a child's picture book. But logic tells us that the result should be a smooth curve, so we can produce a better result by drawing a smooth curve through the dots instead. We have now constructed a continuous function from a set of discrete values.

If you know that the result has to be a smooth curve, you can probably manage with fewer points. In plotting the characteristic curve of a film, there are usually 20 readings available from the standard step tablet, but because of the regularity of the curve you can easily manage with only 10. Generally speaking, the more regular the curve the fewer readings you need.

This is an idealized example, as you will know if you have actually plotted curves from measured data. What you have to do in fact is to draw a *best curve*, i.e., the curve that is the best match to all the points. To satisfy yourself that this is indeed the correct curve, there are various devices: invoking the limits of experimental error; deleting the two values that look the dodgiest; finding a polynomial expression that fits the curve (and, if you are lucky, the theory too); and other dubious stratagems you can find exemplified in almost any research paper.

Figure 11.1 The tops of the spikes appear to trace out a curve (broken line).

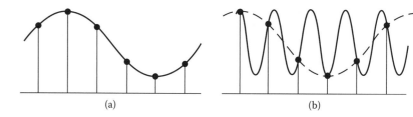

Figure 11.2 Aliasing: (a) Adequate sampling rate; (b) inadequate rate: as shown, reconstruction can produce the wrong waveform (broken lines).

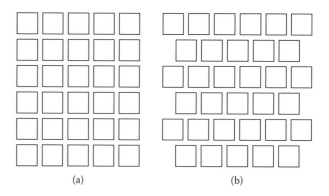

Figure 11.3 Arrays of pixels: (a) rectangular; (b) trilinear.

As we saw in earlier chapters, the most regular curve of all (apart from a straight line) is a sinusoid. If you know that your analog signal is a sinusoid, you should be able to reconstruct it by taking only one sample per cycle, at the peak value. But of course this only works if you already know the frequency, in which case the exercise is trivial. In any case, a wave of exactly twice the frequency would fit the data equally well. It turns out that you have to take samples at not less than twice the maximum frequency at which the signal is operating in order to be able to build a reliable replica of the original waveform from the sample data. This is known as the *Nyquist criterion* (see Box).

We saw in Chapter 6 that *any* signal, however irregular, can be analyzed into a combination of sinusoids. The highest frequency present determines the sampling frequency needed.

The "staircasing" of diagonal edges referred to earlier is another aspect of the same general phenomenon. Harry Theodor Nyquist or Nyqvist (1889–1976) was a Swedish physicist who worked at the Bell Telephone Laboratories. He won several prestigious awards for his research in the field of telecommunications.

THE NYQUIST CRITERION

If the sampling rate is slightly higher or slightly lower than the highest frequency present in a waveform, the samples themselves will fluctuate sinusoidally at a low frequency equal to the difference between the two. With radio or sound signals this will result in beats, and with spatial frequencies it will result in moiré fringes. This effect is known as *aliasing*. Nyquist showed that aliasing would not occur as long as the sampling frequency was at least twice the highest frequency present in the signal (Figure 11.2).

Digital Recording of Luminance

Digital photography involves electronic measurement of the luminance of the optical image, in a closely set mosaic of points all over the image. But how closely set must they be? This is where the Nyquist criterion comes in. The designer has to ask, "What is the highest spatial frequency the device needs to record?" and the Nyquist limit is twice that frequency. So if the requirement is for a resolution of 50 cycles per millimeter there must be at least 100 discrete readings per millimeter, or 2400

× 2600 on a 35-mm standard frame. Each point is called a picture element (*pixel* for short), and this spacing represents a total of 8.6 megapixels. This is the kind of performance one would expect from an ISO 800 film emulsion used in a camera with a high-quality lens. This has long been surpassed. At the time of writing (2009) 12 megapixels on a standard digital SLR (DSLR) format is nothing out of the ordinary. The usual arrangement of the sensitive elements is a rectangular or trilinear (hexagonal) array (Figure 11.3). Each pixel records the luminance level of its own tiny area as part of a continuous function, that is to say, it is in itself an analog recording device, but the light levels recorded are digitized at the storage stage: 4096 levels (around the limit of human visual discrimination) are represented by 12 bits of information.

Exactly the same situation arises in digital sound recording. Here it is usual to store 20 bit levels.

BITS, BYTES, AND BINARY ARITHMETIC

For people who count on their fingers, the decimal or denary (base 10) system is a convenient form of arithmetic, so much so that it is the universal counting system, at least on this planet. But for electronic circuitry it is unnecessarily complicated. There is no reason, apart from the obvious physiological one, for counting in tens: it would be equally easy to use a system of twelves or eights. But the simplest system of all is the binary (base 2) system, where there are only two digits, 0 and 1. The "place" convention is exactly the same as in denary notation, but the number 10 comes round very quickly, as you can see from Table 11.1, which shows the equivalent binary and denary numbers up to 10_{base10}.

Binary numbers are particularly convenient to use in digital calculations and measurement because there are only two digits. They can thus be represented by the "off" (0) and "on" (1) positions of a switch. One switch codes for 0 and 1; two switches code for 0 to 3 (00 to 11); three switches for 0 to 7 (000 to 111); and so on. In order to count up to 10 you need four switches (four binary digits or *bits* give you up to 16 choices). To code all 26 letters of the alphabet you need five switches (up to 32 choices), the leftovers being used either for punctuation marks or a correction code. In binary arithmetic you can carry out all the normal operations of arithmetic such as addition, subtraction, multiplication, division, powers, fractions (vulgar and "bicimal"), and so on. All these operations are greatly simplified in binary notation as compared with denary.

You probably have noticed that each successive place in a binary number represents a step in powers of 2, just as each successive place in a denary number represents a step in powers of 10. To see how the correspondence works, let us look at the number 269.375 expressed in full, first in denary, then in binary notation:

$2 \times 10^2 + 6 \times 10^1 + 9 \times 10^0 + 3 \times 10^{-1} + 7 \times 10^{-2} + 5 \times 10^{-3} = 269.325$ (in denary)

$= 1 \times 2^8 + 0 \times 2^7 + 0 \times 2^6 + 1 \times 2^5 + 0 \times 2^4 + 1 \times 2^3 + 1 \times 2^2 + 0 \times 2^1 + 1 \times 2^0 + 0 \times 2^{-1} + 1 \times 2^{-2} + 1 \times 2^{-3} = 100001101.011$ (in binary)

So, as you can see, the rules for notation are identical. But as you can also see, binary notation is comparatively long-winded, so programs often count in

TABLE 11.1
Binary and Denary Equivalents

Denary	0	1	2	3	4	5	6	7	8	9	10
Binary	0	1	10	11	100	101	110	111	1000	1001	1010

eights (octal notation) instead, leaving the machinery to convert this to binary—something it can do much more easily than converting to binary from denary.

The principle of octal notation gave rise to the concept of "octal bits," which have become known as *bytes*. Originally a byte was just a three-bit binary number, but gradually the meaning of the term has expanded to take in the idea of a single piece of information or "syllable" without a rigid length specification. As it makes more sense to specify the memory of a computer in terms of the information it can hold rather than the amount of binary data it can store, computer memories are usually quoted in megabytes (or giga- or even terabytes!). On the other hand, when discussing the number of distinguishable levels in a digital sound reproduction system, or illuminance levels in a digital photodetector system, it makes sense to talk in terms of bits, for example, 12-bit (4096) levels or 16-bit (65,536) levels.

Extending the Sensitivity

One problem that has bedeviled digital image recording is the short characteristic curve compared with that of a typical silver halide emulsion. In the latter the recordable log-exposure range is extended considerably by coating a second, slower emulsion, which takes over at the point where the fast emulsion reaches the point of overexposure. This arrangement permits the accurate recording of a scene with a contrast of 400:1 (a log-exposure range of 2.6) or more. The digital sensor can only cope with around 100:1 (log-exposure range 2.0), which limits both the latitude of the sensor and the subject contrast it can cope with. Faced with a subject of extreme contrast, a digital photographer may be compelled to make two exposures, one set at normal exposure and the other at one or two stops underexposure, so that the first shot copes with the shadows and the second with the highlights. These two results can then be combined in Photoshop or a similar system. This is plainly a considerable nuisance, and has led one manufacturer to rearrange the pixel array with low-sensitivity sensors inserted in the diagonal links (Figure 11.4). Another manufacturer has simply alternated vertical rows of high and low sensitivity pixels over the array. Of course, this latter kind of interpolation lowers the horizontal resolving power of the sensor, and was not feasible until the available pixel count of a sensor array exceeded 10 million.

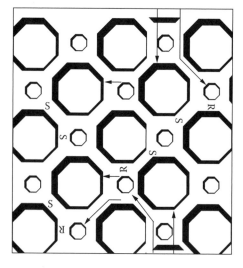

Figure 11.4 Fuji's pixel mosaic. The smaller pixels have a low sensitivity in order to record highlight detail in a high-contrast subject.

Principles of Electronic Information Storage

Just as you can understand the physical principles of what goes on in a car engine without necessarily knowing all the details of its electronic ignition system, so you don't need to know every last theoretical detail about the electronic circuitry of a digital camera in order to understand its principles. Most active electronic components these days are microchips of one sort or another, often containing thousands of elements, and even at an advanced level of understanding one is really only concerned with the relationship between what goes in and what comes out. The great-grandfather of all microchips, the 741 operational amplifier, contains well over a hundred transistor, resistor, and capacitor elements, but when seen from the outside it has just a current supply, a signal input, a gain control, and an output. In effect, it is just a transistor with style.

However, in order to grasp what goes on in electronic image recording, you do need to have some inkling of the principles underlying modern electronic components. They are all based on *semiconductors*. Now, although a semiconductor has an electrical resistance somewhere between an insulator such as porcelain or PVC and a conductor such as copper or silver, it isn't just a partial conductor like wood or damp earth. Conducting materials have many free electrons in their structure, and can thus carry an electric current freely. Insulators have all their electrons bound up with their molecular structure, and there are none free to carry a current.

Semiconductors also have their electrons bound up with their structure, but they can be shaken out of their bonds by the application of sufficient electrical energy (i.e., voltage), so that the material becomes able to carry a current. The most commonly used chemical element is silicon, in the form of a single crystal. This has a structure of atoms with four outer electrons in orbitals locked in with those of adjacent atoms, so that each atom "sees" a full orbital of eight electrons.

An electric current is just a movement of electrons in a conductor from the negative to the positive terminal of a voltage source. A current of 1 ampere (A) represents 6×10^{18} electrons entering and an equal number leaving the conductor per second (not the same ones, of course).

The Semiconductor Diode

A pure silicon crystal is called an *intrinsic* semiconductor. It will conduct electricity only if the voltage and the temperature are high enough. However, if we add a very small controlled amount of the right kind of impurity, its electrical properties change considerably. If we add an element with five outer electrons instead of four, such as phosphorus, there will be extra unbound electrons in the crystal, and it will now conduct electricity when only a low voltage is applied, as the electrons can be readily freed. If instead of this we add an element with only three outer electrons, such as aluminum, there will be spaces in the orbitals that can be filled by any stray electron, leaving a "hole" where it has come from—effectively a *positive hole*, as the donor atom has lost an electron. This crystal will also conduct electricity, by the laborious-seeming method of positive holes moving in the opposite direction to the electrons.

The two types of "doped" semiconductor crystal are called, respectively, *n-type* and *p-type semiconductors*. On their own they are, of course, electrically neutral, as there is no external source of either electrons or holes. The exciting thing is when a *p*-type and an *n*-type semiconductor are in contact (Figure 11.5).

Electrons from the *n*-type material are now available to the *p*-type material, and as there is a powerful urge for the orbitals to be complete, many of the electrons migrate from the *n*-type to the *p*-type, giving the former a positive charge and the latter a negative charge. An electric field is thus set up opposing the drift of any further electrons towards the *p*-type material. If an external positive voltage is now applied to the *p*-side and a negative voltage to the *n*-side, the field will be overcome and a current will flow.

Although a "hole" is just the absence of an electron, it represents a positive charge equal and opposite to that of an electron, and can be thought of in the same terms. A "hole" current is represented by holes moving from the positive to the negative terminal. The process of adding a controlled amount of impurity is called "doping."

The crystal structure has to be continuous across the junction, so the *p*-doping and *n*-doping have to be carried out on a single crystal of silicon.

Remember, the actual flow of electrons is from negative to positive.

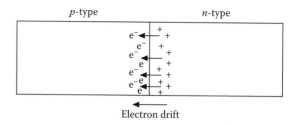

Figure 11.5 A *p–n* junction. Electrons drift from the *n*-type zone to fill the lattice gaps in the *p*-type zone, resulting in the accumulation of a negative charge in the latter and a positive charge in the former. Applying a positive voltage to the *n* side and a negative voltage to the *p* side increases the field opposing it, and no current flows. Reversing the voltage overcomes the field and current flows.

On the other hand, if the voltage is applied in the opposite direction, the field will increase and oppose any initiation of a current. We have a one-way device, a rectifier that will allow electrons to flow only in the direction *n*-type to *p*-type. It is called a *semiconductor diode*.

The Junction Transistor

This is a semiconductor device that is capable of amplification. It consists basically of two *p-n* junctions back to back, in a single crystal. There are two types: *p-n-p* and *n-p-n*. The very thin central region is called the *base*, and the outer regions are called the *emitter* and the *collector*. In an *n-p-n* transistor the emitter is connected to a negative voltage and the collector to a positive voltage. When the base (the *p*-region) is given a small positive voltage ("bias") a large number of electrons will pass into the base from the emitter, and as the base is very thin many of these will be carried over into the collector region, so that a current of electrons flows from the emitter to the collector. The base current has a large influence on the collector current, and a small change in base current can result in a 20- to 100-fold increase in collector current, thus achieving amplification. A *p-n-p* transistor operates in the same way, but with holes instead of electrons. There are several different types of transistors operating on somewhat different principles; this type is known as the *junction transistor*. Figure 11.6 illustrates the principle.

This is, of course, a much-simplified picture. If you are anxious for a more rigorous treatment, you will need to go to one of the sources listed in Digging Deeper.

The Field-Effect Transistor (FET)

A transistor, by its nature, is an analog device. But if the bias is high enough it is effectively a full conductor, and it can be operated as a switch. This is its main

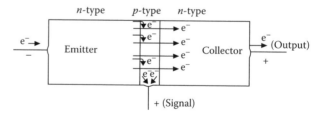

Figure 11.6 Junction transistor (*n–p–n*). When the number of holes in the base is reduced by applying a small positive voltage, a large number of electrons pass from the emitter into the base, and as these are now in a majority over those on the collector, they flow into it, completing the circuit. The base current is very small and the collector current large, providing amplification.

function in digital devices. However, its conductivity is sensitive to changes in temperature, and there is always a leakage current, so for many purposes it has been supplanted by one or other of the forms of *field-effect transistor (FET)*.

The FET fulfils much the same functions as the junction transistor, but is much more efficient. Its construction is different, however. In its basic form, instead of a *p-n-p* or *n-p-n* configuration it has a narrow channel of intrinsic semiconductor material joining two heavily doped layers of *p*-type material and *n*-type material, respectively. These are known as the *source* and the *drain*; the channel is called the *gate*. In one common type, the *MOSFET (metal oxide semiconductor field-effect transistor)*, the gate has a thin layer of an insulator with a conductor attached to its upper surface. When a positive voltage is applied to the gate an electron current flows along its surface. There is no "base current" as such. The schematic construction of a MOSFET is shown in Figure 11.7.

The Charge-Coupled Device (CCD)

A capacitor is a device for storing an electric charge. In its most basic form it consists of two metal plates or *electrodes*, separated by an insulator (the *dielectric*). The very earliest capacitor was called a Leyden jar, and was simply a glass jar lined inside and out with metal foil (Figure 11.8). Application of a voltage between the outside and inside foils caused electrons to flow into one side and out of the other, creating a large potential difference between the two foils. When the applied voltage was removed the charge would remain for a considerable time.

In the mid-nineteenth century, before Faraday's revelations on electromagnetism and invention of the dynamo, the only way to produce high voltages was with electrostatic generators such as Wimshurst's machine. This produced little current, but with a Leyden jar you could store an accumulated charge and obtain quite large currents from it for a brief time. Crude though it is, you can get a very nasty electric shock from a charged Leyden jar.

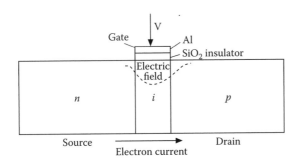

Figure 11.7 Principle of the field-effect transistor (FET). Instead of a base current there is an electric field resulting from a voltage on the gate. The strength of the field controls the current flowing through the substrate.

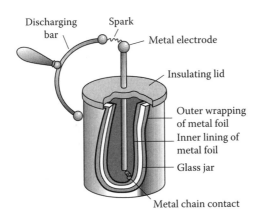

Figure 11.8 A Leyden jar, the earliest capacitor.

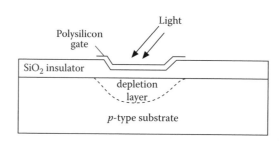

Figure 11.9 Principle of the metal oxide semiconductor (MOS) capacitor. The gate is made of a material that releases electrons into the substrate when stimulated by light. The electron charge remains until it is read out. This device forms one pixel of a CCD array.

Modern capacitors have a very thin dielectric layer and a large area of foil, folded or rolled up; they can store large numbers of electrons in a small space. The thinner the layer of dielectric, the greater is the capacitance (charge-storing ability). A variant of the MOSFET, known as a MOS capacitor, is the basic component in the *charge-coupled device (CCD)* that forms the heart of the digital camera. The MOS capacitor consists basically of a layer of dielectric material (silicon dioxide, SiO_2) only a few molecules thick, deposited on a substrate of *p*-type silicon. The uppermost electrode, or gate, is of transparent polycrystalline silicon, which is an intrinsic semiconductor. When light falls on the device the photon energy transfers electrons into the substrate, where they remain as a surface charge (Figure 11.9).

This type of MOS capacitor represents a single picture element (pixel) of the photosensitive array in the focal plane of a digital camera. Each pixel records the light intensity of the optical image (to be exact, the exposure) at that point of the focal plane, coded in the form of an electric charge. The next stage, of course, is to read those values and store them in such a way that they can be re-created, in the correct sequence, on a screen or printed on paper.

Getting the Image Out of the Camera

Linear CCD Arrays

The simplest kind of CCD array for forming an image is the single-line linear array, recording in one dimension. This type of array is appropriate for scanning and photocopying, where the array moves across the subject matter, or for the equivalent of a slit camera. There are several systems for retrieving the information, all of a similar nature.

The term "charge coupling" refers to the way signal charges are transferred from one electrode to the next. In the three-phase system, at any one instant only one out of every three electrodes carries a high potential, and any charge will be localized under that electrode. If an adjacent electrode has a potential applied and the potential on the original is cancelled, the charge moves to the new position. At each full cycle of clock voltages each charge moves one pixel forward. Charges can thus be passed along the line in what has been called "bucket brigade" action, to be recorded sequentially at the exit end. Three separate but interconnected clock generators control the potentials. (Three are necessary to ensure the charges move in the right direction.) Figure 11.10 illustrates the principle.

More recent developments include "buried-channel" CCDs, which carry the charge at a deeper level. They have a higher transfer efficiency and permit higher

This sounds like a slow business. It isn't: it happens in less than a microsecond.

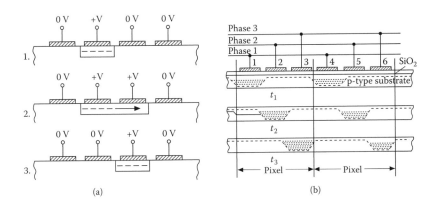

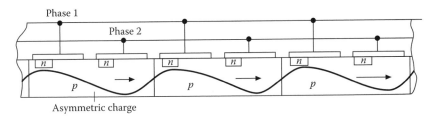

Figure 11.10 Principle of signal transfer. (a) The voltage pulses are on a three-phase system. The charge is always under the voltage, so when the voltage is transferred to the next cell (2), the charge moves with it and (3) becomes established in the next cell, and so on to the end of the line, where it is recorded. (b) "Bucket brigade" movement of charges.

Figure 11.11 Two-phase CCD principle. The asymmetrical doping ensures charges move one way only.

clocking frequencies. By employing asymmetrical doping it is possible to operate a two-phase system that can shift in only one direction, improving resolution by 50 percent (Figure 11.11).

Two-Dimensional Arrays

Linear CCD arrays are employed in flatbed scanners, and for recording moving images as in satellite imagery. But for straightforward photography we need a two-dimensional array. There are several methods for transferring charge in two dimensions, the full-frame transfer array being preferred for still cameras. In this system all the horizontal line readouts are fed to a shift register, where a separate clock performs a vertical shift. The output from the line readout is amplified as it emerges, pixel by pixel, then fed to a 12-bit (or higher) analog-to-digital converter, and its value and location coordinates are stored on the camera's software, which nowadays is a memory card. The image is displayed on a small screen at the back of the camera (on professional and semipro cameras there is usually a conventional viewfinder in addition) and the contents of the memory card are downloaded into a computer, where images can be enhanced if necessary, and printed out. Some enhancements, such as sharpness control, may be available within the camera.

Color in a Digital Camera

Monochrome digital cameras are used only for special applications. General-purpose cameras are always equipped with color systems. There are two ways of acquiring color images in a digital camera. One is to separate the red, green, and blue components of the optical image by dichroic filters, using the kind of beamsplitting arrangement described in Chapter 6 under "separation negatives"; this demands the use of three CCD arrays, and gives the highest resolution, but is expensive and comparatively bulky. A typical arrangement using totally reflecting prisms to orientate the images correctly is shown in Figure 11.12.

A more common method, and one used until recently in all amateur digital cameras, is to employ a three-color filter arrangement over the pixel array. A simple system used in rostrum cameras and scanning devices (including satellite survey cameras) is the triple linear array (Figure 11.13a). General-purpose cameras have a color mosaic filter with one filter element to each pixel. A common arrangement, known as a Bayer mosaic array after its inventor Bryce Bayer of the Eastman Kodak Company, is as in Figure 11.13b. The filter densities need to be selected so that the color sensitivity of the array matches that of the eye. In Figure 11.13b you will notice that there are twice as many green-filtered pixels as red or blue. This helps to make a match to the spectral sensitivity of the eye. In addition, it helps to preserve the sharpness of the image, as it is the green element that contributes most to the luminance component of the image.

You will also notice that in this latter case, when a saturated red or blue is imaged, only one-quarter of the pixels are operational. In practice the spectral transmissions of the filters overlap, so that all the pixels are active to some extent even for a saturated color. In addition, an interpolation algorithm is built into the camera software which, when it detects a uniform area of a saturated color, fills in the lowest pixel outputs so that there are no actual gaps. This does degrade the resolution by about 25 percent, but the result is preferable to a pattern of tiny holes in the uniform color area.

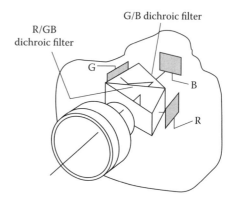

Figure 11.12 Color digital image capture using three CCD arrays.

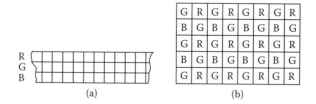

Figure 11.13 (a) Triple linear color CCD array for scanning devices. (b) Bayer mosaic filtered color CCD array for general-purpose digital cameras.

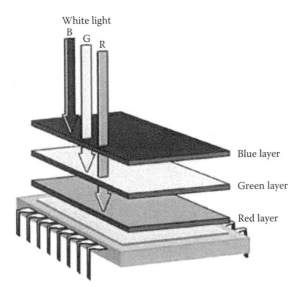

Figure 11.14 Foveon color-recording system (schematic). The three layers are part of the same silicon sensor element.

A recent advance in sensor technology enables every pixel to record all three colors. This depends on the fact that the three wavelength bands penetrate the sensors to different depths.

This system has been developed by Foveon Inc. The recording operates in a way analogous to that of silver halide color film with its three layers having sensitivity respectively to red, green, and blue. The earliest commercial product has a multilayered silicon sensor fabricated on a standard complementary metal oxide semiconductor (CMOS) array. Its operation depends on the fact that blue light is absorbed by the upper layer of the silicon, green by the middle layer, and red by the lower layer (Figure 11.14).

The wavelength absorption coefficient of silicon is shown in Figure 11.15 (color plate). This determines the depths of the electrodes required in order to determine the appropriate spectral sensitivities (0.2, 0.6, and 2 μm in Figure 11.15). This system can in theory produce higher sensitivity and higher resolution than mosaic systems, and has the added advantage that in the image there can be no colored moiré or multicolored staircasing artifacts accompanying fine detail or sloping edges. Such artifacts as remain are in the original optical image, and are not only

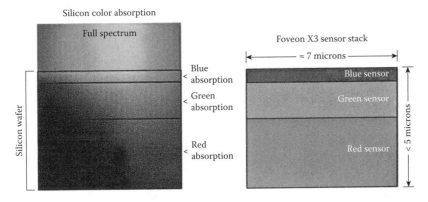

Figure 11.15 (See color insert following page 154.) (a) Color absorption in silicon; (b) relative thickness of subpixel segments.

less obtrusive in the raw image, but easier to deal with without loss of image quality. While it would be foolish to predict that this is the way sensors will inevitably go in the future, the system does seem to hold a good deal of promise.

Compression

If you associate with audio enthusiasts, you will be aware that the data compression that enables a minidisk to play for as long as an audio CD, or an iPod to store a huge amount of music in a very small space, is a contentious matter. Fastidious photographers tend to have a similar attitude towards data compression in visual image recording, and professional photographers often insist on recording completely uncompressed (RAW) data. However, camera software can hold only a limited amount of data, and although the amount of memory that can be held in memory cards increases more or less according to Moore's law, as digital cameras become progressively smaller and lighter the images that can be held in the camera's memory are still limited in number, particularly as so many advanced amateur cameras can now record minutes of continuous movie time. There are several different types of compression techniques, the simplest types of which store only the data that are changing from one pixel to the next. Thus a scene containing large areas of uniform tone and/or color need take up less memory than a complex scene containing a wealth of color and detail.

The compression system for sound recording involves deleting data for any sound that is masked out by louder sounds. As this treatment is likely to wipe out, for example, the harpsichord continuo from loud passages in an early Haydn symphony, perhaps the audio buffs have a point.

The Future for Digital Cameras

At the time of writing, the resolving power of a CCD or CMOS array only approaches the capabilities of a typical fine-grain 35-mm silver halide color film. But there is no theoretical reason why the resolution shouldn't approach the best that conventional films can achieve. The technological problems are being tackled, and there are regular announcements of yet another megapixel being added to the latest sensor array. Transfer efficiency is now at 99.9 percent, and the number of recording levels has in the past few years increased from 8 bits (256 levels) to 12 (4096 levels). Card memories have replaced discs, so that cameras can be made smaller and lighter—and cheaper. The earliest digital cameras were little more than (and often literally) digital backs grafted onto conventional camera bodies. Indeed, this is still the norm for studio cameras. After only a few years the very first purpose-built digital cameras have become dinosaurs. The convenience and cheapness of a camera that uses no film; offers instant viewing and retakes, with cine and sound recording facilities; is able to download directly to a computer; and—for news photographers—affords direct transmission to a central office, seems a sure indication that digital photography must completely take over both the amateur and photojournalist fields before long. And digitally stored images don't fade. Mind you, there are still many things traditional methods can do that digital technology is unlikely to do better or even equally well. But the future for digital cameras is assured.

Scanners and Scanning Methods

The scanning principle has a long pedigree. The earliest scanners appeared in the 1930s, and were used for sending black-and-white pictures by wire (later by wireless telegraphy). A collimated light beam was focused on the surface of a rotating drum bearing the material to be transmitted, and an associated photocell picked up the reflected beam. A lead screw caused the lamp and photocell to scan the material helically at about 80 turns per inch, and the signal went down the line in binary form: "on" for black and "off" for white. Such a system could not cope with continuous tone material, but it could manage fairly well with halftones, provided the spacing and orientation of the dot structure didn't result in moiré patterns. In the 1950s the principle was adapted to produce copies of drawings or typescript as stencils for ink duplicating machines, the stencils being cut by an electric spark.

Today's scanners have come a long way from these primitive beginnings. Modern scanning equipment belongs here for two reasons, namely that it employs a linear CCD array, and that scanners are used routinely to digitize silver halide photographic prints. Apart from their use in moving-film cameras ("slit cameras") such as photo finish cameras, aerial and satellite imaging, rotating panoramic cameras, and peripheral cameras, linear CCD arrays are also useful where the subject is stationary and the detector does the moving. In one type of scanner, the drum scanner, it is the subject that moves under the CCD array, but most scanners are now flatbed.

The handheld scanner resembles a miniature vacuum cleaner head (Figure 11.16). The scanner head contains a row of colored light-emitting diodes (LEDs) to illuminate the original, and a linear color CCD array to record it. To operate the scanner you simply draw it over your document. By sweeping overlapping sections you can scan documents that are wider than the scanner. The accompanying software stitches the swathes together to give a single image. The ability to scan areas of almost any size is one of the main advantages of the handheld scanner.

And a very nasty smell the process produced, too, as typing pools were only too aware.

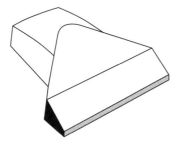

Figure 11.16 A handheld scanner.

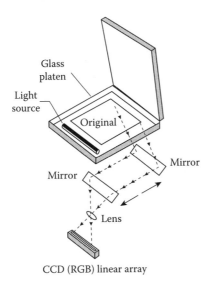

Figure 11.17 Typical flatbed scanner (schematic).

The more familiar flatbed scanner operates more like a photocopier. Typically, the original, positioned face down on a glass platen, is illuminated from beneath by a tubular light source, and a system of mirrors moves down the original, focusing an image on the CCD array via a lens system (Figure 11.17). Color scanners may operate either by making three separate passes under different filters, or by using colored tubes that flash in synchronism with the CCD clock, but the preferred system now seems to be a triple-CCD array with red, green, and blue filters.

Optical resolution is usually quoted in dots per inch (dpi) horizontally, the figure depending on the number of CCD elements per inch, and in lines per inch (lpi) vertically, this figure depending on the pitch of the stepping mechanism. Most flatbed scanners have an adaptor for scanning 35-mm transparencies and negatives, but this represents only a compromise, and a dedicated 35-mm scanner is essential for professionally acceptable results.

The retention of these obsolete measurements is partly a result of the American origin of much of the earlier equipment, but mainly of a stubborn resistance to change in the printing trade.

Frame Grabbers

The frame grabber is the successor to the still picture extracted from a movie sequence, or, more closely, to a photograph of a TV screen. Physically, it is a printed circuit board (PCB), which takes the output from an analog TV picture, digitizes it, then feeds the signal into a computer in a form similar to that given by a digital camera. Presumably, once TV becomes wholly digital the task of the frame grabber will be greatly simplified.

Digging Deeper

If you know little or nothing about electronics, or have had difficulty in following the explanation of CCD mechanisms given in this chapter, try *Starting Electronics*, by Keith Brindley (Newnes, 1999), which begins from scratch. It even teaches you how to solder connecting wires and terminals. A somewhat more advanced book, leaning towards the applied side, is *Electronic Products*, by Barry Payne and Daniel Rampley (Collins Educational, 1997). *High Speed Digital Design*, by Howard Johnson and Graham Martin (Prentice Hall, 1993) is an undergraduate-level text. A more controversial choice is *The Art of Electronics*, by P. Horowitz and W. Hall (2nd edition, Cambridge University Press, 2006). You will either love its friendly, helpful approach or hate the way important concepts seem to be left only half explained. Some electronics engineers swear by it. Borrow it from a library before you make up your mind.

In a subject that is moving so fast, it is difficult to provide a booklist that is fully up to date. The best I have come across is the late Ron Graham's *Digital Imaging* (Whittles Publishing, 1998). The explanations of the workings of CCD arrays are so clear and well expressed that I have found it difficult to avoid quoting at length from them in this chapter. Some of the equipment described is already going out of circulation, but a book as good as this deserves frequent updates in this respect, and no doubt will get them. For a more recent (though also more superficial) treatment of the subject, try *Easy Digital Photography* by Scott Slaughter (Abacus, 1999), which combines basic principles with practice. To find out more about the new generation of sensors, at present you will need to go to the Web, for example, www.foveon.com. All the amateur photographic periodicals carry information on the latest digital equipment, the most informative being the professional *British Journal of Photography*, which carries an in-depth study of a new digital camera almost every week.

Chapter 12 Halftone, Electrostatic, and Digital Printing

Continuous Tones with Printer's Ink

The earliest photographers made all their prints by contact, but as camera formats became smaller, contact prints became less attractive, and as early as the 1860s enlargers were in regular use. An enlarger is basically a camera that produces an image at a magnification greater than unity, and an enlarger lens is basically a camera lens that is optimized for close-up work and used back to front.

A problem arose when there began to be a need for photographs in newspapers and magazines. Certainly, there were methods of producing continuous-tone reproductions on plain paper without the use of silver halide, but they were slow and messy, and usually involved special preparation of the paper surface.

The difficulty is that printer's ink is black and opaque, so that printing is effectively a binary process. Black ink is either present or absent. Making steel engravings from photographs is a skilled and time-consuming business, but before the halftone process was invented this was the standard method of producing pictorial matter for printing in papers and books. By making the engraved lines very fine and close together, a skilled engraver could produce a fairly convincing illusion of the presence of shades of grey in the picture. And that is the clue to the next step.

Pre–Second World War Leitz enlargers were designed to use the standard 50-mm Elmar lens from the Leica camera in this manner. Some models even had autofocus linked to this specific lens.

An exception was *Woodburytype*, a process that involved a gelatin relief image, which produced an indentation in a soft metal plate when clamped to it under heavy pressure, the plate then being used for intaglio printing. With the advent of photogravure the process lapsed, though there are still some books with Woodburytype illustrations to be found in antiquarian bookshops.

The Halftone Principle

The idea that a grid of cross-hatched lines, or a pattern of black dots, would look grey when viewed from a distance such that the pattern would not be discernible led to the principle of the halftone illustration, invented in 1878 by Frederick Ives.

A "halftone," or middle grey, could be achieved by a mosaic of tiny black and white squares of equal size, like a microscopic chessboard. A light grey would result from a mosaic where the black squares were smaller than the white ones and a dark grey where they were larger. A continuous-tone image could thus be reproduced with nothing more than black ink and white paper, provided the original image was coded into black and white dots of varying sizes. Ives's ingenious solution was to use a high-contrast photographic plate and to install an opaque screen bearing a fine mosaic of tiny transparent squares, made by ruling opaque lines on a glass plate, a short distance in front of it. In such a system each of the apertures formed on the screen acts as a pinhole camera, forming a slightly blurred image of the camera lens aperture at the plane of the optical image of the subject. The high-contrast emulsion behaves like a binary switch: below a certain exposure threshold nothing happens; above the threshold there will be total black. Since the pinhole images are blurred, only the centers of the darkest parts of the original receive an

Frederick Eugene Ives (1856–1937) worked for a local newspaper in Litchfield, Connecticut. He was a pioneer of color cinematography.

171

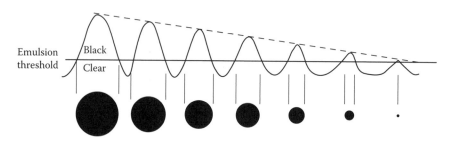

Figure 12.1 Principle of traditional halftone screen system.

over-the-threshold exposure, whereas in the brightest areas of the original subject much more of the dot image will be over the threshold. The final dot size is thus proportional to the brightness of the tiny area imaged (Figure 12.1).

When this negative is printed onto a copper plate coated with a photoresist (the original medium was dichromated albumen), the exposed areas become insoluble.

The unexposed areas are washed clean and the plate etched with acid. The protected dots remain raised above the surface, and once the whole plate has been cleaned it can be wrapped round a printing cylinder and used to print the halftone reproduction. Figure 12.2 shows a continuous-tone scene and a magnified example of halftone equivalent.

Owing to their expense and fragility, separate screens are no longer in use. Contact screens with graduated dots do the same job. They are often magenta rather than black, and are used with an orthochromatic process (green-sensitive, high-contrast) emulsion. If a halftone negative is needed and your lab does not have access to a process camera, you can use *auto-screen* film. This has been preexposed to the threshold point under a contact screen, so that any further exposure will bring up the dot pattern just as though you had used a screen in the conventional way. However, in the computer age it is a simple matter to turn a scanned picture into a halftone with a few keystrokes, and this is now becoming the most common method.

Screen Sizes

The largest dot spacing in regular use is 80 dots per inch (dpi), but such a coarse screen is used only with low-quality newsprint; 125 dpi is more usual for good-quality newsprint, and for the highest quality coffee-table books 250 or more dpi may be used. Owing to the high sensitivity of the human eye to vertical and horizontal lines, the dot alignment on a black-and white halftone is usually tilted to 45°.

A (negative) photoresist is a substance that becomes insoluble when exposed to light. A number of colloids—such as gelatin, gum Arabic, and albumen—behave in this way after having been treated with a dichromate.

This method of printing is called *letterpress* and is still used for very long print runs such as those of national newspapers, though even here it has been largely supplanted by photolithography. This is a process in which the dots are made ink-attractive and the spaces ink-repellent. The inked image is transferred to paper via a soft rubber roller, so the printing plate itself is under little stress and can be made of thin aluminum sheet or even tough paper.

Sorry about the inches but, as I said earlier, printers are a conservative lot.

(a)

(b)

Figure 12.2 (a) Halftone print; (b) part of print, magnified.

Printing in Color

If you want to print in color (and these days even local free newspapers carry color pictures), you need a basic three printings—cyan, magenta, and yellow—in register. The screens have to be set up at different, preferably unrelated angles; otherwise, there will be blotchy colored moiré patterns. To minimize the risk of this, the yellow printer may be made with a somewhat coarser screen than that of the other colors. (Yellow is chosen because it contributes little to the overall resolution.) The printing plates are made in the usual way from separation negatives, which may be masked as described below.

Because some of the colored dots appear alongside one another and some of them overlap, the resulting image is partly additive and partly subtractive. This places restrictions on the spectral content of the inks, and inevitably these colors together do not produce a good black, or even a neutral grey. It has therefore been the practice for many years to include a "black" printer (actually grey) made from the color original using a medium yellow filter. A more sophisticated system uses *undercolor removal* (see below). The four-color combination (cyan, magenta, yellow, and black) is given the abbreviation CMYK, the "K" being used for "black" rather than "B," to avoid possible confusion with "blue."

Color Masking

In Chapter 7 there was a box explaining the way that built-in color masking is used in color negatives to compensate for inherent errors in hue and saturation of the cyan and magenta dyes. It would be a good idea for you to go back and look at this box again, as I am going to analyze this problem more closely. A color negative consists of three color-separation negatives superposed in register, and it is easier to see exactly what is needed if we think of three actual color-separation negatives made on black-and-white film and then printed in printers' pigments (cyan, magenta, and yellow).

In a three-color separation printing process the red-filter negative (the red record) is printed in cyan ink, the green record is printed in magenta, and the blue record in yellow. But printers' inks suffer from the same errors in hue and saturation as the dyes in a color film, and to a much greater extent. In particular, the magenta ink absorbs blue light that it should reflect, and the cyan ink absorbs both blue and green. The error with standard printing inks is some 30 percent for each of these. Considering first the magenta ink: this absorbs around 30 percent of blue light, and thus behaves as if it was mixed with 30 percent yellow. So whenever magenta and yellow occur together, we must compensate for this error by removing 30 percent of the magenta value *from the yellow printer*. The way to do this is to make a positive of 30 percent contrast from the green record (the magenta printer) and bind it in register with the blue record (the yellow printer). That takes care of every situation where both green and blue records bear an image.

The cyan ink is both desaturated and bluish. It behaves as if it contained some 30 percent magenta and 15 percent yellow. So, in a similar manner to the magenta correction, we make a 30 percent positive mask from the cyan printer (red record) and bind it in register with the green record, to reduce the magenta content where cyan is present, and a 15 percent mask from the same cyan printer, to bind in register with the blue record, to reduce the yellow content by this smaller amount where cyan is present. This corrects the errors in hue and saturation sufficiently for most purposes. However, if you are to obtain full correction for every color, you need

It doesn't take care of a pure magenta where no yellow is present. Photographs of magenta flowers with highly saturated colors *do* look reddish, and there is nothing you can do about this.

to take into account the small absorption of red by the magenta dye and the small absorptions of blue and green by the yellow dye; so in theory you need six masks, two for each separation negative. As if this wasn't enough, none of the three inks can produce a fully saturated color, so black or neutral grey areas where the three dyes are printed in equal amounts still don't add up to a true black, only to a dark muddy brown. So a grey printer has to be added, just to give a good black, and this is traditionally made by a fourth negative taken using a yellow filter, as explained earlier.

If you think this sounds like an immense amount of work you are not far wrong. In practice, masking has seldom been taken to these lengths. For the finest quality color work most of the final correction was entrusted to the skill of a fine-etcher, who worked from the proofs and made small corrections to the printing plates with a fine paintbrush and nitric acid. Today nearly all printing is carried out either by offset lithography or by electronic or electrostatic printing methods, where fine corrections by hand are not possible. Fortunately these can now be made electronically with great accuracy. There are dedicated programs that can carry out both masking and undercolor removal matched to the specific printers' inks (see below).

Undercolor Removal

Three-color inks are expensive and, as we have seen, they are not 100 percent efficient. Wherever cyan, magenta, and yellow appear together, we could substitute grey. So why not do so? With electronic analysis and control it is easy. All the computer has to do is to subtract equal quantities of cyan, magenta, and yellow from any area where use of all three colors is indicated, until only two colors are left present, and then print those two colors in appropriate proportions, plus the required amount of neutral grey. This usually does away with the necessity for the fourth (K) exposure.

Electrostatic Copying (Xerography)

Electrostatic printing depends on the properties of certain types of semiconductor material (such as amorphous selenium) that become conductive when illuminated. The main application of the process is in xerographic document copying.

The word *xerography* comes from two Greek words meaning "dry writing." "Amorphous" comes from Greek words meaning "shapeless" and means that the material has a glass-like texture as opposed to a crystalline one.

The xerographic process was invented by Chester Carlson in 1938 and patented by him in 1942. A typical xerographic photocopier uses a large metal drum coated with a layer of selenium as the light-sensitive medium. A strip light illuminates the original document, and a lens focuses the strip image onto the drum; the image is moved in synchronism with the rotation of the drum.

There are seven steps in the production of an electrostatic photocopy:

1. The drum passes under a grid of fine wires charged to a potential of 5–10 kV in the dark. The resulting corona discharge puts a uniform electrostatic charge of several hundred volts on the drum surface.
2. The drum passes under the exposing slit and the projected image discharges the electrostatic charge in proportion to the illuminance. The drum now carries a latent image in the form of an electric charge.
3. Electrostatically charged toner is introduced to the drum surface, where it accumulates on the areas that have retained their charge. The toner usually

consists of particles of carbon black about 5–20 µm mean diameter, in a resin binder.

4. The receiving paper is charged to a high potential, and fed into contact with the drum surface. The toner particles are now attracted to the paper surface.
5. The receiving paper passes out through heated rollers, which press the toner particles into the surface, fuse the image into the paper, and eject the finished print.
6. Any remaining toner particles are removed from the drum by a brush.
7. The drum is exposed to light to dissipate any remaining charge.

All these steps take place in a single revolution of the drum. There are several types of optical systems that may be employed to obtain the synchronized moving image, involving the movement of the illuminating lamp or the lens system or the platen, or all three. A schematic diagram of the system is shown in Figure 12.3.

For color copying, three or four toners are needed to produce a CMY or CMYK image. The exposing light can be either three or four successive strip lights with appropriate filters, or a row of RGB light emitting diodes (LEDs). In the simplest version there are three or four separate passes, one for each toner.

Copiers intended for rapid runs of several hundred copies use a charged belt instead of a drum. As pure selenium is brittle, organic photoreceptors with a degree of flexibility are used instead for the belt coating. These consist of strong electron donors (or acceptors) in a polymer substrate, usually containing a pigment to even out the spectral sensitivity. This type of photoreceptor gives better reproduction of continuous-tone material than selenium, and it is now becoming universal in both color and black-and-white photocopiers.

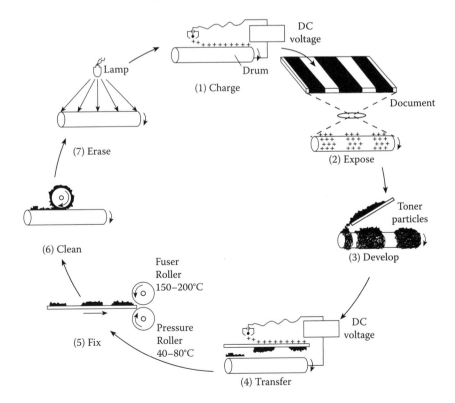

Figure 12.3 Cycle of operations of an electrostatic photocopier.

Xerographic photocopiers have now become standard equipment in printing sections in universities and other institutions where runs of several hundred copies are needed in a hurry. Recent models can turn out up to 100 copies a minute.

Printers

A printer takes the stored image from a computer, pixel by pixel, and translates it into digitally coded hue, density, and saturation values, which it then turns into a printed image.

There are four basic methods of achieving this: direct mechanical contact, controlled deposition, xerographic, and quasi-photographic.

Direct Mechanical Contact

The oldest and simplest of these is suitable only for text and simple line decorations (as for menu cards). It uses a standard typewriter format and a daisy wheel. The *dot matrix* printer grew up alongside the daisy wheel. It uses a block of up to 24 pins filling a rectangle the size of a typescript capital. Like the daisy wheel printer it uses an ink ribbon, but can operate at more than 10 times the speed. It can be used for pictures, but the quality is poor, and its main advantage is its simplicity. It is generally used for calculator printouts and cash register receipts, but even for these purposes it is being phased out.

Controlled Deposition

The most commonly used system is the ink-jet system. The ink jets themselves are near microscopic in size, up to 256 of them fitted into a space the size of a typescript capital. They operate by forcing out a tiny drop of ink either by heating the nozzle (which is finer than a human hair) or by squeezing its upper part with a piezo element. To make up for the deficiencies in the standard CMYK dyes, high-quality printers have additional dyes, sometimes as many as seven.

A piezo element is a crystal or ceramic block that expands when a voltage is applied across it.

The two other systems are *dye sublimation* and *thermal wax*. Both systems use a contact ribbon, and transfer is by the application of heat. The printer head resembles the dot matrix head, except that instead of 24 pins it has several hundred microscopic needles, all of which are temperature controlled. "Sublimation" means "vaporization without melting," and in the dye sublimation process the application of a hot pin to the ribbon causes the pigment to vaporize into the paper, the density of the dot depending on the temperature of the needle. (As the needles are so fine their temperature can be varied very rapidly.) The thermal wax system is similar, but at a lower temperature. The dyes are in a wax base, which under a hot pin melts into the paper. Both systems require four passes to produce a CMYK image. Some printers can be adjusted to take either dye sublimation or hot wax ribbons.

Ink-jet printers are now capable of giving very high quality results, and the alternative processes seem to be waning in popularity; a number have already ceased production. The market in printers for home use is now exclusively ink-jet, with machines capable of results that are fully up to professional quality.

Xerographic Printers

The sole representative is the laser printer. It is a logical extension of the electrostatic copier. Instead of exposing the drum to an optical image, the drum is scanned across its width by a tightly focused diode laser beam. The laser is fixed, and the scanning

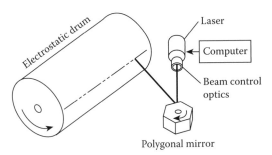

Figure 12.4 Principle of laser printer.

is accomplished by a rotating polygonal mirror. The beam intensity is varied as appropriate to each pixel during the scan. Figure 12.4 shows the schematic layout. Many so-called laser printers don't use lasers at all, but a row of hundreds of near-microscopic LEDs instead. This arrangement does away with the need for a rotating mirror. The diodes are selectively switched according to instructions fed into the printer from the computer.

To produce CMYK prints from a laser printer it is necessary to use four passes, one for each color. This can be done on a single drum with four changes of colored toner or, in a printer intended for fast production runs, with four drums.

Quasi-photographic Printers

At the time of writing there are two printing systems that use photographic-type material, both originating from Fuji. In the *thermo-autochrome* process the print material is somewhat similar to traditional color print paper, with RGB-sensitive layers containing CMY color-generating material. The scanning is by a three-color LED row, and the print is developed by the application of heat and fixed by UV radiation. The *Pictrography* process (the name is Fuji's coining) covers printers that use red, green, and blue diode lasers to make a single pass over a dye donor material, the dyes of which are transferred to a final base by contact, in a manner somewhat similar to Polaroid professional color material. Both of these systems yield prints of photographic quality, and it seems likely that variants of these techniques are likely to be introduced by other manufacturers, as they have the advantage over other processes of not using ribbons or printing inks. In addition, the systems operate directly from RGB files without the necessity for CMYK conversion.

It has to be said, however, that, in a similar manner to sound and video-recording methods (see Chapter 14), one system inevitably comes to dominate the market, and not necessarily the best one. Both of these systems seem to be on the way out in the face of competition from the ubiquitous, and constantly improving, ink-jet techniques.

Digging Deeper

There is a vast amount of information on conventional halftone printing. Perhaps the best is U. M. Adams's *Printing Technology* (Delmer, 1996). This also contains a full account of the digital technologies outlined in this chapter. If you are interested in printing per se, you couldn't do better than hunt for old copies of *The Penrose Annual*, which chartered the progress of traditional printing, and always contained breathtakingly beautiful examples of the printer's craft. Bamber Gascoigne has written a splendid survey of printing up to 1860 in *Milestones of Color Printing* (Cambridge University Press, 1997). Frank Romano's *Dictionary of Digital Printing* (Delmer, 1997) steers you through the jargon surrounding the subject, and Ron Graham's *Digital Imaging* (2nd revised edition, Whittles Publishing) takes you up to 2002 in technology and equipment. If you are interested in the way printers managed to produce pictures before photoengraving was invented, there is a new and splendid book on the subject called simply *The Printed Picture*, ed. Richard Benson (Museum of Modern Art, New York, 2008). As far as new equipment is concerned, you will need to consult the *British Journal of Photography* and the *Professional Photographer*, which regularly publish detailed surveys of equipment. Best of all, visit the annual Focus on Photography exhibition, which takes place at the National Exhibition Centre in Birmingham every February, for a full update on equipment.

Chapter 13 Television

Beginnings

The concept of television as a practical technology began with the work of a Scot, John Logie Baird, in the early 1920s. His work built on an idea of Paul Nipkov, a German engineer who in 1883 had taken out a patent on a scanning disc that contained a spiral of holes positioned so that when it was spun the holes would scan an entire image, line by line, in one revolution. The brightness of the image at each point would be transmitted by wire and reconstructed by a similar disc synchronized to the disc in the transmitter. Nipkov did not exploit his patent, and it was left to Baird to produce a practical system, using a radio link. In Baird's transmitter the luminances tracked by the rotating Nipkov disc were recorded and coded as modulations of a radio signal (Figure 13.1).

This signal was transmitted to the receiver, which was equipped with either another synchronized Nipkov disc or a rotating polygonal mirror, which generated a flying spot that covered the screen area and matched the luminances of the original subject.

In 1887 Karl Ferdinand Braun developed the cathode ray oscilloscope in order to study high-frequency alternating currents.

In this device the scene is focused by a lens on a plate, and the image is scanned by an electron beam, the forerunner of the image orthicon tube. The receiver used a synchronized flying spot to reconstruct the image on a fluorescent screen. A number of workers in Germany and Russia played with the idea of transmitting pictures using this device, but the first practical demonstration of a working electronic television system was by another Scotsman, Archibald Campbell-Swinnerton, in 1908. This led to collaboration between the two companies Marconi and EMI (Electrical and Musical Industries) to produce a commercially viable system. By 1936 the two rival systems were sufficiently developed for the BBC to begin television broadcasts, transmitting

Karl Ferdinand Braun (1850–1918) held positions in a number of German universities before settling in Strasbourg, where he founded, and became director of, the Institute of Physics. He pioneered the operation of the crystal rectifier ("cat's whisker") used in early wireless receivers ("crystal sets"), and developed the principles of antennas, before inventing the cathode ray tube. He shared the 1909 Nobel Prize for physics with Marconi (see margin note below).

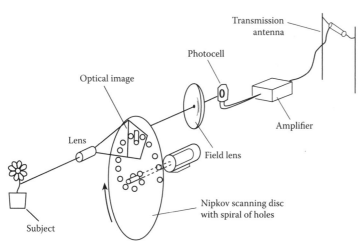

Figure 13.1 Baird's original television system. The receiver used a similar system in reverse, with a lamp replacing the photocell.

Baird died in comparative obscurity in 1946, his contribution to the birth of television almost forgotten. The Irish-Italian inventor Guglielmo Marconi (1874–1937) managed to retain a higher public profile. He exploited Hertz's discovery of radio waves commercially, setting up a profitable establishment in London, and in 1901 succeeded in sending a Morse code message across the Atlantic. Soon after this he founded a worldwide telegraph network for the British Government. Unlike Baird, he died loaded with honors, and has a lavish tomb in Florence Cathedral.

When illuminated on one surface, a photovoltaic material produces a potential difference between the light and dark surfaces. A photoconductive material is a semiconductor that becomes conductive when light falls on it. Both types of material have a broadly linear response.

by the Baird and the Marconi-EMI systems on alternate days. The following year the Baird system transmissions were abandoned—a curious decision, under the circumstances. At that time Baird's system was visually superior to the Marconi-EMI system, and the receivers were simpler. Unfortunately, at about the same time Baird lost much of his equipment, including his first electronic camera, in the disastrous Crystal Palace fire. But Baird didn't give up. At that time he was already working on a color camera. Early in 1940 he was seconded to the War Office, and spent the whole of the Second World War working on radar and other advanced electronic equipment. Because of the secret nature of his work, few people knew about it at the time.

The EMI transmissions continued until the outbreak of the Second World War, then ceased for six years. By the time they resumed the cathode ray tube was king, and remained so for more than 60 years. It has now been superseded in both cameras and receivers by semiconductor sensors and displays, which are lighter and smaller, and require neither a high vacuum nor a high voltage.

The Television Camera

The optical part of a TV camera is similar to that of a conventional still camera, i.e., a lens equipped with an iris diaphragm focuses an optical image on a light-sensitive surface. In the earliest TV cameras the recording device was called an *image orthicon tube*. The light-sensitive part was a plate coated with photovoltaic material.

The optical image produced a negative charge on its rear surface in proportion to the illuminance at each point. This plate formed the "screen" for an electron beam tube, a scanning device that needed a high vacuum for its operation. The scanning electrons were repelled in proportion to the image illuminance and were caught by an "electron trap," which thus recorded the illuminance. It was cumbersome, and was prone to image artifacts, especially in scenes with bright lights. It was gradually replaced by the *vidicon tube*, which used a photoconductive plate. This allowed the electron current from the scanning beam to be read directly from the plate (Figure 13.2). The scanning rate was (and remains) the same as that of the TV receiver. (The scanning mechanism is described under The Cathode Ray Tube below.)

A few cameras used in professional broadcasting still employ vidicons, but these have now been largely supplanted by the smaller and more convenient CCD chip, which also is faster, has a more linear response, and does not require a vacuum tube. Amateur camcorders are now little bulkier than a traditional 35-mm still camera, and even these are now being supplemented, if not entirely replaced, by smaller dual-purpose still/cine cameras (see Chapter 10). TV cameras can now be miniaturized to such an extent that they can be inserted into human body cavities

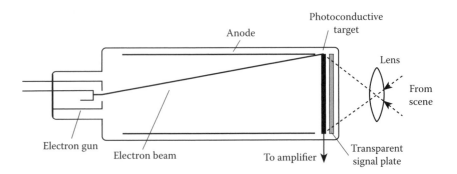

Figure 13.2 Principle of vidicon tube.

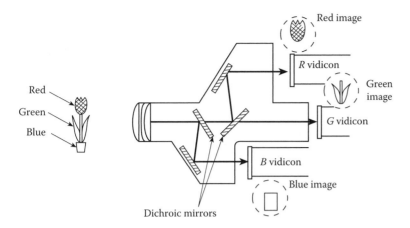

Figure 13.3 Simplified diagram of a modern professional TV camera with three CCD arrays.

for diagnostic purposes, and to monitor the progress of surgical procedures. These tiny cameras find their way into all sorts of unlikely places such as bird boxes and termite mounds, even a cricket stump.

As with still cameras, TV cameras need to record three images, respectively, red, green, and blue. The introduction of color presents extra problems over monochrome, as the responses of the three receptors need to be an exact match, and the color balance of the lighting has to be adjusted to give an acceptable white. (This can nowadays be achieved automatically by circuitry within the camera.) You have already met the CCD chip as used in the digital still camera. Professional studio TV cameras generally have three similar chips for the red, green, and blue images, using dichroic mirrors or a prismatic construction (Figure 13.3). Portable TV cameras employ a mosaic filter similar to those used in still cameras in front of a single CCD array.

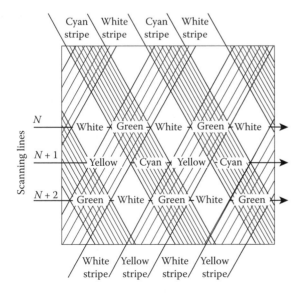

Figure 13.4 Example of a mosaic filtering system for recording RGB signals economically.

However, the Bayer mosaic arrangement, once standard in such arrays, is inefficient, as no more than one-third of the light energy can reach any element in the array. A more economical scheme criss-crosses cyan/white diagonal filter stripes with yellow/white stripes, giving a pattern of cyan, green, yellow, and white (Figure 13.4). The green content is recorded directly, the red by electronically subtracting the cyan signal from white, and the blue by subtracting the yellow signal from white. This system allows at least twice as much light to reach any element in the array, at the expense of a small loss of color quality. It seems likely that a development of the Foveon system (see Chapter 11) may overtake this technology, too.

Data Storage Methods

The traditional storage device for both analog and digital TV cameras has been a tape cassette. Successive reductions in size have brought these (for amateur and small field professional cameras) down to about the size of an audiocassette, while still permitting up to four hours of recording time with compression. Disc and memory card storage are now becoming the norm.

The luminance (local image brightness) and chrominance (hue and saturation) signals are separated electronically for storage. The chrominance record requires only the red and green components, as these can be subtracted from the luminance record to give the blue component, thus reducing the storage content.

Transmission and Reception of a TV Signal

The full theory of TV transmission and reception lies beyond the scope of this book. There are whole books on the subject (see Digging Deeper for Chapter 14), and the technology varies to some extent in different countries. In addition, with the advent of digital high-definition television (HDTV), new standards have been evolved. As far as analog TV is concerned, there are at present two main systems, differing only in detail. The system used mainly on the American continents is called NTSC, which stands for National Television Systems Committee. The NTSC analog system operates on 525 lines with interlaced frames at 60 pps. The PAL (phase alternating line) system has been adopted by most of the rest of the world. Using 625 lines at 50 pps (also interlaced) it averages out color errors by reversing the phase of one component of the chrominance signal on alternate lines (hue depends on phase).

"Interlaced frames" means that the odd and even numbered lines are shown on alternate scans.

This enables material shot at 25 (30) pps to be shown at effectively 50 (60) pps, thus minimizing the flicker that is sometimes obtrusive at high light intensities. Some analog receivers show each complete frame twice, producing effectively 100 (120) pps, a rate that eliminates flicker entirely.

Color control with the NTSC system has always been difficult, and in the early days the popular interpretation of these initials was "Never The Same Color."

In France and Russia, and parts of Africa and East Europe, the SECAM (séquence couleur à memoire) system was used until 1983. This was an 825-line system operating at 50 pps, using separate chrominance signals for alternate lines. After 1983 the system reverted to 625 lines, in which form it is still in use (most SECAM analog receiving sets are now dual-system with PAL as an alternative). However, with the advent of digital TV the discrepancies between analog systems have become irrelevant.

The Signal

In analog TV the luminance and chrominance signals are encoded onto the carrier as *amplitude modulation* (AM); the sound is transmitted as *frequency modulation* (FM) on a separate carrier at a slightly different frequency. FM gives better quality

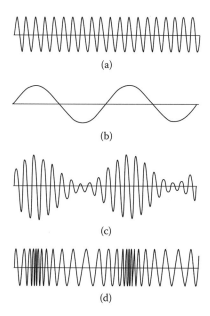

Figure 13.5 Modulation of a carrier wave. (a) Carrier; (b) waveform to be encoded; (c) amplitude modulation (AM); (d) frequency modulation (FM).

sound, free from interference. The reason AM is used for the picture signal is that any visible "echoes" are stationary (with FM they would fluctuate).

Figure 13.5 shows the way a simple sinewave signal is coded onto a carrier by AM and FM. The decoder for AM can be a simple rectifier to cut out the negative-going part of the signal, plus a smoothing circuit. The FM decoder is more complicated. In practice both decoders are quite sophisticated, as luminance, chrominance, and two or more channels of stereo sound also need to be retrieved from the complete signal.

In digital transmission the carrier is sampled at a higher frequency, in accordance with the Nyquist requirement. The luminance signal is produced by matrixing the RGB components to form a monochrome (Y) signal. The chrominance signal, which uses much less bandwidth, is coded as two components, namely R-minus-Y and G-minus-Y, from which the RGB components can be teased out electronically. As chrominance signals do not need to contain the same amount of detail as the luminance signal, the digital sampling for chrominance can be at a lower rate than that of the luminance.

Transmitting Antennas

The transmission of a signal depends on the energy present in the electromagnetic field that exists between the transmitter and the receiver. The device that sets up this field is called an *aerial*, or more correctly, *antenna*. In order to set up the field with maximum efficiency, the transmitting antenna has to be matched to the wavelength of the signal it is required to transmit. Radio waves are electromagnetic radiation, just as light is (see Chapter 1). At the low and medium frequencies used for sound radio (i.e., 0.15–1.5 MHz, or 2000–200 m wavelength), diffraction enables the radiation to follow ground contours, and though higher frequencies may be blocked by tall buildings and hills, they are reflected by the ionosphere at a height of around 300 km. The transmission antennas for these frequencies need to be

183

The ionosphere consists of a layer containing large numbers of charged particles, and is situated between about 50 and 500 km above the Earth's surface. The Appleton layer, at about 280–304 km, is particularly rich in free electrons, and acts as the main reflector for short-wave radio.

All TV transmitters in the United Kingdom (except for a few relay stations) transmit a horizontally polarized signal. It is the electric component that is horizontal. The magnetic component is thus vertical, which it needs to be in order to create the maximum electric potential in the antenna elements.

matched to the corresponding wavelengths, and accordingly are very large. Such huge "aerial farms" can be seen at the long- and medium-wave transmitting stations at Droitwich, Daventry, and Rugby in the United Kingdom.

Because the receivers are all around it a transmitter antenna needs to be *omnidirectional*. Low-power transmitters use a ring of dipoles (see below), their outputs overlapping so that the power distribution is uniform in all directions. Individually they need to be fairly large in area, as they have to dissipate a good deal of heat, so the greater the surface area the better. Because of this they may be taller than their width, even though the polarization of the transmitted beam is horizontal.

If this rectangular emitter is higher than it is wide, diffraction in a vertical direction is restricted and power output is confined to a comparatively narrow horizontal region. Most cellphone relay towers have antennas designed on this principle, as you will probably have noticed. A single antenna can be made omnidirectional by bending it into a cylinder (Figure 13.6). This type of antenna also has good heat dissipation. Figure 13.7 shows a reception antenna on the same principle, matched for VHF radio reception.

Receiving Antennas

Dipole Antenna

When the radiation encounters a conductor it sets up an electric potential across it, and this can be used as the input to a receiver. At the frequencies used for TV broadcasts the input signal received will be very weak unless the conductor (the aerial, properly known as the *receiving antenna*), is facing in the right direction and is of the right length to match the transmitted wavelength. There are several ways of achieving this match, the simplest of which is the *half-wave dipole* (Figure 13.8a). It consists of two conducting rods (usually aluminum alloy tube), each one-quarter of a wavelength long, orientated perpendicular to the direction of propagation and parallel to the direction of polarization. The effect of matching the wavelength in

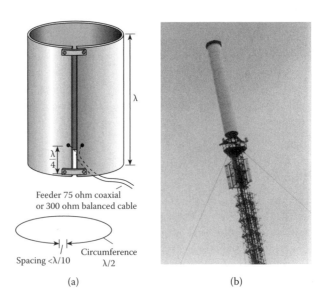

Figure 13.6 (a) Proportions of a typical cylindrical transmission antenna. (b) TV mast showing several different types of transmission antenna. The cylindrical ones are at the top, protected by a fiberglass shield.

Figure 13.7 A domestic TV antenna consisting of a pair of Yagi arrays, plus an FM radio antenna consisting of a folded dipole bent into a circle for omnidirectional reception.

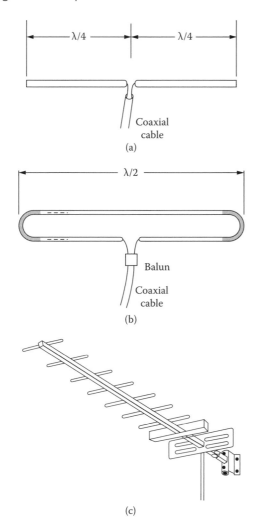

Figure 13.8 (a) Dipole antenna. This consists of two quarter-wavelengths of conducting metal rod aligned perpendicular to the line joining the antenna to the transmitter. (b) Folded dipole antenna. (c) A nine-element Yagi array.

185

this manner is that at the tips of the antenna the oscillating voltage is a maximum and the current is close to zero, whereas at the innermost ends the voltage is close to zero and the current a maximum (see Figure 13.9).

A strong signal produces an output of 1–2 millivolts (mV) from the antenna, and this is led via a cable to the input of the TV receiver. In the United Kingdom coaxial cable is used, which minimizes interference from outside sources, as the outer conductor is connected to earth and shields the inner conductor. Mainland Europe mostly uses twin or *balanced* cable. The two conductors are then at opposite potentials. The input from the antenna is fed to a transformer, which passes it to the decoding circuit within the receiver. A coaxial input has one end of the transformer primary at earth potential; a balanced input has the earth in the middle.

Folded Dipoles

The output of a dipole antenna is approximately doubled by joining the ends of the dipole to form a loop (Figure 13.8b). Note that this does not constitute a short circuit, as you might suppose, because at the ends of the dipole the current is zero.

IMPEDANCE

Whenever electrical power is transferred from one system to another there is always a loss. This is reduced to a minimum when the two systems are matched for *impedance*. Impedance is the a.c. equivalent of resistance, and like resistance it is measured in ohms. It is derived by dividing the a.c. voltage by the a.c. current. The two quantities are not necessarily in phase, and in a dipole element they are 90° out of phase, as can be seen from Figure 13.9.

A simple dipole has an impedance of 75Ω, which is the same as the impedance of a coaxial cable with the outer element earthed, so it can be connected directly to the cable going to the receiver. A folded dipole has an impedance of 300 ohms, which is the same as that of a twin (balanced) cable. To obtain maximum power transfer with a coaxial cable it is necessary to incorporate an impedance matching device between the dipole and the cable. This is called a *balun* (*bal*-anced to *un*balanced), and is a simple combination of resistors and capacitors. It is usually installed in the antenna head mounting.

The symbol for ohms is Ω, the Greek capital letter omega. If you want a practical example of impedance mismatch, try connecting a pair of hi-fi speakers to the headphones outlet of an audio amplifier. You will get hardly any sound out of them, because output impedance of the headphones is typically 100 Ω or more, and the speakers have an input impedance of only about 5Ω.

A balanced cable, with an impedance of 300Ω, has the advantage that it can be fed directly from a folded dipole without the need for a balun, so the losses are lower than with a coaxial cable. On the other hand, it is more subject to interference.

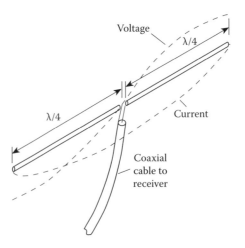

Figure 13.9 At the outer ends of the dipole the voltage is a maximum and the current is zero; at the center the voltage is zero and the current is a maximum.

Multi-element Antennas

There are two main factors that make a single dipole antenna inefficient. The first is that it acts itself as a radiator. This means that much of the incoming energy is reradiated into space. The second is that it can receive an equal amount of radiation from behind, which means that any signals coming from behind the dipole can cause serious interference. A single dipole also picks up signals within an angle of around 30° to either side, so it is liable to receive further unwanted signals from other transmitters.

Reflector Element

This is an unconnected dipole about 5 percent longer than the driven element, positioned one quarter-wavelength behind it. Its reflected signal is in phase with, and reinforces, the main signal, while its transmitted signal is in antiphase with the onward transmission and cancels it, and at the same time it assists in blocking signals from unwanted transmissions coming from behind the active element. Figure 13.10a and b show the radiation acceptance pattern for a dipole antenna without and with a reflector element. Some reflectors are in the form of a slotted plate, which increases their efficiency (Figure 13.8c).

Parasitic (Director) Elements

The signal strength can be further augmented by adding unconnected *parasitic* (or *director*) *elements*. The addition of these also further reduces both of the above problems. When an antenna array has parasitic elements, the dipole that actually

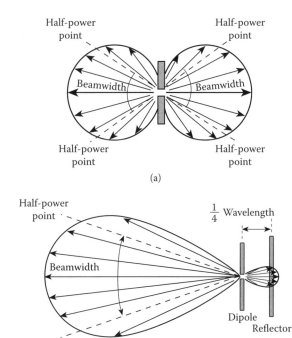

Figure 13.10 (a) Radiation acceptance pattern for a simple dipole. (b) Pattern for a dipole with reflector element.

You may feel that there is something wrong here: the spacing should be a half-wavelength, not a quarter-wavelength. The latter is, however, correct: in the process of reflection the wavefront undergoes a phase reversal, and this results in the phase required for destructive interference.

If you deduced that the parasitic elements were operating in an analogous manner to the quarter-wave antireflection coating of a camera lens, full marks. The principle is indeed the same. However, if you look carefully at an actual Yagi TV antenna array you will see that the parasitic elements are in fact nearer to each other than a quarter-wavelength, and become still closer together towards the front end, as well as becoming shorter. This is to obtain uniformity of signal capture over a fairly wide band of frequencies.

This type of transmission is also used for cell phone and VHF radio relays, which is the reason so many tall buildings sport forests of antennas and dishes on their roofs.

feeds the signal to the receiver is called the *driven element*. The parasitic elements are slightly shorter than the driven element, and are spaced approximately one quarter-wavelength apart. Each element passes the signal onward to the next, and the overall gain in a multi-element setup can be as much as 15 decibels (30 times). Figure 13.8c shows a typical nine-element UHF array. This is known as a Yagi array, after one of its inventors.

Antenna design is almost as much a craft as a science, and many variants are around. Some less orthodox designs have been found to give maximum gain for weak sources; others give maximum directionality where there is external interference. Some aerials can be very complicated indeed (Figure 13.11).

Microwave Relay Transmission

If you want to send very high frequency (VHF) or ultrahigh frequency (UHF) signals over distances beyond the line of sight you will require some kind of relay system. The VHF band, nominally 97–105 MHz, is used for analog sound broadcasting using frequency modulation (FM). The UHF band, 300–600 MHz, is used for analog and digital TV broadcasting, as well as digital sound. In the past, relay stations were fed from aerial farms or by landline, but today all UHF signals to be transmitted over long distances employ microwave links. Microwaves have wavelengths of only a few centimeters, and behave very much like light waves. They undergo refraction, diffraction, and reflection, and they travel in straight lines. If you raise the carrier frequency to several gigahertz (1 GHz = 10^9 Hz) you can focus the signal into a tight beam with a paraboloidal reflector, like a searchlight (Figure 13.12), and direct it to a receiver (also equipped with a paraboloidal reflector, which focuses the beam onto the receiving antenna) within the line of sight. The receiver amplifies the signal as necessary, and transmits it onwards to the next relay point, and so on to the final station, where it is converted back to the original carrier frequency and radiated to local TV receiver antennas in the usual way.

Satellite Transmission

The ultimate in line-of-sight relays is satellite transmission. At a height of 36,000 km above the equator, a satellite's orbital period is exactly 24 hours, so that it remains stationary relative to a given point on the Earth's surface. A large number of satellites occupy this *geostationary orbit*, and many of them belong to the

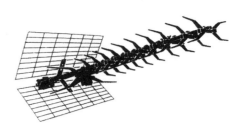

Figure 13.11 This complicated array has a gain of more than 20 dB, the equivalent of some 30 dipoles.

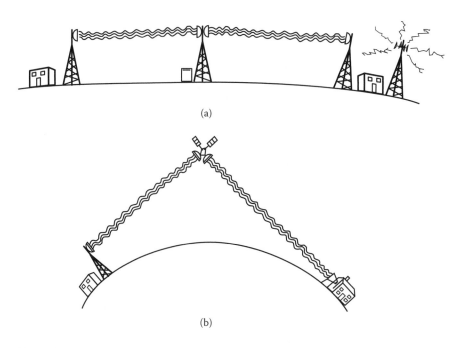

(a)

(b)

Figure 13.12 Microwave links: (a) ground relay; (b) satellite relay.

Direct Broadcast Satellite system (DBS). The earthbound transmissions use frequencies of around 14–14.5 GHz, and are retransmitted from the satellite at downlink frequencies between 11.7 and 12.2 GHz. The satellites are powered by solar energy, and radiate sufficient power to be picked up by quite small (45-cm diameter) dishes.

Owing to problems with the reception of linearly polarized signals in this mode of communication, satellite signals are circularly polarized. They are thus received in equal strength no matter what the polarization requirement of the receiving station may be.

In order to be able to mount receiving dishes flush to a vertical wall, they are usually designed as off-axis sections of a paraboloid. The antenna at the focus is known as a *horn antenna* from its shape; it collects the signal and feeds it to circuitry that amplifies and frequency-converts it before passing it to the TV receiver (Figure 13.13).

You may have noticed that when a newscaster on TV talks to a correspondent in some distant country, there is a noticeable pause before the correspondent is seen to reply. This is because it takes a measurable time for the signal to be converted from analog to digital, and then to travel to a distant satellite and on to the correspondent; the reply then takes the same time to return.

Figure 13.13 Wall-mounted off-axis paraboloidal satellite receiving dish with a horn antenna.

Cable Transmissions

Poor weather conditions can interfere with airborne signals. In particular, heavy snow can have a devastating effect, especially on an analog signal. Many of the more difficult areas are now spanned by underground (and underwater) cable. Copper is no longer used, owing to its adverse electrical properties at high frequencies. Instead, the transmission is by means of *single-mode optical fibers*. These are constructed of low-loss silica glass drawn into fibers that are only a few (optical) wavelengths in diameter, and convey the signal in the form of digital pulses of coherent light, which travel along the fiber with very little loss over large distances. Because of the very short wavelength of light compared with that of UHF radio waves and the correspondingly large bandwidth available, it is possible to transmit a large number of different signals along an optical fiber simultaneously. Optical fibers are also used for so-called cable TV.

The TV Receiver

The receiver is in principle just a captor of electromagnetic signals, equipped with a tuner, a decoder, and an amplifier, with its output connected to a picture display system and one or more loudspeakers. The beam scanning system is synchronized to the incoming signal by pulses added to the signal at the transmitter.

The Cathode Ray Tube

Until comparatively recently the cathode ray tube (CRT) dominated the field of television receivers, computer monitors, and visual display units, but other methods of display are now superseding it. Properly called an electron beam tube, the cathode ray tube is admittedly a cumbersome and potentially dangerous piece of equipment, with a susceptibility to screen burnout if left on for long periods with a static image; but for many decades there was no substitute, and even now it has few rivals for clarity, color saturation, and contrast.

The Electron Beam

If you heat up a metal filament in a vacuum and give it a negative charge it will emit electrons. If you leave the filament itself neutral, and instead surround it with a negatively charged metal cylinder coated with a powerful electron donor, this will emit electrons much more efficiently. This arrangement is known as a *hot cathode*. If you now position a positively charged metal screen some distance away (still in the same vacuum chamber) the electrons will fly towards it at high speed. If its surface is coated with a *phosphor* (a substance that emits photons when hit by electrons) the screen will glow. The pioneer experimenter William Crookes discovered this *electron current*, and called the effect "cathode rays"; the name has survived in the term "cathode ray tube."

You can focus the stream of electrons into a narrow beam by surrounding it with a negatively charged cylinder (the *focusing electrode*), so that the beam reaches the positive surface (the *screen*) as a small spot. A second cylinder close to the cathode (the *control grid*), carrying the signal in the form of a variable negative charge, repels the slower electrons and is used to control the intensity of the beam.

We've already seen something of scanning, but it is worth looking at more closely here, as it is a fundamental technique in television and video. In a TV display the field is applied as a sawtooth waveform, rising linearly and returning abruptly.

For many years the only available type of display was the cathode ray tube (CRT), and as a large minority of extant TV receivers still employ this method of display, I make no apology for discussing it in some detail. Besides, although production of TV receivers with this type of display has now all but ceased, cathode-ray oscilloscopes (CROs) are still in universal use for fault diagnosis and waveform display.

Sir William Crookes (1832–1919) was an English physicist and chemist who carried out all his researches in his own laboratory. He discovered the element thallium, and invented the Crookes radiometer, a scientific toy, the working of which was eventually explained by Maxwell. In the 1870s Crookes investigated the passage of electric current through gases at low pressure, and described the behavior of "cathode rays" in electric and magnetic fields. He was awarded the Order of Merit in 1910.

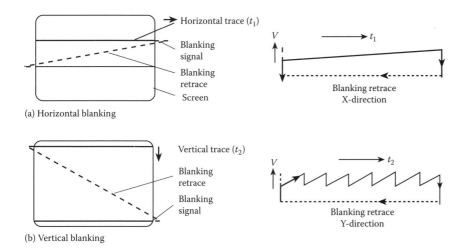

Figure 13.14 Waveforms in X and Y directions for generation of a TV raster.

This causes the spot to repeatedly traverse the screen at a constant speed and then flick back quickly to its starting position. The downward scanning movement is controlled by a second magnetic field at right angles to the first, with a much slower sawtooth, so that it returns to its starting point only once per complete scan. In order to avoid leaving a streak across the screen where the spot has to fly back at the end of each traverse, the beam is deflected off the screen area by a *blanking pulse*. Figure 13.14 shows the way the field varies with time.

The Phosphors

To create a picture, the electron beam has to generate light at the screen face, which forms the positive electrode. Monochrome display screens are coated with a phosphor that emits white light when it is energized by a beam of electrons. (Phosphors for CROs are usually green rather than white.) Figure 13.15 shows a cutaway view of a TV tube.

In Chapter 8 I discussed the phenomenon of persistence of vision, and explained that when individual pictures were presented successively at a high enough rate the sensation of flicker would be absent, and movement within the frame would appear to be continuous. In television, unlike motion pictures, we don't have a whole picture presented at once: instead, the flying spot traces out the luminances of the image, line by line. At any instant only the spot is present, but it moves fast enough to appear to fill the whole frame, on account of the persistence of vision. In the standard (SDTV) analog format the spot completes 625 traverses of the screen in 1/25 second (525 in 1/30 s in the United States).

We don't see all 625 lines, though, as a few lines at the top and bottom are reserved for synchronizing pulses and teletext. Also, in analog TV only alternate lines are presented at each scan, with the frames interlaced on alternate scans: in one scan, odd-numbered lines are presented, and on the next, even-numbered lines. This means that the rate of picture presentation is 50 a second instead of 25 (60 in the United States), so that flicker is much reduced. Some modern TV receivers generate each frame without interlacing, showing each one twice, i.e., a frequency of 100 pps, to eliminate flicker altogether and improve the resolution. The intensity of the spot is continuously modulated to match the picture signal at that point, i.e., it is lowered for a dark area and raised for a light area. The color signal is also varied between the three electron guns to match the hue of the area being scanned.

All phosphors glow for a time after stimulation ceases, and this helps to minimize flicker. In a CRO this "decay period" may be several seconds, but for television only a short period is appropriate, about 20–40 milliseconds.

These figures do not apply to high-definition transmissions, which are digital, and are considered later in this chapter.

191

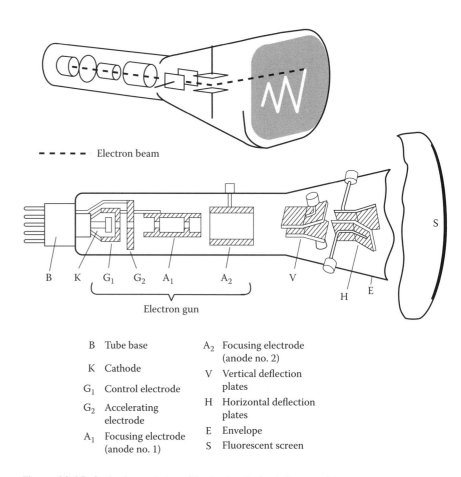

Electron beam

B	Tube base
K	Cathode
G_1	Control electrode
G_2	Accelerating electrode
A_1	Focusing electrode (anode no. 1)

A_2	Focusing electrode (anode no. 2)
V	Vertical deflection plates
H	Horizontal deflection plates
E	Envelope
S	Fluorescent screen

Figure 13.15 Cathode ray tube with electrostatic deflector plates.

If you compare a television tube with an oscilloscope tube you will notice that the latter is much longer and thinner. This is because CROs operate with electrostatic deflectors, and an electrostatic field has much less deflecting effect on an electron beam than a magnetic field (which can be made very strong indeed). The magnetic coils on a TV tube turn the electron beams through nearly 90°, and because of this, and the pinpoint accuracy that is necessary to obtain a sharp and undistorted picture, the magnetic coils have to be designed with a cunning that is near to witchcraft.

One recent attempt to reduce the bulk of the tube was to route the beams from the side, turning them through a right angle before scanning began. This reduced the depth of the receiver, but did not solve other problems such as the necessity for a vacuum and high voltages.

Beam Deflectors

How is the spot steered? There are two possible methods. The first is by electrostatic deflection. If the beam passes between two plates that bear opposite charges it will be deflected towards the positively charged plate and away from the negatively charged plate. The other method is magnetic deflection. A beam of electrons in a magnetic field is forced to travel along a circular path as long as it remains in the field. By using specially shaped electromagnets the beam can be deflected through a large angle. Figure 13.16 shows the principle. Electrostatic deflectors have less effect on the beam than magnetic deflectors, and are found only in CROs.

CRT Display

As explained earlier, the luminance and chrominance signals are transmitted on the same carrier. In the receiver the image hues have to be separated into red, green, and blue (RGB) components. The picture tube has a triple electron gun providing one beam for each of the three primaries, red, green, and blue. A perforated metal screen aligns the beams with phosphor dots that respond with light of the appropriate hue (Figure 13.17). (In a monochrome receiver the chrominance signal is ignored.)

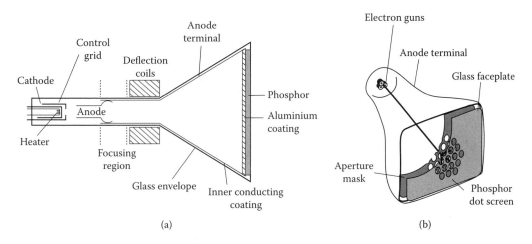

Figure 13.16 (a) Schematic layout of a TV tube; (b) cutaway view.

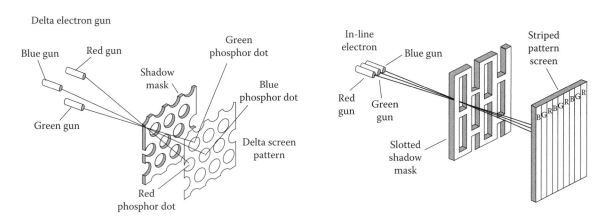

Figure 13.17 Operation of a color mask: (a) delta format; (b) in-line format.

Liquid Crystal Display Screens

These were developed largely owing to a growing market for laptop computers, which needed flat screens and low power consumption. They are becoming common on portable TV receivers, are standard on mobile phones and amateur digital camera viewfinders, and are appearing regularly on the streets in the form of animated billboard posters. "Liquid crystals" are optically active gelatinous substances that rotate the plane of polarization when a voltage is applied across them. A *liquid crystal display* (LCD), also known, more logically, as a *spatial light modulator*, is an array of pixels that can be stimulated separately by a series of contacts along the horizontal (X) and vertical (Y) edges of the array. The optically active material is sandwiched between crossed Polaroid sheets, so that until it is stimulated no light is transmitted. When an electric field is applied, the plane of polarization is rotated to a greater or lesser degree, allowing light to be transmitted in proportion to the applied voltage. For a color image the screen is covered by a mosaic of primary color filters in groups of three subpixels, respectively, red, green, and blue. LCDs can operate by either reflected or transmitted light. For TV and other monitor displays they are backed by a uniform white light source (Figure 13.18).

One manufacturer now includes yellow subpixels using a signal derived from a combination of red and green signals.

193

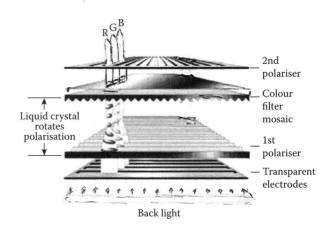

Figure 13.18 Schematic of LCD screen. Applying a voltage to the liquid crystal rotates the plane of polarization, allowing light to pass through the second polarizer when a pixel is stimulated.

Plasma Screens

In a plasma display the pixels are self-luminous. Each pixel consists of a cluster of three tiny fluorescent tubes containing a mixture of neon and xenon gases at low pressure, coated with phosphors that emit, respectively, red, green, or blue light when stimulated by voltages across them. The luminance of each subpixel is proportional to the applied voltage, so for each whole pixel a full gamut of hue, lightness, and saturation is possible. Plasma screens are manufactured only in comparatively large sizes, and have a shorter life than LCD displays. Also, like electron beam tubes, they are liable to local screen burnout when a stationary image is left displayed for a long time.

OLED Displays

Organic light-emitting diodes (OLEDs) are LEDs that can be incorporated into very thin substrates, as they are self-luminous. At present the problem is a limited life, but much work is in progress, and at least one manufacturer has put a TV receiver with an OLED screen on the market. Once the various production difficulties are overcome, there is little doubt that they will be an important part of receiver technology.

Projection Systems

There are two types of projection: "front" and "rear." Front projection is onto a white screen, as used with slide or PowerPoint projection systems. Rear projection has the image source situated behind the screen, which needs to be fully diffusive in order to obtain even illumination and a wide viewing angle. This is usually achieved by a Fresnel lens mounted behind and in contact with the screen, which itself is made of diffusing material.

CRT projection systems employ three separate small tubes, one for each primary, each with its own lens system, giving precisely superposed images. The tubes are designed to give a very bright image, using red, green, and blue phosphors, respectively. Each tube may have its own optics, or the three beams may be combined by a prism system (Figure 13.19).

The earliest projection system used *eidophor tubes*, which (improbably) used the electrostatic distortions of a thin film of oil on the inside face of the tube to produce diffraction effects that modulated the beams to produce the image.

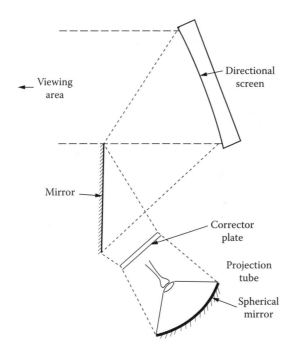

Figure 13.19 Typical arrangement of front projection TV optics (CRT display).

Plasma systems are not in general suitable for projection TV. On the other hand, an LCD system can be mounted for either front or rear projection, as the brightness of the image depends only on the intensity of the illumination behind it. Individual manufacturers use several variants of the technology, but they all operate on similar principles. However, a more recent system operates on a different principle. Known as *digital light processing* (DLP), it can be used for front or rear projection, and is bright enough to illuminate a full size cinema screen. The image is created by a matrix of microscopically small mirrors on a semiconductor chip. These mirrors can be altered in orientation so that they reflect light either through the projection lens or into a heat sink or "light dump." Although they are essentially binary devices (i.e., either on or off), their reaction speed (of the order of microseconds) is such that by controlled fluttering they can produce what are effectively grey scales, the depth of grey governed by the on/off time ratio. There are two methods of obtaining the color. In the first the colors are produced by a rotating segmented RGB filter wheel (which may also include segments of secondary hues and white) synchronized with sequential color signals. In the second the projector uses a prism system with dichroic filters to direct the RGB components to three DLP chips; the component signals are recombined and directed out through the lens (Figure 13.20). The latter system is the one generally used for large screens and movie theatres. Another method for projection that is beginning to come to the fore is laser TV. Three lasers are used to scan the screen from behind. This method is promising for large public TV displays, as it produces images with high contrast and color saturation, which are bright enough to be shown outside in full daylight. Like OLED technology, laser TV is still under active development, and the first models are becoming commercially available.

Digital Television: The Advantages

Analog television has a number of drawbacks. It is prone to interference from outside sources. Bad weather and solar storms can cause patterns, distortion, and

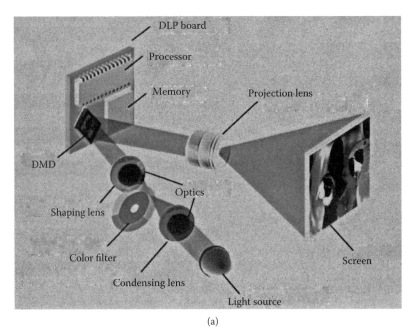

(a)

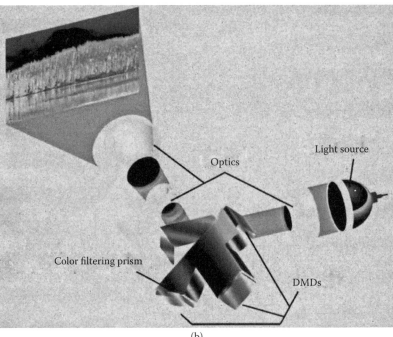

(b)

Figure 13.20 Two versions of a digital light processing (DLP) display using digital micromirror devices (DMDs). (a) Rotating color wheel version. (b) Combining prism version.

fading, and transmission of the signal takes up a large bandwidth of frequencies. In addition, video recording by analog techniques results in some inevitable deterioration in quality with succeeding generations of imagery, and the videotape itself becomes degraded with repeated use. Digital television operates on similar lines to digital still and movie photography, except that the images are transmitted via a carrier beam rather than being downloaded into a computer. The carrier beam itself is much as I have already described, and the only additional task of the decoding system is to transform a set of spikes into its envelope, as indicated in Figure 13.1.

So where are the advantages of digital TV? Quite a few, as it happens.

- Narrower bandwidths allow more channels in a given bandwidth allocation.
- Higher resolution, especially with the introduction of 1250-line HDTV.
- Simpler color coding, hence more reliable color (this applies in particular to NTSC).
- Repeated transfer to recording systems over any number of generations of copies without any deterioration.
- Tape wear, where used, is much less of a problem.
- Easy combination with data such as subtitles.
- Simplification of the mechanism of frame grabbers, download to computers, etc.
- Easy compression of signal without loss of information; variable levels of compression available, depending on required image quality.

Perhaps the only snag (apart from the user's need to buy new equipment or ancillaries) is that when the noise becomes excessive the signal disappears suddenly and totally. But by this point an analog signal would also have become incomprehensible.

Aspect Ratio

A perennial problem with the screening on TV of commercially produced films is that they often don't fit the format. Older films made with a 4:3 aspect ratio fit fairly well on the old standard TV screen format of 5:4, but almost all films made for showing in cinemas are now made in wide-screen format, with an aspect ratio of from 8:5 to 5:2, depending on the process.

Aspect ratio is the ratio of width to height.

To deal with the problem a wide-screen film has either to be shown with black bands at the top and bottom of the screen ("letterboxing"), or to be given the Procrustean treatment of lopping off the extremities of the image. Sometimes a wide-screen presentation may get away with squeezing the picture a little (this is common with introductory titles and final scrolls), but on occasion whole scenes may have to be reshot with television in mind. A new generation of TV receivers with a screen aspect ratio of 16:9 has now superseded the older format, so the problem is considerably lessened, though now the difficulty lies with old films, where black bars appear at both sides of the picture ("pillarboxing"), or alternatively the tops of heads are cut off (you have a choice).

High-Definition Television

The concept of high-definition television (HDTV) arose out of the need for an image that would match the quality of existing motion picture films. It was clear that analog broadcasting would use far too much bandwidth at the required resolution, and contemporary digital experiments gave results little better than standard transmissions. However, the evolution of MPEG systems of data compression promised a solution, and in 1993 the MPEG system became accepted as the standard method.

MPEG stands for Motion Picture Experts Group. The evolution of HDTV depended on the development of a suitable compression technology, as otherwise the necessary bandwidth would be far too great.

A 16:9 aspect ratio, a specific colorimetry regime, and scan modes of 1080i (interlaced lines) and 1080p (progressive lines) were agreed. The display was standardized at 1920×1080 (square) pixels. The refresh rate remained governed by the local a.c. mains frequency, i.e., 60 Hz in the United States, Japan, and some South American countries, and 50 Hz elsewhere.

HDTV displays have twice the resolution of SDTV displays. When standard 35-mm film is shown, it is scanned in PAL at 25 pps, about 4 percent faster than

197

the standard cine filming rate of 24 pps, each frame being displayed twice for a 50-Hz refresh rate, to avoid flicker. For NTSC countries with a 60-Hz refresh rate, alternate frames are shown twice (1/30 s) and three times (1/20 s), thus achieving the correct film rate. Video recordings are made as determined by the broadcaster, and converted electronically between the two systems as required.

To achieve the highest quality picture and sound information from an HDTV signal, the ordinary coaxial or balanced cable connecting the antenna to the receiver is usually inadequate. It is necessary to use a high-performance HDMI cable, as this has to carry not only the RGB components of the picture signal, but also the coded 5.1 channels needed for surround sound broadcasts.

Compression

The whole purpose of a signal is to convey information from the transmitter to the receiver. By definition, "information" means something that is unpredictable. But if you consider an actual TV image, say a landscape containing a moving figure, you will appreciate that the image contains a great deal that is *not* information in this strict sense. For example, a uniform blue sky contains only one piece of information, and that one piece is repeated throughout the whole area; it therefore needs to be transmitted only once. There is also little change from one frame to the next, so there is a similar redundancy with respect to time.

Cartoon animators have long been familiar with these concepts, and from the earliest days have superposed moving subjects on stationary cels, saving a lot of redrafting in the process.

Compression systems are designed to eliminate as much of these types of redundancy as is practicable. An ideal compression system would keep all the information and discard all the redundancy. In practice such a system does not exist, though the MPEG-2 system comes close.

MPEG-2 defines the nature of the bitstream between the encoder and the decoder, i.e., the transmitted signal. This defines the nature of the decoder precisely, but allows scope for different encoder designs. It covers all types of video screens from cell phones to full surround sound electronic cinema, including HDTV. The compression of individual frames closely follows the JPEG standards used for compression in digital stills photography, and temporal compression is analyzed in terms of what moves between frames—only the difference signal being stored. The analysis of this is complicated, and leads to various partial solutions: some actual information may have to be discarded, and the various systems are configured to discard only that which is likely to pass unnoticed by the viewer (such as very low contrast detail).

Digging Deeper

I haven't yet discussed the way the signal is recorded on tape or disc. That belongs to the next chapter. As most of the texts on television also deal with video recording, I have left the recommendations until the end of Chapter 14. One book does deserve mentioning here, though. *Vision Warrior*, by Tom McArthur and Peter Waddell (Scottish Falcon, The Orkney Press, 1990), is a biography of John Logie Baird that sorts out many of the myths about the man and his work. Tom McArthur is a freelance journalist, and Peter Waddell is former Reader in Engineering Science at Strathclyde University, Glasgow, where there is a small but fascinating museum chronicling the history of television. If you get the opportunity, it is well worth a visit. The National Media Museum in Bradford also has a section devoted to the history of television, with a working model of Baird's original television apparatus.

Chapter 14 Video Recording and Replay Systems

Sound recording is now well over a hundred years old. On the earliest surviving cylinders are preserved the voices of Tennyson and Gladstone and of Edison himself, even the piano playing of Brahms. Motion pictures are about the same age: we have cine records of Queen Victoria's Jubilee and funeral ceremonies, and of city life in Paris and London in the 1890s. The addition of sound to the visual record came a good deal later, with the advent of the "talkies" in the 1920s. In contrast, sound recording preceded by four decades the addition of visual material to its medium. Video recording technology was initially based on sound recording, and to a large extent still is. So it is logical to begin with a discussion of sound recording.

Magnetic Tape Recording

Recording of both sound and television is necessarily a one-dimensional business. The vinyl LP disc, with its single groove more than 0.6 km long, is an obvious example. Audiotape, which grew up around the same time, is another. It was born, as were the first experimental 33⅓ rev/min discs, in the 1930s. One of the earliest tape machines was the Blattnerphone, which used a 6-mm steel ribbon passing over a magnetic head at five feet per second, no less. The tape reels were enormous, and editing involved the use of tin snips and a spot welder. In the 1940s steel tape was replaced by fine steel wire, and the recorders accordingly shrank to suitcase size. Steel wire was itself soon replaced by plastic tape coated with a layer of magnetic iron oxide particles.

Analog Sound Recording

In an analog audiotape recorder the signal is fed, in the form of an electric current, to a head consisting of an electromagnet with a very small gap between the poles (Figure 14.1). As the tape passes over the head in contact with it, it becomes magnetized in proportion to the signal, retaining its magnetism as it is wound

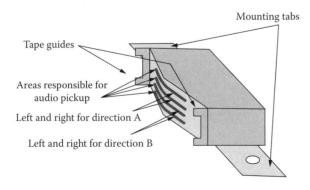

Figure 14.1 Layout of a tape head for audiocassette. The magnetic gap is at the apex of the shallow angle. Note that in an open-reel analog recorder tracks 1 and 2 (direction A) are not adjacent, but are interlaced with tracks 3 and 4 (direction B).

onto a receive spool. When the magnetized tape is run through the same equipment, the fluctuating magnetic field preserved in the track induces a current in the head that duplicates the signal, and can be amplified to produce a replica of the original sound.

FERROMAGNETISM, HYSTERESIS, AND A.C. BIAS

Ferromagnetism

Elements and compounds are made up of molecules that tend to face in a particular direction when immersed in a magnetic field. The extent to which this happens in response to a given field strength is called the *permeability* of the material. A few substances such as iron and some of its compounds have a very high permeability. They are said to be *ferromagnetic*. In such substances groups of molecules about 0.1–1.0 mm in diameter form *domains* aligned in a single direction. Under the influence of a strong magnetic field the domains that are aligned with the field grow in size until, if the field is strong enough, they all become aligned, and the material is said to be "saturated." When the external field is removed, some ferromagnetic substances, notably compounds of iron, nickel, and chromium, as well as some lanthanides such as samarium, retain their magnetism.

Compounds suitable for recording have a high permeability, and can be readily demagnetized by means of an alternating magnetic field that decays to zero over a short time. Suitable materials for coating on recording tape are iron (III) oxide, Fe_2O_3; chromium (IV) oxide, CrO_2; and pure iron, Fe.

Hysteresis

When stimulated by a fluctuating magnetic field at audio or higher frequencies, the magnetization of the tape coating lags behind the stimulus. This effect is called *hysteresis* (Figure 14.2). In analog recording this causes a particularly obnoxious type of distortion called *crossover distortion* (Figure 14.3a).

a.c. Bias

Soon after tape recording was introduced, it was found that the distortion caused by hysteresis could be eliminated by superimposing a high-frequency signal (around 50 kHz) on the audio signal. This is called *a.c. bias*. Figure 14.3b shows the effect on the signal of adding bias.

Lanthanides are a group of elements with atomic numbers from 57 (lanthanum) to 71 (lutetium), possessing different electronic structures but similar chemical properties.

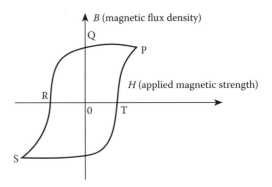

Figure 14.2 Hysteresis. The magnetic flux density (i.e., the degree of magnetization) of a ferromagnetic substance lags behind the magnetic field strength producing it. When the applied external field falls to zero, some magnetization is retained.

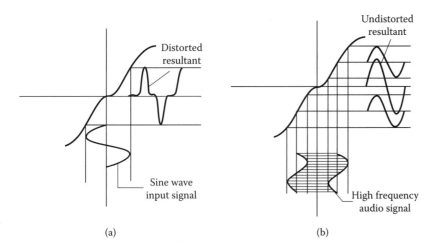

(a) (b)

Figure 14.3 (a) A signal supplied directly to the tape suffers crossover distortion due to hysteresis. (b) The addition of a high-frequency a.c. signal (a.c. bias) allows the output to be derived only from the linear regions of the transfer characteristic, and eliminates the distortion.

Tape Recording Mechanisms

Videotape was itself originally analog, and was developed from audiotape. So to begin with let's look at the way an analog audiotape recorder is put together. In a professional sound recorder there are three magnetic heads (Figure 14.4). The head gap is narrow, 3 μm or less (about one-twentieth of the thickness of one of these pages). This gap is filled with nonmagnetic material such as polytetrafluoroethylene (PTFE), which forces the magnetic flux to leave the gap and penetrate the recording medium.

As the tape moves from left to right, the first head it encounters is the *erase head*. This applies a high-frequency alternating magnetic field to the tape, which fades to zero as the tape moves on, wiping out any previous magnetization. Next comes the *recording head*, which carries the audio signal plus the a.c. bias, and impresses this on the tape in the form of a varying magnetization. The third head is the *replay head*, which monitors what is now recorded on the tape. This head is only seen in open-reel recorders and professional and semiprofessional cassette recorders. In cheaper recorders monitoring is directly from the incoming signal.

The standard writing speeds of audiotape show its history. They are all submultiples of five feet per second (152.4 cm/s). Professional audio analog recordings are made on open-reel recorders at 15 or 7½ in/s (38 or 19.5 cm/s). Sources with less stringent

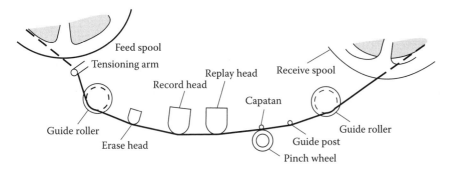

Figure 14.4 Head layout in a professional audiotape recorder.

The recording machines notoriously used by the Nixon administration were open-reel "crawlers" operating at 15/16 in/s (2.4 cm/s), and the sound quality was correspondingly low. However, a modern analog cassette recorder, with improved magnetic heads and better tape coatings, can at $1^{-7}/_8$ in/s equal the performance of what were once top-of-the-range open-reel recorders.

White noise is a sound containing all frequencies in equal amounts, with random phases. It sounds like escaping steam.

Although tape hiss is not a problem with digital equipment, noise reduction systems are still necessary, as noise of various kinds is inevitably generated in the electronics of the decoding and amplifying equipment.

requirements, such as speech, can be recorded at 3¾ in/s (9.5 cm/s). Cassette recorders record at 1⅞ in/s (4.75 cm/s).

Professional open-reel digital sound recorders operate at 30 in/s (76.2 cm/s) because of the higher frequencies involved, but digital audiocassette recorders operate at the same speed as analog ones. Rotary Head Digital Audio Tape (RDAT) uses larger cassettes with helical scanning similar to the VHS videocassette system described later in this chapter, and low linear tape speeds of 8.15 or 12.225 mm/s.

NOISE REDUCTION SYSTEMS

When cassette recorders were first introduced, with their narrow tape tracking at a relatively low speed, the relatively large size of the domains produced an effect analogous to the graininess of silver halide photographic negatives. It manifests itself in the form of white noise, and is known as "tape hiss." Ray Dolby, one of the pioneers of high-fidelity sound reproduction, evolved a system for reducing its effect.

He reasoned that as this hiss became obtrusive only in quieter passages of music, and its upper frequencies were the most obtrusive, it was necessary to suppress the higher frequencies present in the hiss wherever the music was quiet. He designed circuitry that would, in recording, emphasize all the higher musical frequencies in the quieter passages (the hiss level would of course remain the same), and de-emphasize them (now including deemphasizing the hiss) on replay (Figure 14.5). This system, known as *Dolby B noise suppression*, is still standard in the better models of cassette recorders and players, and in all VHS videocassette sound systems. A more complicated system, *Dolby A*, is used in professional analog sound recording. Most modern, high-quality semiprofessional equipment now uses a simpler version of this, called *Dolby C*.

Ray Dolby continued to be involved in noise suppression, in video as well as audio contexts, and a number of refinements in video recording and reproduction also bear his name. Various manufacturers of audio and video equipment have found other methods of noise suppression, and these may be offered as alternatives in high-end equipment.

Videotape Recording Techniques

The first videotape recorders were also open-reel, but the inconvenience of their complicated lacing system led to their being supplanted commercially by several types of videocassette, the lacing being carried out automatically within the machine. These different types were mutually incompatible, and eventually the struggle for commercial survival was won by the Video Home Service (VHS) format.

The recording of an analog video signal presented no theoretical problems that had not already been solved by audiotape technology. There was a practical problem, though. Whereas audiotape needs to record frequencies only up to 20 kHz (the limit of normal human hearing, about 2½ octaves above the topmost note of a piano), video frequencies go from around 25 Hz to more than 4 MHz. This means that the writing speed on the tape has to be nearly six meters per second. The solution is a classic example of lateral thinking: move the recording head as well as the tape. All modern videotape recorders operate on this principle. The two magnetic recording heads are installed flush with the surface of a slightly tilted cylinder (Figure 14.6).

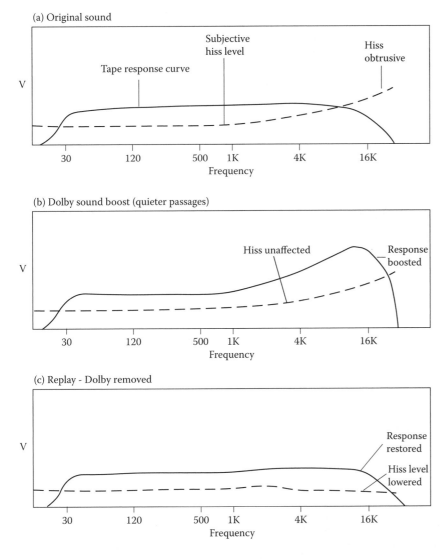

(a) Original sound

Subjective
hiss level

Hiss
obtrusive

Tape response curve

V

30   120   500   1K   4K   16K
Frequency

(b) Dolby sound boost (quieter passages)

Hiss unaffected

Response
boosted

V

30   120   500   1K   4K   16K
Frequency

(c) Replay - Dolby removed

Response
restored

Hiss level
lowered

V

30   120   500   1K   4K   16K
Frequency

Figure 14.5 Principle of Dolby B noise reduction system.

The tape is wrapped round one half of the cylinder, which rotates at 25 rev/s in the direction of the tape movement. The information is thus recorded helically, as a succession of long thin oblique bands, crossing the width of the tape over a distance equal to half the circumference of the cylinder. Each traverse records one complete TV line. At the "standard play" setting the tape itself moves linearly at about 33 mm/s. On the "long play" setting the tape speed (and the image quality) are halved.

The VHS Format

The advent of the compact disc (CD) and, in particular, the ability to record on a disc format directly from a home computer, has spelled the death knell of the VHS system, and VHS recorders are no longer made. However, as with audiocassettes, large numbers of VHS recorders (and tapes) are still in daily use, so the system cannot be ignored here.

The standard layout of a VHS recorder/player transport mechanism is shown in Figure 14.7.

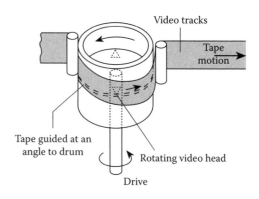

Figure 14.6 Record/replay head in a VHS recorder.

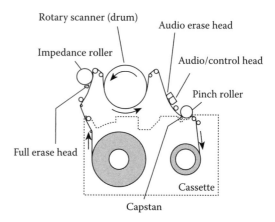

Figure 14.7 Tape path in a VHS recorder.

When you insert the cassette into the recorder the internal mechanism opens the lid, unlocks the spools and pulls out a loop of tape, which it wraps round the cylinder. The rest of the guidance system then moves into place. Once threaded, the tape passes first over an erase head, which removes any previous tape signal. It then passes over an "impedance roller," the action of which is to smooth out any possible juddering of the tape. Next, the tape passes over a guidepost, round the scanning drum in a semicircle, and past a second guidepost. Two video/sound heads are built into the scanning cylinder. There may also be a linear sound head that can be used for a subsequent voice-over commentary. The sound and video signals are recorded on the same part of the tape, but because the video signal is at a much higher frequency than the audio signal the video recording is confined to the surface of the tape coating, erasing the surface part of the audio signal in the process, but leaving it intact in the deeper part of the coating.

The depth of penetration of a signal is inversely proportional to its frequency, due to self-inductance in the medium, and is very small at megahertz frequencies ("skin effect").

Digital Recording

In digital recording the shape of the characteristic curve is irrelevant, as the pulses are an all-or-nothing event; the magnetizing current is set to almost saturate the recording medium. The amplitude is constant, and the pulses are created by reversals of its direction (Figure 14.8a). On replay, owing to the inductive nature of the replay head, an output is produced only as the flux in the head changes from positive to negative and vice versa (Figure 14.8b).

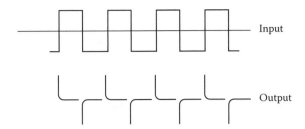

Figure 14.8 Digital input and output in a magnetic head.

The polarity of the resultant pulses alternates as the flux changes from positive to negative and back. What is needed now is a circuit that can locate the timing and direction of each pulse, and restore the square pulses. To reduce noise, only the peaks of the pulses are detected.

Inductive replay heads are at a disadvantage in both analog and digital recording systems as the output is low at low frequencies (it is zero for d.c.), and also falls off at high frequencies owing to the finite gap width of the magnetic head. A useful replacement has been the *magnetostrictive head*, which measures the actual flux density rather in than its rate of change. As this device is not sensitive to current direction, a positive d.c. bias is applied to bring the negative pulses up to zero, so that the output readings are either 1 or 0, as required of a binary code.

As I explained in Chapter 11, digital broadcasting has the advantage over analog broadcasting in that it is almost totally free from both interference and signal corruption; the same applies to digital recording. When digital signals are recorded, there is a further advantage: digital information can be compressed. In television pictures (and in sound) much of the information remains the same over a period of time. So you don't have to keep sending the same information repeatedly over this period. You need only send a new signal when the information changes. Of course, the compression that is possible varies a good deal. In rapid action sequences little or no compression may be possible, but in static titles (or silence in audio recording) the compression ratio may be upwards of 40:1. A further benefit is a simpler interface between video and computer, and precise editing is easier, too. Professional studio digital sound and video recording does not use compression, though field recording equipment does. Any loss in quality is virtually undetectable by eye or ear. What is important to the field worker—and the amateur—is the saving of space.

If you are mathematically minded, you will have spotted that the head output is the derivative of the input, and the detector needs to be designed to perform an integration on the signal.

Digital Videotape

With the runaway success of compact audiodiscs (CDs), which are encoded digitally (though not compressed), right from the time of their introduction it was clear that digital recording would also come eventually to dominate the videotape field. Mechanically, there is little difference between analog and digital video recording, but with the latter the tape width can be halved, and the cassettes can be made smaller and lighter than analog recorder cassettes.

The advantages of digital video recording have also become more obvious as digital TV becomes universal: for one thing, it is possible to record a digital TV program without any loss of quality, or deterioration with repeated use. Tape is likely to continue to be an important vehicle for storing both audio and video images, though its role is already being taken over by discs, largely owing to their ease of rapid access. This, of course, was the reason computer memories moved from tape to disc many years ago.

Hard Disc Recorders

The hard disc is the heart of a computer's memory system. It uses magnetic recording in the same manner as audiotape, but with much higher resolution, as the magnetic domains are far smaller. The magnetic material is coated on an aluminum or glass disc and polished to an optically flat surface. The read/write head is designed aerodynamically so that it hovers a few nanometers above the disc instead of making contact with it. This avoids wear, and allows rapid movement between tracks to locate information virtually instantly at any point on the disc. Typical storage capacity is between 10 and 40 GB. The advantage over CD-type discs is that a hard disc is fully rewritable—as it needs to be when it is installed in a computer. Video recording equipment using hard discs has recently begun to appear, and this permits instant choice of programs from a huge library of sound and visual recordings with full surround sound. Such equipment can record a TV program and replay an earlier part of it while a later part is still being recorded.

Magneto-optical Discs

Magnetic tape has the disadvantage that the tape head comes into contact with the tape itself during both recording and replay, and this causes both tape and head wear, as well as a buildup of oxide particles on the heads. With the magnetic hard disc the air gap between the recording/replay head and the recording surface limits the maximum recording frequency. In the magneto-optical system there is no physical contact with the recording medium, and the resolution is limited only by the size of the focal spot, which depends on the numerical aperture of the focusing lens and the wavelength of the laser.

When a magnetic material is heated to a specific temperature, known as the *Curie point*, it loses all its magnetism, and if it is in a magnetic field when it cools again, it takes on that magnetization. In the magneto-optical disc system the material is initially magnetized in one direction, and below it is a coil providing a small steady magnetic field in the opposite direction, which, however, is too weak to affect the existing field. But when the track is heated by the laser to the Curie point, as it cools it takes on the reversed magnetic field.

The readout system employs the Kerr effect, in which a magnetic field rotates the plane of polarization of a polarized light beam. The replay head contains a polarizing filter, and variations in polarization produce variations in transmitted light intensity. The recording can be erased by reversing the current in the coil and running the disc with the laser operating continuously; the disc can then be reused.

Camcorders

The earliest camcorders showed their amateur cine ancestry in their shape. They were necessarily rather large, as the sensors were miniature vidicon tubes and the recording material bulky reels of half-inch wide magnetic tape. Some of today's camcorders still use tape cassettes, though in a much smaller format, but CCD sensors as used in digital still cameras are now becoming ubiquitous. More recent versions (Figure 14.9) employ either a minidisc (a small version of a CD) or, increasingly commonly, a memory card.

Numerical aperture is related to *f*/no, but is only used for diffraction-limited optics. It is discussed in Chapter 18.

Memory cards, like discs and tape, carry their information in the form of aligned magnetic domains, but in the case of cards the data are arranged in a rectangular pattern, an X–Y coordinate system providing access as in digital still photography.

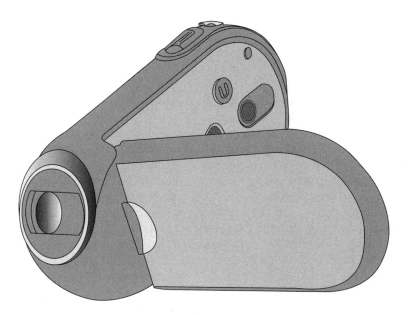

Figure 14.9 Typical modern camcorder.

As I mentioned in Chapter 11, still cameras and camcorders are currently undergoing a kind of convergence, and the digital amateur camera seems to be steadily taking over the role of the camcorder, as the data capacity of memory cards continues to increase. With suitable compression, a card can now hold several GB of information.

CDs and DVDs

CDs first appeared in the late 1960s. Since the videodisc evolved using the same technology, we can look at both together. The digital videodisc (DVD) was developed using the technology of CDs via the CD-ROM, a random-access memory disc for computers that supplanted the floppy disc with its limited capacity. (A CD-ROM can hold around 45 MB of information, as against 1.25 MB for a floppy disc.)

BEFORE THE CD

Until the advent of the CD, all record players used an analog signal in the form of a spiral groove hot-stamped into the surface of a shellac disc (later PVC). This groove actually duplicated the waveform of the sound energy. This principle goes right back to Thomas Edison, inventor of the cylinder phonograph, and Emil Berliner, inventor of the flat disc gramophone record. By 1960 the technology had reached a high level of sophistication. In order to give the effect of a fully spread sound stage, two channels could be coded into the groove, one for a left and one for a right speaker, forming a right-angled V at ± 45°.

Various designs were later used to encode two further channels into the PVC groove in order to provide surround sound, but without much commercial success. Another venture was binaural sound recording, using microphones in the ears of a dummy head, to be listened to through headphones. This produced startlingly realistic results, but never caught on. By the mid-1970s analog sound recording had reached its limit, and the vinyl LP disc, delicate, noisy, and prone to the effects of wear, dust, and electrostatic charges, was being seen as inadequate.

This stereophonic principle is much older than you might think. The conductor Leopold Stokowski was experimenting with stereo recording in 1934, and there are reports of demonstrations of binaural sound (using two telephones) in Paris around 1900. But the real leap forward was the 45/45 principle patented by Alan Blumlein, a genius of electronics, who lost his life tragically in an air accident in 1943 while working on airborne radar development.

The Digital Principle

First, a recap on the principle involved. To describe a waveform in digital terms you measure its amplitude at a number of closely spaced sampling points. Then you draw the curve that passes through all the values. But can you really get the true waveform from those samples? What if there were some kinks in the waveform in between your measurements? The answer is that if you know the highest frequency present in the waveform (and you always do know it) you simply need to sample at a minimum of twice that frequency (the Nyquist criterion). The waveform then *has* to be smooth between the sampling points. The next thing is: how accurate does the measurement have to be? Digital measurements operate only in whole numbers. Suppose you sample at intervals 1, 2, 3, ... 10. Any of these values could be as much as ± 0.5 in error, and that might not—would not—be accurate enough. You would get large steps in the reconstruction. But binary digital measurements can work in much smaller steps: you can sample in as many small steps as you need. Sixteen-bit accuracy is common: this gives 65,536 steps; 20-bit accuracy gives just over a million. For an audio CD the sampling rate is 44.1 kHz, which is well above the limit of human hearing. (This somewhat odd figure was arrived at from the conversion of TV signals from analog to digital, the same equipment being used to convert analog audio signals to digital.)

Structure of CDs

CDs have a diameter and thickness of 120 mm and 1.2 mm, respectively, and are fabricated by polycarbonate injection molding. The master (original) recording is made as a spiral groove, tracking from inside to outside at constant linear speed. An infrared laser of wavelength 780 nm is focused on the data surface, which is protected by a thin layer of polycarbonate, and follows the track as the disc rotates. The digital encoding is in the form of microscopic raised areas (bumps); these are 0.12 μm deep, and in a CD 0.4 μm wide and a minimum of 0.4 μm long. This pattern is coated with a thin layer of reflective aluminum, so that the light from the laser is reflected back to a beamsplitter, and thence into a monitoring photocell, which passes the digital information to a decoder and amplifier. The spacing between adjacent tracks in a CD is 1.5 μm, which matches the size of the focal point of the laser. Figure 14.10 illustrates the operation of a CD player.

This spacing is so close that the grooves form a diffraction grating, which is why you see spectral colors in the light reflected from the surface of a CD.

The original recording is burned (literally) by a laser into a plastic master disc, which is then duplicated in metal by electroforming to produce a stamping "shim" in the same way as an LP record. It bears the digital information in the form of pits, so when the CD is formed it is a negative of the master, and the pits become bumps. Their height is around one quarter-wavelength of the tracking laser, so that, in playing the disc, as a bump comes under the focused spot the reflected light from the bump interferes destructively with the light from the surround, thus registering a 0. In between the bumps the light is fully reflected, and registers a 1; the result is thus a binary signal. The disc, running at a constant linear speed, as it runs from inside to outside slows from some 600 rev/min to about 130 rev/min. The total length of a CD track is some 6 km, and playing time is up to 74 minutes (exceptionally, 80 minutes). There is no compression.

Structure of DVDs

DVDs have the same diameter and thickness as CDs, but differ from CDs in having two layers of information, and sometimes by being double-sided, the two sides having been formed separately and glued together to make a final disc of the same thickness as a CD. The upper layer bears the video information; in this case the

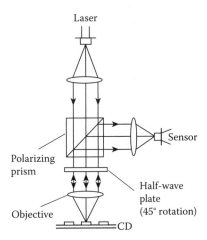

Figure 14.10 CD information retrieval. The reflected light has its polarization turned through 45° twice, and is thus wholly reflected at the polarizing prism on its return. There is destructive interference between the light reflected from the bumps and their lands, and this is used to distinguish the bumps from the lands.

bumps are a minimum of 0.2 μm wide and a minimum of 0.4 μm long. The lower layer is a normal CD track, which allows the disc to be played as an audio CD on a DVD player, the upper layer in this case being ignored. The laser for the DVD track has a shorter focal length, 650 nm, in the red region of the visible spectrum, and the tracks are correspondingly narrower, with 740 nm separation. The total track length is around 11 km. This allows the DVD to contain roughly four times the (raw) information of a CD. However, with compression it can actually contain about 20 times as much information as a CD, which allows it to contain a full three-hour film with 5.1 surround sound, optional subtitles in several languages, commentaries, and extra film snippets appropriate to the material. A DVD used as a computer memory can store 4.7 GB of information in each layer, more than 18 GB altogether. This is more than 25 times the capacity of a CD, which itself has more than 60 times the capacity of a floppy disc.

"5.1 surround sound" caters for a central speaker, two side front speakers, two rear speakers and a subwoofer (the ".1" of the 5.1.).

The Blu-ray System

Further development of the DVD was held up for some time, as the comparatively long wavelength of the laser limited its resolution. However, the 1980s saw the evolution of laser material that would operate stably at 405 nm in the far violet region of the visible spectrum. For a time there was fierce competition between rival commercial systems, but eventually (as had happened with VCRs) one system won. This was Sony's Blu-ray system, which has a groove pitch of 300 nm and the dimensions of the bumps reduced accordingly. A Blu-ray disc can hold some 25 GB of data, which is comparable with the capacity of a hard disc (see below). Because of the very short focal length of the laser lens the data layer needs to be much closer to the surface of the disc, although the overall dimensions of the disc remain the same as those of a CD.

Figure 14.11 compares the three systems.

Blu-ray discs can also be fabricated as a double layer, giving double the playing time. With such discs, as the playing of one layer is completed the laser switches focus onto the second layer. To avoid the laser's having to backtrack, the second layer may run from outside to inside of the disc.

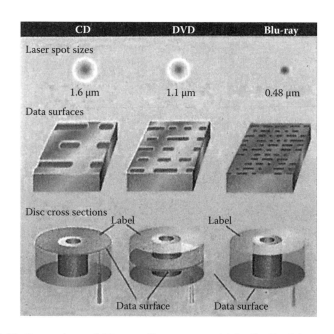

Figure 14.11 Comparison of CD recording systems. (a) Audio CD; (b) DVD; (c) Blu-ray.

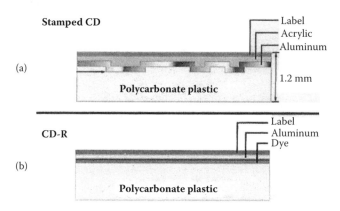

Figure 14.12 Construction of (a) a commercial CD, (b) a CD-R. The CD-RW has a similar structure to the CD-R but with a different type of light-sensitive layer.

Burning a CD

CD-R Systems

All modern computers are equipped with a facility for producing your own CD or DVD, or indeed any data you need to save permanently. Plainly, it would be impracticable for a home computer to produce a recording, with its bumps and lands, in the conventional manner. In the CD-R disc system the aluminum layer is flat and remains so after the recording process. It has been coated with a layer of dye that is initially translucent to light from the laser. For recording ("burning") purposes the laser power is increased so that when it acts on the dye it renders it opaque, so that although the physical shape of the dye layer is unchanged, the opaque areas appear as bumps and can be read in the same manner as a CD or DVD. This type of disc is not rewritable.

CD-RW Systems

CD–RW (rewritable) discs are constructed differently, using what is called "phase shift technology."

In this system a special compound is sandwiched between two inert layers. The laser actually heats the compound to about 600°C, at which point its structure changes from crystalline (translucent) to amorphous (opaque). The disc plays in the same manner as the CD-R, but its data can be erased. To overwrite a section the "write" laser heats the track to 200°C, which causes the compound to revert to the crystalline form, then it writes the new information onto the (now blank) area. Figure 14.12 compares the two types of coating with a standard CD.

Don't confuse the term "phase shift" with the phase of a wave. It comes from a different branch of physics, and refers to two different crystalline states of a material (or, in this case, crystalline and amorphous.

Digging Deeper

Television and video recording is a huge subject, and I am well aware of having dealt somewhat superficially with both the science and the technology. I have tried to avoid cutting theoretical corners, though. Finding really good books on the theory of television and video systems isn't easy. For one thing, in this area there is no such thing as science without technology, and it is difficult (and pointless) to try to discuss scientific principles without also discussing both the technology and the hardware to which it refers. The best books are nearly all American in origin, and quite a few details differ from those of British and European technology. Most books on television and video deal with either the creative side—production, editing, and so on—or with the equipment itself: servicing, fault finding, etc. Some of the best of these are to be found in the Focal Press list (there are more than 100 titles!).

The standard work in this field is indeed published by Focal Press: *The Art of Digital Video*, by John Watkinson, regularly updated and splendidly comprehensive. It is a model of clarity, too. The most recent edition at the time of writing is the 4th, dated 2008. But you need a fairly deep pocket to buy it, and a spell of weight training before you carry it home. There is a companion volume, *The Art of Digital Audio*, also by John Watkinson, if you are interested in this aspect, too (it is even heftier). Watkinson has also written two other books, *An Introduction to Digital Video* and its companion *An Introduction to Digital Audio*, also for Focal Press. These eschew the tougher mathematics and stick to run-of-the-mill equipment. Make sure you buy (or borrow) the most recent edition.

One of the best accounts of television theory is in *Basic Television and Video Systems*, by Bernard Grob and Charles Handon (6th edition, McGraw-Hill, 1999). In spite of its title it is very comprehensive, though the section on videodiscs shows its age. Surprisingly for a textbook at this level, it is short on diagrams, though long on somewhat muddy photographs. There are self-testing questions at the end of each chapter, but they are all of the true/false type and not very mind-stretching. Like the other books it is American in origin, but the authors do deal fairly thoroughly with both (traditional) PAL and SECAM technologies. In contrast to most other manuals, Charles Poynton's *A Technical Introduction to Digital Video* (John Wiley, 1996) looks at the subject from the perspective of computing and communications. It assumes a good knowledge of electronics and computing principles, and is aimed at computer system designers and television engineers seeking to broaden their horizons. Not a book for beginners, but chock full of information.

Any book that takes more than a year in the writing and publishing can't hope to be right up to date, so fast do things move in this field. For information on the most recent equipment you have to go to the specialist magazines. You can find other valuable sources on the principles of television and video recording on the Internet. The much-maligned Wikipedia is fairly sound when dealing with this subject matter. Another useful site that gives a great deal of background information is http://electronics.howstuffworks.com, and its content is pretty reliable.

Chapter 15 Three-Dimensional Imaging

How We See Depth

Monocular Clues

Looking out of a window, you can see that your image of the world has depth. You don't need to move around to see this. Even if your window is only a single peep-hole, so that your viewpoint is one-eyed and fixed, the world still doesn't look flat. Why should it? After all, it doesn't look flat in a photograph, or even in a sixteenth-century painting. In fact, with a fixed *monocular* (one-eyed) viewpoint there are four clear clues to depth, and you perceive these quite automatically.

1. *Relative image size.* Near objects appear larger than far objects, and parallel lines appear to converge towards distant points.
2. *Image overlap.* Near objects overlap far objects.
3. *Aerial perspective.* Haze causes far objects to have a lower contrast and color saturation, and to appear more bluish, than near objects.
4. *Modeling and texture.* Highlights and shadows indicate the characteristic shapes and surface textures of objects.

Figure 15.1 (color plate) shows two examples of these clues. The ancient Greeks and Romans were familiar with them, and according to contemporary accounts they produced *trompe l'œil* murals of startling realism. The laws of perspective were mislaid during the Dark Ages, but were rediscovered in the fifteenth century by Italian painters.

Figure 15.1b shows a view of the harbor entrance at Rhodes. All these clues can be seen in this photograph.

There are nevertheless important clues to depth that occur in real life but are necessarily missing from a painted picture or a photograph. In real life, if you change your viewpoint, the relative position of near and far objects changes, too. When you look at a scene your eyes are moving constantly, and so, probably, is your head. A change of viewpoint, however small, produces these positional changes in the visual image: near objects appear to move in the opposite direction to the way your head moves, while far objects appear to remain stationary. This effect is called *parallax.* So we can add to our static clues:

5. *Parallax.* The effect of parallax is most marked when you watch the countryside go by through the window of a train or car. The nearest objects flash past, middle distance objects move more slowly, and features in the far distance seem to scarcely move at all. Parallax probably gives the most powerful sense of depth in real situations. A static picture does not possess this property, of course, but a cine or TV image does. And the camera need not move, either: the subject can move, or it may simply rotate—a property often exploited in computer imaging.

A *trompe l'œil* painting deceives the eye into believing that the painted perspective is real depth. In Italian churches there are many examples of false alcoves and domes painted on flat surfaces (Figure 15.1a). The rediscovery of the laws of perspective is usually attributed to the painter Brunelleschi, but this is an over-simplification: many other painters were involved, particularly Uccello and Piero della Francesca. Renaissance painters achieved some remarkable effects with wall and ceiling paintings. In conventional paintings the rendering of light and shade on objects, and the texture of fabrics, reached a peak with the Dutch and Flemish masters of still life. It was said at the time that their images were so realistic that birds would peck at the grapes, and bees would visit the painted flowers. (The birds I can accept; I'm not convinced about the bees, though: see Chapter 19.)

Figure 15.1 (See color insert following page 154.) (a) A famous seventeenth-century *trompe l'œil* painting on the barrel (cylindrical) ceiling in the church of Sant'Ignazio in Rome by Fra Andrea Pozzo.

A further clue to depth is the necessity to refocus your eyes between near and distant objects. This is called *accommodation*. This is relevant to objects nearer than about two meters from the viewer. So we can add this too.

6. *Accommodation*. This complements the other clues, but as the action of accommodation is automatic and largely unperceived, it probably contributes little to the sensation of depth. In any case, most people lose much of their ability to accommodate by the time they reach the age of around 45.

Binocular Clues

There are two further clues to the perception of depth. These depend on the possession of two forward-facing eyes, and this is important in more than one way.

Figure 15.1 (See color insert following page 154.) (b) A photograph showing clues 1–4. Notice (1) the relative size of the two pillars, (2) the overlap of objects, (3) the slight haze reducing the contrast of the distant fort, and (4) the modeling and texture resulting from the side light on the nearer pillar.

Whereas the first set of clues to depth was based on geometry, these two (along with accommodation) are physiological.

7. *Ocular convergence.* As an object approaches, your brain tells your eye muscles to turn the optic axes of your eyes inwards so that they meet at the surface of the object. The feedback from the muscles of your eye tells your proprioceptive system how much convergence has been necessary, and this information is matched to a preexisting model in your brain, to keep tabs on the distance of the object.

8. *Stereoscopic fusion.* This is a quite complex concept, as it is associated intimately with the visual cortex of the brain. Because your eyes are around 6–6.5 cm apart, the images on your two retinas are slightly different, owing to the difference in viewpoints (i.e., the parallax). By a process

Proprioception is the sense of position. It tells you, for example, whether your left arm is bent or straight. It tells you too, even with your eyes shut, whether your left forefinger is about to scratch your nose rather than miss it and scratch your chin instead. Eye convergence is one of its more subtle perceptions.

215

that is still only imperfectly understood, your visual cortex compares the two images and interprets the difference between them in terms of depth. Physiologists call this facility *stereopsis*.

Stereoscopic perception varies greatly between individuals. Probably only about 5 percent of the population has the sense developed to the highest degree. This perceptual elite no doubt includes top-flight cricketers, baseball players, and others whose talent demands the instantaneous and exact assessment of the position of a fast-approaching object. Of the remainder of humanity, about four out of five have normal stereoscopic vision; around one out of five have only feeble stereoscopic perception, while out of these 1 in 20 have effectively monocular vision, usually owing to an eye injury or a development fault in childhood.

Stereopsis begins to develop in human babies at the age of a few months, and is usually fully developed between one and two years, though there is some variation. What is certain is that if something prevents its development before the age of about four to five, it will fail to develop fully. This problem was at one time common in individuals who had had an uncorrected squint as small children because of unbalanced eye muscle development. Nowadays this condition is quickly diagnosed and rectified, but if it is ignored it leads to lazy eye syndrome, with vision in the nondominant eye disappearing almost completely.

Although the unfortunate one in five may be cut off from the appreciation of stereoscopic pairs of images they are by no means deprived of three-dimensional vision, as the first six clues to depth are still available to them. After all, many people with only one fully functional eye can still play a decent game of tennis.

The Limits of Stereo Pairs of Images

Let's recap those clues to depth:

1. Relative image size
2. Image overlap
3. Aerial perspective
4. Shadows, modeling, and texture
5. Dynamic parallax
6. Accommodation
7. Ocular convergence
8. Stereoscopic fusion

Not all of these are equally important, but they do all contribute to depth perception. It is worth noting, though, that a stereoscopic pair of photographs lacks a number of these clues, in particular numbers 5, 6, and (to a large extent) 7.

Early Stereoscopic Images

We don't know exactly when the first stereoscopic images were made. There have been theories about drawings dating back to the fourteenth century and earlier, though such drawings may just be different views of an object. It is in fact quite easy to draw stereoscopic pairs of simple geometrical objects (see Digging Deeper). You need, though, to develop some ocular skill to be able to fuse two such images without some sort of optical aid. A pair of stereoscopic images appears in Figure 15.2 (color plate) for you to try later.

Left   Right

Figure 15.2 (See color insert following page 154.) A stereoscopic pair. To view these images stereoscopically without an optical aid, hold the book about 40 cm away from you just below eye level, while you look at a distant object in line with them. Then bring the book up into your line of vision, keeping your eyes relaxed. You will see one blurred image in the middle, flanked by two other (also blurred) images. Now slowly bring the middle image into focus without losing fusion. (This may take some practice.) You may find it easier to deliberately cross your eyes, in which case the stereoscopic depth will be reversed.

Probably the first model of a working stereoscope was that of Charles Wheatstone in 1860. This has survived virtually unchanged to the present day, and is still used for entertainment by enthusiasts, and (not for entertainment) by photo-interpreters of aerial reconnaissance and survey photography. (More about this latter usage later.)

Wheatstone's stereoscopic principle is simple and elegant. Two photographs taken of the same scene from viewpoints the same distance apart as the distance between the eyes are mounted with their optical centers at the interocular distance, and viewed through lenses that present a magnified image at infinity (Figure 15.3).

Viewing the images with the eyes relaxed (i.e., not converging) produces binocular fusion and the sensation of three-dimensionality. Accommodation and dynamic

Sir Charles Wheatstone is probably better known by generations of physics students for the Wheatstone bridge (which he didn't invent, but merely improved; cf. Robert Bunsen and the Bunsen burner). Wheatstone made numerous contributions to optics, electrical theory, and acoustics. He was also an accomplished musician, and invented the concertina.

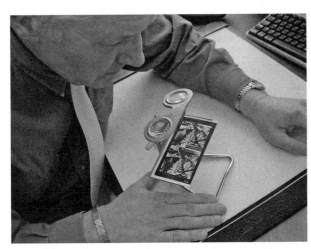

Figure 15.3 Wheatstone's stereoscopic principle in use. His original design included mirrors (see Figure 15.9).

217

These developments include variants on Wheatstone's original stereoscope, from the 1860s onwards, the Viewmaster transparency viewer of the 1940s, the polarized demonstration movies of the 1950s, the anaglyphic presentations on TV in the 1960s, the Nimslo camera of the 1970s, and the interlaced computer-drawn images of the 1990s. A new generation of stereoscopic movies is even now appearing. Each of these systems has had its merits, and I shall be explaining how they work.

parallax are not involved, and though there may be some variation in convergence when examining objects in the foreground, it does not necessarily correspond with that involved in viewing the original scene. In fact, out of the eight clues listed earlier, only the last one (stereoscopic fusion) is added to the monoscopic clues I began this chapter with, and this sense is weak in a large minority of people. So it's not surprising that an interest in stereoscopy by the general public only appears as a brief burst of enthusiasm when a new development comes along. And each previous system in turn becomes a case of "Whatever happened to … ?"

Stereoscopic Camera Formats

Conventional Cameras

You can easily make stereo pairs with a conventional camera. All you need is a small platform fixed to a tripod, with a batten along which you can slide the camera. A piece of boxwood meter rule makes a good batten, as you can use the graduations to adjust the distance between the two viewpoints. For general stereoscopic photography you simply slide the camera horizontally some 6.5 cm between the two exposures. If you are using a film camera with the conventional left-to-right film wind, and you want the two negatives to be in the correct order, you need to take the left eye view first (remember, the images on the film are inverted) (Figure 15.4). For a 35-mm negative, you will need to enlarge the image by a factor of a little less than ×2 to have the optical centers of the images about the same distance apart as the distance between your eyes. For a 4.5 × 6 cm format you will need about ×1.4 enlargement, and for a 6 × 7 cm (landscape format) or 4 × 5 inch (portrait format) you can simply set up contact prints side by side, and view the results directly through a Wheatstone stereoscope.

Cameras and Camera Conversions

In the early days of stereoscopic photography there were quite a few cameras designed especially for the purpose, and many still survive. They were effectively two cameras built into one body. Some took double-width plates, and others used standard film stock. At the outbreak of the Second World War there were at least 10 models in production, and several more appeared after the war. These cameras

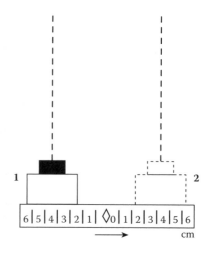

Figure 15.4 Setup for a stereoscopic pair for a conventional camera, using a calibrated slide bar.

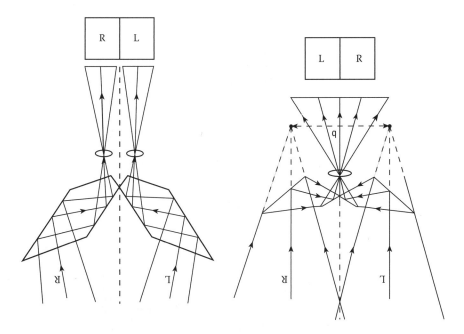

Figure 15.5 Two optical adaptors for making one-shot stereophotographs with a conventional camera.

are no longer made commercially, but a number of modern versions are available, and some camera manufacturers still offer stereo converter kits for conventional cameras, prismatic systems with double lenses to give portrait-format stereo pairs on 35-mm film. The principles underlying the two most common optical arrangements are shown in Figure 15.5.

Some specialists make stereoscopic cameras to special order. These may be near-replicas of earlier cameras, but more often they are standard 35-mm cameras that have been turned into Siamese twins (see Figure 15.6). In order not to waste film the exposures are usually interleaved, and the right and left images are in reverse order. With digital cameras, of course, this is of no importance.

Some cameras have more than two lenses, and these produce a different type of stereogram, which I will come to later.

Changing the Base Length

Stereoscopic imaging has far wider applications than simply providing a quasi–three-dimensional experience. In particular, it earns its keep in applied photography. Almost all practical applications, though, involve changing the base length from the normal interocular distance.

Stereophotography in Aerial Survey

When an aircraft is employed to take a series of vertical photographs in order to prepare a map of a ground area, it makes a series of parallel runs, taking photographs that usually have a side overlap of around 10–20 percent, to avoid the possibility of missing any detail. But the longitudinal overlap is much larger, typically 60 percent (Figure 15.7). This is to allow stereoscopic viewing. When you view two prints taken with this kind of overlap through a suitable stereoscopic viewer, you can see the image in three-dimensional detail. You can actually see more detail this way than you can from either of the two photographs viewed singly.

I remember, many years ago, a photo-interpreter colleague spotting a kite being flown in one such pair of aerial images, a detail virtually invisible in either of the single prints.

(a)

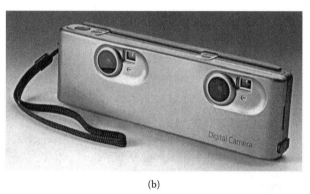

(b)

Figure 15.6 (a) Two standard digital cameras converted for stereophotography; (b) compact stereoscopic camera. (Photographs courtesy of David Burder, 3-D Images Ltd.)

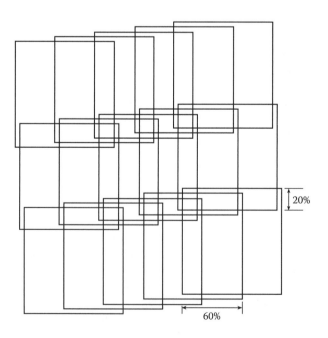

Figure 15.7 Overlap in aerial survey photography.

Hyperstereoscopy in Aerial Photography

However, when you view the pairs with a simple viewer, the apparent size and distance of the perceived image is much reduced, so that it appears like a Lilliputian model viewed from a comparatively short distance. The reason for this is that the base is far larger than your interocular distance: anything from a couple of hundred meters to several kilometers. This effect is called *hyperstereoscopy*. It can seem a bit odd the first time, but you soon get used to it. There is a special viewer with magnifying lenses that puts the image back to infinity and eliminates the "model" effect, but it also exaggerates the relief, so that objects appear about four times as high as they should be. The actual factor can be calculated and optimized using a rather complicated piece of geometry, but for most purposes the most satisfactory rendering is when the ratio of base length to aircraft altitude is about 0.6. The vertical exaggeration of long-base stereoscopy is in fact useful in the interpretation of aerial reconnaissance photography. Incidentally, all this is true of satellite photography, too: the distances are greater, but the angular proportions remain the same.

Most aerial photography is for survey purposes and map making. Stereo pairs of photographs are particularly useful here. Photo-interpreters often work directly on the negatives rather than on prints, as the resolution and tone separation are better. Equipment for plotting contours directly from stereo pairs existed as early as the 1930s. The Zeiss Stereoplanigraph did this semiautomatically, requiring the operator only to steer a pair of floating spots so that they coincided on adjacent overlapping images. Modern computer-driven equipment now does this job with much less fuss, and at far less cost.

The Stereoplanigraph was said to be the most complex, and the most accurate, piece of optical equipment ever built. As far as I know, only three were ever made.

Effect of Incorrect Lens Separation

Some stereoscopic adaptors for cameras have lenses that are closer together than 6.5 cm, and when you examine the results in a viewer in the usual manner they show diminished depth. The familiar term for this effect is "cardboarding," from the cutout appearance of objects. Even when the inter-lens separation is the full 6.5 cm, the effect can be present in mid-distance shots. Close objects may have the opposite appearance, with exaggerated depth. As a general rule, for a satisfactory three-dimensional reconstruction the subject matter should lie at a distance of between 30 and 50 times the base, i.e., 2 to 3.25 m for a 6.5-cm base. For landscapes it is advisable, if possible, to increase the base by about 50 percent. The reverse is true for extreme close-ups and macro work, where the parallax is too great for a normal base, and the camera lenses have to be shifted or their optic axes toed in (Figure 15.8). Shifting is preferable to toeing in, as the latter results in keystoning, and there may be difficulty in visually fusing the images unless the effect is corrected in the computer.

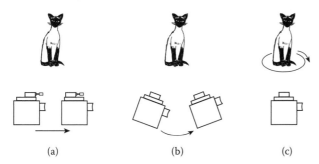

(a)   (b)   (c)

Figure 15.8 Hypostereoscopy. (a) Parallel camera movement using shift lens; (b) swinging the camera; (c) rotating the subject. The parallax angle should not exceed about 7° (the diagram is exaggerated for clarity).

Hyposteroscopy in Microscopy

In microscopy, particularly electron microscopy, it is not usually possible to take two suitably spaced views of the subject. Even in binocular microscopes with double objectives, which do permit live stereoscopic viewing, the photography port will be monocular. To obtain stereo pairs of images you have to make two exposures, rotating the specimen by a few degrees between them. Rotating the specimen is exactly equivalent to using a pair of toed-in cameras; the keystoning effect will be minimal as long as you keep the parallax angle down to no more than about 7°. You are effectively making the stereo pair with a much reduced inter-lens distance (*hyposteroscopy*). There will now be a Brobdingnagian effect: the image will seem to be some distance away, and very large. This is usually unimportant, as viewers, aware that they are observing something greatly magnified, will be expecting such an impression.

Viewing Methods for Stereo Pairs

The inventor of stereoscopic photography was, as I said earlier, Charles Wheatstone (he of the Wheatstone bridge). His stereoscopic viewer appeared in 1838, more or less coinciding with the first public appearance of photography itself. At the time there were no enlargers, and camera formats were sizeable, so the images were viewed via mirrors. The prints had to be made laterally reversed (Figure 15.9). Wheatstone's viewer was designed for large prints, but the mirrors made it cumbersome.

When cameras became smaller, the viewer became less popular than the simpler one that had been introduced by David Brewster (he of the Brewster angle). This could be used with the popular quarter-plate size prints (3¼ × 4¼ inches).

There are two versions of the Brewster stereoscopic viewer (Figure 15.10). The split-lens version allows the viewing of somewhat larger prints and a small amount of eye convergence, which makes for more comfortable viewing. Both types are still extant.

The Viewmaster stereoscope of the 1940s and 1950s (Figure 15.11) was an enclosed Brewster stereo viewer that held a commercially produced disc bearing stereo pairs of transparencies made on 16-mm Kodachrome film at opposite points on its diameter, with a lever advance mechanism. It proved immensely popular, and the catalogue of discs ran into many hundreds—not just of scenery, but often also of serious educational material. For a time you could even obtain a kit for making your own discs.

As a result of the introduction of his viewer, it was claimed on behalf of Brewster that he had himself invented stereoscopy. Brewster did nothing to counter this story, and this resulted in some acrimony between him and Wheatstone.

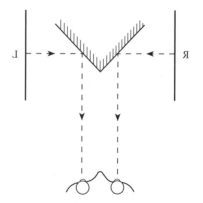

Figure 15.9 Wheatstone's original stereoscopic viewer for large prints.

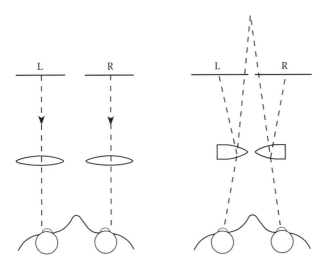

Figure 15.10 Two versions of Brewster's stereoscopic viewer. The version on the right permits some eye convergence and can be more comfortable for viewing.

Figure 15.11 The Viewmaster stereoscope.

A type of viewer employed for large aerial survey photographs (250-mm square) uses a double mirror. Its design is due to Hermann Helmholtz (he of the coil and the resonator), later modified by Caze (who he?). A more recent version employs 45° TIR rhomboids and binocular eyepieces to obtain a highly magnified image (Figure 15.12). The rhomboids can be swivelled to align the images accurately.

Viewing without Optical Aids

Practiced viewers can view a stereo pair without optical aids, by controlling their ocular convergence. This is a physical trick that has to be learned, like wiggling your ears or raising just one eyebrow. The method usually suggested for this is to look at a distant object, then to bring the stereo pair up in front of the eyes without at first refocusing them. With a little practice you can now focus sharply on the

223

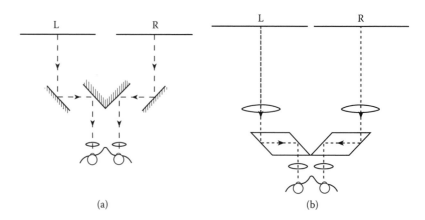

Figure 15.12 (a) Helmholtz's stereoscope for viewing large prints. (b) Modern stereoscopic viewer with telescope optics and TIR rhomboids (adjustable for alignment).

images without your eyes converging and losing binocular fusion. You will then see a three-dimensional image in the center of your field of vision, with a vague two-dimensional image on either side. Try this now on Figure 15.2.

Coincident Image Stereograms

If we could print a stereo pair of images one on top of the other and still direct each image only to the appropriate eye, we could make more or less full use of ocular convergence. In fact, there are several methods of achieving this. An important consequence of this technique is that we can arrange any desired plane of the image coincide with the plane of the print itself. This will be the plane where the two images are in exact register on the print; then, instead of the image appearing at infinity as it does in conventional optical viewing, it will lie across the plane of the photograph itself. For landscapes and distant objects the most realistic effect is when such images are projected on a distant screen; this also results in the greatest perceived range of depth. There are two ways of making coincident image stereograms: anaglyphs and Vectographs.

Anaglyphs

This is the better-known method. One image is printed in red and the other in blue or cyan. You view the result through color filters—red for one eye, blue or cyan for the other. The red-filter eye sees only the blue image, and the blue-filter eye sees only the red image. The two images fuse to give a three-dimensional image in something approaching neutral grey. There may be a small amount of color fringing for near and far objects, but your perceptual mechanism will probably ignore this. David Burder has patented a system for color stereoscopic photographs whereby a color photograph is separated into red for one eye and green plus blue for the other. When you view the compound image through cyan/red spectacles you see the scene stereoscopically in full color. There is still some color fringing, and saturation is compromised to some extent, but the effect is convincing, and the system has been piloted on British television with some technical success. One of its merits is that it also permits monocular viewing without glasses, as you can see from Figures 15.13 and 15.14 (color plates), albeit with some color fringing and loss of resolution.

If you can't manage this no matter how hard you try, don't feel you are a failure. At least a quarter of the population can't do it either. It *is* an unnatural muscular action, and if you practice it persistently for some time (as many RAF photo-interpreters did during the Second World War) you may finish up with episodes of double vision.

With red/cyan glasses, you can get the full stereo effect from these two illustrations. The red filter must be on the left. David Burder has generously offered to supply red/cyan to any reader who contacts his website, Burder3D@aol.com and mentions this book.

Figure 15.13 (See color insert following page 154.) Monochromatic anaglyph. This is an example of hypostereoscopy. To view this illustration and Figure 15.14 in stereo you need a pair of red/cyan glasses, with the red filter in front of your left eye. (Red/green will do for this figure, but will not give accurate colors on Figure 15.14.)

Vectographs

A somewhat more satisfactory system is based on the optical phenomenon of polarization. In the simplest method of presentation the two images, which may be in full color, are projected through two linearly polarizing filters set orthogonally at ± 45° to the horizontal, onto a metallic screen. The viewers wear polarizing spectacles also set at ± 45°, so that each eye receives only the appropriate image. This is probably the most satisfactory system to date, as there is full color to each eye and negligible crosstalk (negligible, that is, until you tilt your head). The Vectograph system was evolved during the Second World War by the Polaroid Corporation. Vectograph film was a double-sided film with the two sides polarized orthogonally, each side bearing one-half of a stereo pair. It was viewed through polarizing spectacles. Vectographs could be made using a process similar to that used for making matrices by the dye-transfer color process (dye transfer in red/blue was also used to prepare anaglyphs). This system can be used for both reflection prints and transparencies, but it involves some technical expertise in the processing. The name is now often applied carelessly to any polarizing system for stereoscopic viewing.

Interlaced Images

An adaptation of the unaided viewing principle for viewing larger prints without eye strain has been made possible by computer graphics. The two stereo images are interlaced in an intricate pattern (any pattern will do, as long as it is too complicated to be obtrusive once the hidden image has revealed itself). Figure 15.15 is an illustration. When you look at this in the same way as with an unaided stereo pair—that is, you relax your gaze to infinity and then bring the picture up into your line of sight—the image appears in depth. The resolution is somewhat limited, and at the present stage of development the technique can be regarded as little more than a plaything. But it could conceivably be used as a serious form of illustration in, for example, a textbook on stereochemistry, or a thesis on Renaissance sculpture, given some properly funded development. It is an example of an *autostereogram* (see below).

Figure 15.14 (See color insert following page 154.) Color anaglyph. The red and blue/green images are in register in the plane of the photograph, so that the tiger's eyes appear to be in this plane.

Single Image Auto Stereograph. David Burder

Figure 15.15 An interlaced stereogram. To view the image, hold the book about 40 cm away from you and stare steadily at the figure. The stereoscopic image will jump out after a short while.

Autostereoscopic Systems

An *autostereoscopic* system is one in which the viewer can perceive the image with full parallax without any kind of optical aid or visual training. The supreme example of this is, of course, holography, and its principles are so important that they have a whole chapter to themselves. But there are several systems that are purely photographic, and others, some of them waiting in the wings, that depend on computer-stored video. There are also hybrids of holography and photography, which are also dealt with in Chapter 16.

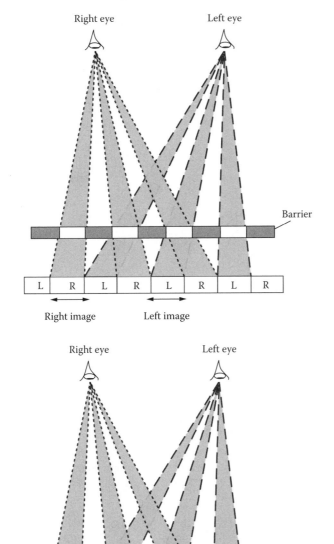

Figure 15.16 (a) Barrier stereogram. Each eye sees only its appropriate image. The barrier may be an LCD array activated specifically for viewing a pair of computer images stereoscopically. (b) When an array of lenticules is substituted for the barrier, the directed light makes the image brighter.

Parallax Barrier Stereograms

In its earliest form, dating back to the early twentieth century, the left eye image was printed through a fine screen, or barrier, of opaque vertical bars. The screen was then moved through the width of one bar and the right eye image printed in the interstices. The composite print was then mounted under a transparent sheet about 0.5 to 1 mm thick, with the screen mounted on top. When viewed from an appropriate distance one eye would see only the left eye image, and the other eye would see only the right eye image (Figure 15.16a).

This system has recently been revived commercially for displaying stereo pairs on a suitably modified home computer monitor, as well as on cell phone displays. In this system the barrier is generated by an LCD screen positioned in front of the image display screen. This barrier can be switched off for 2-D displays. With this simple system the image is somewhat dimmed owing to the presence of the bars, and unless their pitch is closer than the resolution of the eye they can appear like a miniature picket fence. In addition, the horizontal resolution of the image is only half that of a 2-D display. In practice these problems can be dealt with by rapidly switching the spatial phase of the barrier throughout the display of each image, so that the entire image is displayed in each case, and the picket fence effect disappears.

Lenticular Stereograms

A more modern version of the parallax barrier system uses a lenticular screen instead of the bars. The cylindrical lenses that make up the screen are focused on the surface of the print. This comparatively crude system provides only a single stereoscopic image within certain well-defined angles, and the stereo effect reverses to a *pseudoscopic* (inside-out) image at points halfway between. It is used nowadays mainly for advertising cards that switch between two images, usually with a horizontally orientated screen for vertical switching: this avoids fortuitous double images appearing as the viewer walks past (Figure 15.16b).

Modern lenticular stereograms use a large number of images instead of just two, anything from 6 to 200. These are originated by either a battery of cameras or a single camera moving along a rail, recording an image at predetermined intervals. A computer-controlled printer interlaces these images so that each lenticule has a full set of right-to-left viewpoints behind it. The view is now effectively continuous, and you can perceive depth in the image even with one eye closed. Because of the large number of component images there can be a small amount of animation if the exposures have been made successively rather than simultaneously.

Parallax is limited by the focal length of the lenticular elements: the greater the focal length the better the resolution, but the less the parallax angle. The optimum is around 10°–15°. There is still a brief jump into pseudoscopy as you pass the "join," but you get several reruns as you continue to move past (Figure 15.17).

Photographic stereograms have the advantage over holograms that they can be made in any size and at any image magnification; they can be made out of doors, too, and can be viewed by a number of people at the same time, in ordinary lighting. However, the limited parallax (no more than about 15° at most) can be a disadvantage, and the horizontal resolution is limited, too, as only a small fraction of the area of the screen is visible at any one time.

A system capable of full horizontal parallax is also in an early stage of commercial development for home computers and cell phones. Instead of an opaque LCD

In the previous edition of this book a small illustration of this was included, complete with screen. Unfortunately it has not been possible to include this in this edition, but the format appears frequently in advertisements and in cereal packet giveaways, so you may already be familiar with its appearance.

Even in a stationary portrait, the slight movements of the sitter during exposure give a "living" effect absent from that produced by simultaneous exposure by a battery of cameras.

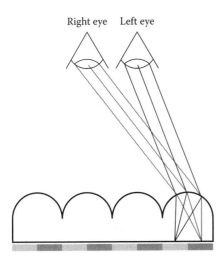

Figure 15.17 Lenticular stereogram with full horizontal parallax. Because the lenticules are retroreflective, each viewpoint sees only a narrow strip behind each. A large number of photographs can thus be printed as narrow interlaced strips, with the lenticular array mounted in register.

barrier, an active lenticular array is generated. The principle uses the property of liquid crystals under electrical stimulation to change not only the plane of polarization but also their refractive index. The lenticules are filled with liquid crystal medium, and the space is filled with a medium with the same refractive index as the unstimulated lenticules. When the electrical field is applied the refractive index of the LC medium increases and the lenticular pattern appears.

A lenticular stereogram does not have vertical parallax, but as a rule this is of little importance. It is, of course, possible to add vertical parallax by making the original series of photographs using a vertical bank of cameras rather than a single camera; you then need to use a mosaic of microlenses instead of cylindrical lenticules to display the composite image. This has recently been tried out commercially, and could be easily reproduced on a home computer system with a suitably modified LCD screen. The barrier system in its more sophisticated form need not necessarily be confined to the small screen. In an example quoted by Javidi, Okamo, and Son (see Digging Deeper) the images were displayed on a screen 3.5 m wide, using 518×288 pixels with 6 mm pitch and a parallax barrier of pitch 11 mm. A suitable viewing distance for these parameters would be 15–16 meters, with a large number of positions available for stereoscopic imaging.

One of the problems posed by barrier and lenticular systems alike is that a small horizontal movement on the part of the viewer switches the image from orthoscopic

The Nimslo camera of the 1980s operated on the lenticular principle. The dissection and mounting of the images was carried out using dedicated machinery at special processing stations. But with only four lenses the results were jerky and the parallax limited, and as the lenses were very close together the final images showed marked cardboarding. David Burder has produced a 12-lens version of the Nimslo camera for producing handheld stereograms (Figure 15.18).

Figure 15.18 A multilens camera for parallax stereograms constructed from three Nimslo cameras. (Construction and photographs by David Burder.)

229

to pseudoscopic (i.e., the image appears inside out). A simple remedy is to space out the groups of pixels with dark bars, so that a large sideways movement simply causes it to disappear.

Stereoscopic Cinema and Television

There are two well-tried methods for screening stereoscopic images, namely anaglyphs and orthogonal polarization. These methods are the same as for still photography, using two cameras set up side by side and synchronized. Anaglyphs do not need any modification to the projection system if the two images are printed on the same film in cyan and red and viewed through red and cyan spectacles. You can show a full-color anaglyphic film in the same manner, using red and cyan images as described earlier for still photographs, but results are better with two separate projectors. The same applies to the polarizing system: with two projectors you don't need special film or overlays, just two polarizing filters, one over each lens; the viewers wear polarizing spectacles. You do need a special metallized screen, though, as ordinary matte screens depolarize the light.

There have been demonstrations of full-parallax cinema, using a battery of projectors and a retroreflective screen, and these appear to have worked well, but such an arrangement is enormously expensive, and would not seem to have a commercial future. In fact, stereo cinema itself has so far never really thrived. Every 10 years or so, some film company makes a film in stereo, usually without much box-office success. The highly successful film version of "Kiss me, Kate" was originally made in stereo, but was only distributed in mono. You can tell it was originally made in stereo from the way the actors frequently throw things towards the camera, presumably to make the audience duck.

As I write, there has been another big push toward 3-D cinema and TV, spearheaded by the film 'Avatar.' Whether the vogue for 3-D presentations will list this time around remains to be seen.

Polarization isn't really feasible for TV stereo presentations, though there was a somewhat impracticable design by Pye television in the 1950s, combining the images from two TV screens set at right angles, with a 45° polarizing mirror as a beam combiner, and polarizing filters on the screens. It was never marketed. Anaglyphs can readily be shown on an ordinary TV screen in either monochrome (red and green merging to brown, as above) or full color (cyan and red). Some years ago the BBC carried out a weeklong series of public experiments using the anaglyph method, including an open-air screening in Covent Garden Market. Technically it was a success, but the public was generally apathetic, and the experiment was quietly dropped.

Further Developments in Stereo Projection

A more recent technical development is the showing of left and right eye images alternately, the viewer wearing spectacles with synchronized LCD shutters. Helmet-mounted projection systems, one for each eye, are now being used in simulators; these may be connected to sophisticated equipment that detects head movements and changes the scene appropriately.

A more practical and much more user-friendly approach is now becoming available. It employs a single projector that needs a detachable addition, and passive reusable viewing glasses. The right- and left-eye images are displayed alternately, and oppositely polarizing filters are alternated between frames. To avoid crosstalk (which occurs with linear polarization if the head of the viewer is tilted) the polarization is circular, clockwise for one eye and anticlockwise for the other. A metallic screen is still required, but this is already becoming common in cinemas. The system is already installed in many cinemas, and promises to become a standard fitting.

Simulated Stereopsis

A much simpler approach produces movies in what appears to be 3-D by deceiving the brain's perception mechanism. It employs nothing more complicated than a pair of spectacles with one glass clear and the other heavily tinted. Perception through the darker glass is fractionally delayed compared with perception through the clear glass. This produces a strong impression of depth with horizontally moving images on the screen, when the direction of movement is from clear glass towards dark glass. Movement in the reverse direction should produce reversed stereoscopy, but it seems that most people don't notice. You can try this out for yourself, using an old pair of sunglasses with one glass removed. It works best with action movies. It is an optical illusion, of course, not genuine stereoscopy.

Integral Photography

I mentioned in Chapter 4 that a camera lens produces an image that is three-dimensional. Not many photographers seem to be aware of this, though you can easily confirm it by setting up a camera with the back open and a small object fairly close to the front of it, with the lens at full aperture. If you focus your eyes on the film gate, you will see the image of the object floating there. It is inverted, of course, and you can only see it with one eye at a time, but if you move around a little you can see that it has parallax. And the image is orthoscopic. Could we make use of this property?

There are two difficulties. The first is that to get a reasonable amount of parallax your lens needs to have a very large diameter. Merely to see the image with both eyes at the same time demands an exit pupil diameter of more than 6.5 cm, for obvious reasons. The second is that the three-dimensional modeling of the image is correct only when the object and image are the same size, that is, the magnification is unity. This is because the longitudinal magnification (i.e., along the optic axis) is the square of the lateral magnification. What is more, for a deep object with its center line at unit magnification, the part of the object nearest the lens will have exaggerated depth ($M > 1$), and the part farthest from the lens will have diminished depth ($M < 1$). This limits the permissible depth of the object to about plus or minus 10 percent of the focal length of the lens. So can there be any optical system that will produce an aerial image with full depth and wide parallax that is also erect, orthoscopic, and undistorted?

Gabriel Lippmann, no less, was the first to investigate this problem, a century ago. In 1908 he proposed a method for producing full-parallax stereoscopic images using microlens arrays in front of the photographic plate and, after processing, reversing the light rays to re-create an image in the position of the original object. Unfortunately, when the image is viewed in the usual way, from the object side, it is pseudoscopic (Figure 15.19).

In a modern version, a "macrolens" array collects the light from each point on the object and collimates it so that the microlens array focuses it on the sensor array, using graded refractive index (GRIN) elements (see p 000) to focus the image. These

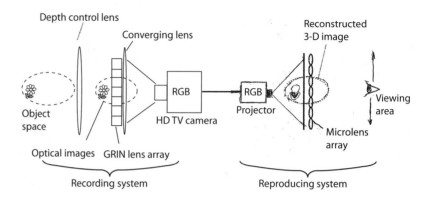

Figure 15.19 Optical principle of integral camera.

produce erect orthoscopic images, so that the overall image is orthoscopic too. This complex image is transmitted to a receiver, which possesses what is in effect a reversal of the recording optics, and produces an image that is geometrically exact, with full parallax and depth. The amount of parallax available depends on the diameter of the arrays, and (unlike a single lens) these can be made as big as you like. At present the system has not been developed commercially, but it has generated a large number of research papers, and promises interesting, if limited, applications.

Digging Deeper

For a full account of the geometry of stereoscopy, the best source is Sidney Ray's *Scientific Photography and Applied Imaging* (Focal Press, 1999), Chapter 16. Chapters 17 and 18 (Photogrammetry, Aerial Photography) also contain useful information on stereoscopy as a measuring tool. Ray gives a huge number of references to research papers and textbooks. The best of the latter are *Stereoscopy*, by N. Valyus (Focal Press, 1966) and *Three-Dimensional Imaging Techniques*, by T. Okoshi (Academic Press, 1976). These are out of print, but aren't hard to find, particularly if you have access to an academic library. Keith Henney and Beverly Dudley's *Handbook of Photography* (McGraw-Hill, 1939) has a chapter on stereoscopic photography, but the chapter on aerial photography is more interesting, because it gives a full account of the method of making maps from overlapping aerial photographs (still in use today); it also has a photograph of the Zeiss Stereoplanigraph.

For a comprehensive review of research up to 1997 you could dip into SPIE Proceedings Vol. 3023 (1997) *Three-Dimensional Image Capture* (SPIE stands for Society of Photo-Instrumentation Engineers), but give yourself plenty of time. The most up-to-date account is the source book *Three-Dimensional Imaging, Visualisation, and Display*, edited by Javidi, Okano, and Son (Springer, 2009). You can keep up with commercial trends by using one of the Internet search engines. The Web site www.stereoscopy.com is particularly helpful. A useful account of stereoscopic vision appears in Richard L. Gregory's *Eye and Brain* (5th edition, OUP, 1998). There are some good examples of 3-D illusions here, but the whole book is well worth acquiring. There is rather more detailed information on binocular perception in N. J. Wade and M. Swanton's *Visual Perception* (Routledge, 1991), but it is a less easy read.

Chapter 16 Holography

Holography is a unique method of producing three-dimensional images. It did not evolve from photography, but from a number of ideas associated with electron micrography and radar, and it was developed not by artists or photographers, but by physicists. Consequently, many people don't associate holography with creative image making, and think you need a PhD and a laboratory full of expensive equipment before you can think about making a hologram. In fact, anyone can make a hologram in a cupboard under the stairs with no more equipment than a glass plate, a laser pointer from a souvenir shop, a piece of holographic film, and some processing chemicals. You don't even need a lens. Holography produces its images by using the principle of optical interference, which we first met in Chapter 1.

Coherence

The examples of interference we have seen previously (Lippmann photography, antireflection lens coating) operate only over very short distances, because after one or two wavelengths the light waves get jumbled up, like army recruits getting out of step. To make a hologram you need a beam of light that stays in step (in phase) over the whole depth of the object you are recording. This property is called *coherence*. A beam of light is coherent if it satisfies three requirements:

1. *Temporal coherence*. The source must radiate at a single frequency.
2. *Spatial coherence*. The light must appear to have been emitted by a point source.
3. *Phase coherence*. At any point where interference takes place, the interfering beams must have a constant phase relationship.

The well-disciplined beam from a laser satisfies the first two conditions. The third is satisfied if the two beams are derived from the same source.

Denisyuk's Hologram

When two mutually coherent light beams meet head-on, they produce stationary interference planes (Bragg planes) one half-wavelength apart. You may remember from Chapter 6 that this was the principle of Lippmann color photography. Lippmann generated his two beams by placing his emulsion in contact with a mirror, but the coherence of the imaging light was so low that it created at most only a few Bragg planes. When the laser was invented in 1961, Yuri Denisyuk substituted a reflective object for the mirror, and shone a beam of laser light directly onto the photographic plate, with the object behind it. The two beams formed a stack of more than a dozen interference planes throughout the thickness of the emulsion. This was the first reflection hologram. This type of hologram is now universally known as a Denisyuk hologram.

At this point it is worth revisiting Chapter 1. Have another look at Figure 1.10a. This is what happens within the thickness of an emulsion when you have a mirror in direct contact with it and expose it to a beam of monochromatic light. When you develop the emulsion you get a stack of planes of metallic silver locating the crests of the standing waves (the *antinodes*), each separated from its neighbor by exactly half

Yuri Nikolayevich Denisyuk (1927–2006) developed his holographic principle while working at the Vavilov Institute in St. Petersburg (then Leningrad), where he was head of optical physics. His research was inspired by Lippmann's interference-based imagery, and he was unaware at the time of the work of either Gabor or Leith. With characteristic modesty Denisyuk called his imagery a Lippmann hologram.

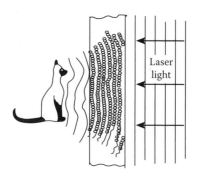

Figure 16.1 Denisyuk's hologram. The object and reference beam are on opposite sides of the emulsion and (somewhat distorted) Bragg fringe planes are formed by the uneven reflecting object.

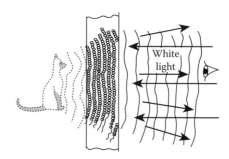

Figure 16.2 The distorted fringe planes of Figure 16.1 reconstruct the beam that was reflected into the emulsion by the object.

a wavelength. Now suppose you substitute a shiny object for the mirror. The interference layers will still be formed, but they will be warped by the uneven contours of the reflecting surface. In effect, they will form a distorting mirror (Figure 16.1)

But this is a very special distorting mirror. When you illuminate it with the original beam, or even with white light, it reflects back a replica of the beam that came from the object! And if you look down this reflected beam you will see an image of the object, just as if it were still there (Figure 16.2).

A few names are in order here, as you will meet them a number of times in this chapter. The undisturbed beam that illuminates the emulsion directly is called the

Denis Gabor (1900–1979) was a Hungarian-born physicist and electronics engineer. He was trying to improve the resolution of electron microscopes when he conceived the idea of a hologram, for which he was eventually awarded the Nobel Prize for Physics in 1971.

234

GABOR'S HOLOGRAM

It may seem a little mysterious that the "distorting mirror" should produce an image in this way, but it is quite logical, really. The first person to form the concept of image formation by interference was Denis Gabor in 1947, though under somewhat different conditions. He did not have a coherent light source, as there were no lasers at the time. The best he could manage was filtered mercury light, with a coherence length of less than a millimeter. So he had to restrict his hologram (a transparency) to pinhead size, with the reference beam traveling along the same axis as the object beam. The result wasn't very successful, as there was a spurious real image that got in the way of the genuine image, but the result was enough to prove his ideas correct. His first publications on the subject used rigorous mathematical arguments that are well beyond the scope of this book. But a simple proof (see Appendix 2) needs no more than an ability to manipulate some basic trigonometrical formulae.

reference beam. The distorted beam that returns to the emulsion from the object is called the *object beam*. The beam that is used to form the image is called the *reconstruction beam* or, more often, the *replay beam*, and the beam that carries the image of the object back to your eyes is called the *image beam*.

Off-Axis Holograms

One difficulty with Denisyuk's first reflection holograms was that with the reference beam falling perpendicularly on the emulsion, the replay beam had to do so as well, so your head got in the way when you tried to view them. But he soon found that the holograms worked equally well when the reference beam was at an angle. This doesn't affect the creation of Bragg planes: it simply tilts them a little. Most reflection holograms are made today with the reference beam above the viewing line by either 45° or 53° (the Brewster angle).

Denisyuk holograms, and other types of reflection hologram, don't have to be viewed by laser light. This is because of the Bragg condition: if you illuminate a Bragg mirror with white light it will reflect only light that has a wavelength close to a match for the separation of the Bragg planes. Other wavelengths will interfere destructively and won't be reflected. In fact, they are transmitted. If you look at the wall behind a red-image reflection hologram, you will see a cyan-colored shadow.

Denisyuk had trouble getting his work accepted by the Soviet authorities—and by his peers. (This situation may have been aggravated by his refusal to join the Communist party.) Consequently, his work went unnoticed in the West for several years.

If the reference beam is at 45°, the Bragg planes will be at 22.5° in the interference space. But within the emulsion they are at a shallower angle (about 15°) and closer together by about one third, because of the refractive index (~1.5) of the emulsion. Laser beams are in general linearly polarized, and the use of the Brewster angle helps to suppress unwanted reflections.

Leith's Hologram

In the meantime, Emmett Leith, working with his research assistant, Juris Upatnieks, behind closed doors for the Department of Defense at the University of Michigan's Willow Run Laboratory, read Gabor's papers and immediately saw the connection between his work and the fundamental theories of synthetic aperture radar, on which they were working at the time.

They saw that Gabor's problem with the spurious real image could be overcome by offsetting the reference beam, which would move the offending image out of the way. Their holographic setup, unlike Denisyuk's, had the reference beam and object beam incident on the emulsion from the same side. (This method also produces standing waves, but instead of being more or less parallel to the emulsion surface, the Bragg planes are approximately perpendicular to it.) In both cases the direction of the fringes bisects the angle between the two beams (Figure 16.3).

Whereas Denisyuk's technique produced Bragg planes that resembled the pages of a book, Leith's produced them in the form of a venetian blind, i.e., more or less perpendicular to the emulsion plane. The results are the same: illumination of the final hologram with a duplicate of the original reference beam produces a virtual image in the precise position of the object. The Denisyuk and Leith-Upatnieks holograms represent examples of two broad classes of hologram, known respectively as reflection and transmission holograms. You can replay a reflection hologram using white light, but a transmission hologram is much less selective, and you have to use monochromatic light, preferably laser light, to display it. If you illuminate a transmission hologram with white light, all you see is a rather lumpy spectrum.

Emmett Leith (1947–2005) spent his entire career at Ann Arbor University, Michigan. He was seconded to the Willow Run Laboratory to work on the optical interpretation of radar images, and Juris Upatnieks (b.1947) was assigned to him as a research associate.

It was something of a surprise to the scientific community that neither Leith nor Denisyuk were to share the Nobel Prize with Gabor. But at least they both later received the Royal Photographic Society's Progress Medal, its highest honor.

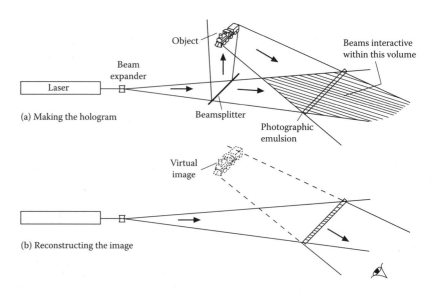

Figure 16.3 Leith and Upatnieks's hologram. The object and reference beam are on the same side of the emulsion, and the fringe planes are formed roughly perpendicular to the surface.

Processing a Hologram

So far, I haven't said anything about methods of processing holograms. In the early days we simply developed and fixed them like photographic negatives. This worked reasonably well for transmission holograms, but the images from reflection holograms were feeble and greenish blue, although the laser light had been red. The main cause of the problem was shrinkage of the emulsion. It turned out that the way to improve the diffraction efficiency and avoid emulsion shrinkage was not to remove any material from the emulsion, but instead to bleach the developed silver fringe planes back to silver bromide, creating alternate layers of high and low refractive index (pure silver bromide has a very high refractive index, more than 2).

There are two possible bleaching techniques. The first, called *solvent bleaching*, resembles the bleaching stage of reversal processing of black-and-white photographic films. The developed silver fringes (the "negative" fringes) are dissolved away, leaving the remaining silver halide (the "positive fringes") unchanged. With this method you have to use a tanning developer such as pyrogallol with metol or phenidone, which cross-links the collagen fibers and prevents the gelatin from collapsing. The second method is subtler. It is called *rehalogenating*. The negative fringes are converted back into silver bromide, which, as it is formed, is deposited on the positive fringes. No material is lost, and the fringe structure remains undistorted. The diffraction efficiency is greatly improved: it can approach 35 percent for a transmission hologram and 95 percent for a reflection hologram.

Other Types of Sensitive Material

Silver halides are not the only materials you can use to make a hologram. In fact, the granular nature of a silver image is a serious disadvantage, as the resolution of the image—indeed, the ability to form an image at all—depends on the fineness of the developed grains, which have to be smaller in mean diameter than half the distance between the fringes (the Nyquist condition again). This means less than 1 μm for a transmission hologram and less than 0.2 μm for a reflection hologram.

Diffraction efficiency is intensity of image beam divided by intensity of (laser) replay beam, expressed as a percentage.

And since a developed grain is something like 10 times the size of a single silver halide crystal, the original crystals need to have a mean diameter of no more than about 20 nm. This makes the emulsion very inefficient, with an ISO speed of less than 1 and a tendency to reciprocity failure at the long exposures necessary when using a continuous-wave (CW) laser. However, it is difficult to find any light-sensitive material that comes anywhere near the photon-capturing efficiency of silver halides, and much research has gone into hunting for such materials. On the plus side, though, the materials we do have are all grainless. Some of them are used commercially, though necessarily with more powerful lasers than for silver halide materials.

Dichromated Gelatin (DCG)

Most colloidal substances, when treated with a dichromate, become insoluble in water when exposed to short wave light. The old "gum-bichromate" process in photography is an example.

The "gum" was gum arabic or gum tragacanth, and "bichromate" is the old term for dichromate.

It is possible to sensitize dichromated gelatin to red light, just as it is with silver halides. After exposure, the unexposed regions of the fringes are washed away with warm water, and the remaining gelatin is coagulated by treatment with methylated spirit. DCG holograms are very bright, but are fragile and have to be protected by a cover glass.

Photopolymer

Polymers are long-chain molecules produced when certain organic substances, called monomers (e.g., ethylene), are subjected to some form of energy in the presence of a catalyst. In the case of photopolymers, the energy is supplied by exposure to light. Photopolymer holograms are usually fixed by a further exposure to light of a different wavelength. Like DCG, photopolymers produce very bright results, but the image is much more robust than with DCG.

Photoresists

These are materials that become either insoluble (negative) or soluble (positive) in certain solvents when exposed to light. Their main use is in the production of embossed holograms, as they produce fringes that consist of ridges raised above the surface of the substrate, and these can be duplicated mechanically.

Other Light-Sensitive Substances

Certain substances, such as the dyes methylene blue and bacteriorhodopsin, can record holographic images, and can also be erased by chemical or further light action. They have been used where holographic images need to be produced and erased repeatedly.

Spatial Light Modulators (SLMs)

At present, CCD and CMOS pixels are still too large to produce holographic images unless the angle between the object and reference beams is reduced to only a few degrees. Nevertheless, they can be employed for in-line (Gabor) holograms, which are sometimes used in scientific and industrial research. It seems unlikely that SLMs will be practicable for conventional holograms in the foreseeable future—which is a pity, as their speed is comparable with that of a high-speed photographic emulsion.

A spatial light modulator (SLM), as noted earlier, is what we used to call a liquid crystal display (LCD). "Liquid crystal" is a poor name for these substances, and new developments in SLMs do not in fact employ them.

Nils Abramson (b.1931) was formerly Professor of Industrial Metrology at the Royal Institute of Technology, Stockholm. He has been responsible for many advances in industrial holography, and was the first person to capture an image of a light pulse in flight.

The Real Image

Viewing a holographic image is rather like looking at the object through a window. Indeed, Nils Abramson once described a hologram somewhat poetically as "a window with a memory."

Because a hologram replicates the entire wavefront reflected from the object, as you change your viewpoint your eye intercepts different parts of the wavefront, and you see the image from the different points of view. Each point on the hologram codes the view of the object from that point, so the larger the hologram and the nearer it is to the object, the greater the angle of parallax (Figure 16.4). The parallax is both horizontal and vertical: if you move up or down, you see more of the upper or lower surfaces of the object.

If you followed the mathematical analysis in Appendix 2, you will see that in addition to the virtual image there is a spurious real image. This is what gave Gabor his trouble. With his in-line setup it was directly in front of the virtual image he required. By having the reference beam off-axis you swing the real image out of the

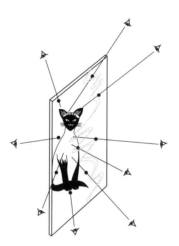

Figure 16.4 A hologram "freezes" the entire wavefront reflected from the object. Each point on the emulsion encodes a different viewpoint.

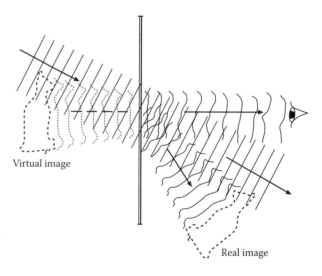

Virtual image

Real image

Figure 16.5 The spurious real image. It is the (–1) diffraction order. It is in fact largely suppressed, as it does not meet the requirements of the Bragg condition.

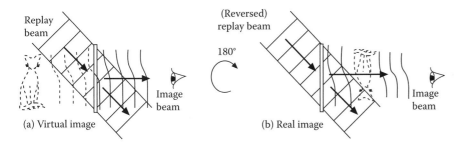

Figure 16.6 Flipping the hologram so that the direction of the replay beam is reversed, reverses the phase of the wavefront and creates a real image in the same position as the virtual image. It is pseudoscopic, i.e., has reversed parallax.

way: it moves to the other side of the reference beam (Figure 16.5). It is usually difficult to see, because it is at a steep angle; if the angle of incidence of the reference beam is more than 45° it won't be there at all. It is weak, too, because this image beam doesn't satisfy the Bragg condition.

However, if you flip the hologram, so that the replay beam is effectively from the reverse direction, you reverse the phase of all the wavefronts coded in the hologram, and the image beam will converge to form a real image on your side of the hologram (Figure 16.6) As the Bragg condition *is* satisfied for this beam, the image will be bright and clear. It is also inside out or *pseudoscopic*. For example, if you move to the right, instead of seeing more of the right side of the image you see more of its left side, and if you move upwards you see more of its underside. This makes the image of, say, a table tennis ball look like the inside of an eggcup, and vice versa.

If you position a ground glass screen (or tracing paper) in the image space, you will see the image projected on it, and if you move the screen in and out you will bring different planes of the image into focus. As Nils Abramson might have said, a reversed hologram is a lens with a memory. You can show that each point on the hologram codes a particular viewpoint and contains the whole image from that viewpoint, by illuminating the reversed hologram with an unspread laser beam, and watching the screen as you move the beam around: all the different viewpoints appear in turn.

If you look at the pseudoscopic image of a holographic portrait you get the full treatment of the visual illusion I mentioned in Chapter 3. The head seems to swing violently forward and back as you move your point of view.

Transfer Holograms

You can generate this pseudoscopic image with either a transmission or a reflection hologram. The exciting thing is that you can make a second hologram using this image as the object. You simply illuminate the (reversed) primary or *master hologram* with a laser replay beam, so that it forms the real image straddling the second or *transfer hologram*. This, of course, needs its own reference beam derived from the same laser. Figure 16.7 shows a typical table layout. As the image is pseudoscopic, you have to have the transfer reference beam coming from "below" (the setup is usually on its side, for simplicity), because you will have to flip the finished hologram to undo the pseudoscopic effect. Made in this manner, the image will straddle the plane of the hologram.

The transfer principle has a number of advantages:

- You can make as many copies as you like, long after the original set-up has been dismantled.
- An image-plane reflection hologram will replay as a sharp image even when the replay light is not a very good point source (i.e., has poor spatial coherence).

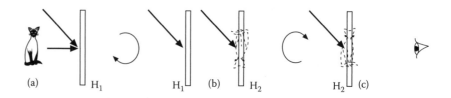

Figure 16.7 Principle of the transfer hologram. (a) Master hologram (H_1). (b) H_1 is flipped, and the real image is projected across the plane of a second film to act as object for the second, transfer hologram (H_2). (c) After processing, H_2 is flipped, and the image is now orthoscopic and across the plane of the hologram.

- You can make a good reflection transfer hologram from a transmission master hologram (it isn't easy to make a reflection transfer from a reflection master).
- By increasing the distance between the master and transfer hologram so that the image lies between them, you can produce a final image that is wholly in front of the hologram.
- You can view an image-plane transfer transmission hologram by white light, provided the image is shallow.
- By using a converging or diverging transfer beam you can reduce or enlarge the image by a limited amount.

Contact Copies

You can make a replica of any hologram by contact copying, using a single-beam transfer process. You simply clamp the master and transfer films in close contact and aim the replay/reference beam at the sandwich. For a transmission copy the transfer film goes on the side away from the laser; for a reflection copy it goes on the same side. Figure 16.8 shows the arrangement for both types of copy.

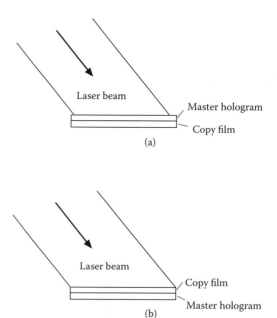

Figure 16.8 Making a copy hologram: (a) transmission; (b) reflection.

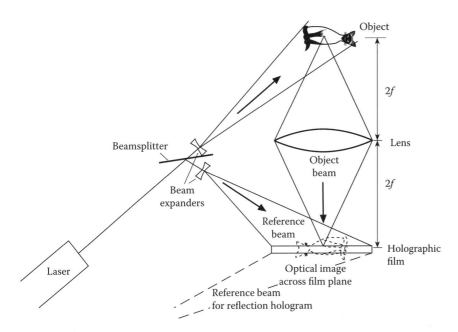

Figure 16.9 Layout for a focused-image transmission hologram. For a reflection hologram the reference beam is from below left (broken lines).

Focused-Image Holograms

A reversed master hologram produces a focused image. As I said earlier, you can think of the master hologram as a kind of lens, though a somewhat peculiar one. Now, a conventional lens can produce a real image, as we saw earlier. We can use this real image as the "object" for a hologram. And if we arrange this image to straddle the plane of the film, we can produce an image-plane hologram in one step. A notional setup is shown in Figure 16.9. The image is inverted, but it is orthoscopic, so it doesn't need flipping. The lens needs to have a diameter at least as big as the diagonal of the proposed hologram, and such lenses are very expensive. The Fresnel lens out of an overhead projector makes a good substitute, if you can avoid stray reflections from the lands.

Rainbow Holograms

Rainbow holograms (also called Benton holograms, after their inventor) are transmission transfer holograms that can be replayed using white light, even though the original subject matter may have been quite deep.

The technique relies on the fact that when you view the image in a transfer hologram you are viewing it through the real image of the frame of the master hologram. Benton masked his master hologram down to a narrow horizontal slit. This meant that a viewer of the transfer image would seem to be viewing the image as though through a narrow letterbox. With such an arrangement you would get full horizontal parallax, but as soon as you moved your viewpoint up or down the image would disappear. That is, if you were replaying the hologram with a laser beam. If, on the other hand, you were to replay such a hologram using a white spotlight, the result would be quite different. In this case, as red light is diffracted more than blue, you get a whole vertical spread of "letterboxes" going right through the spectrum (Figure 16.10). So as you move your viewpoint upwards from below you see the image in pure colors that range from violet to deep red—hence the name "rainbow hologram." Because

Stephen Benton (1942–2003) was Head of the Spatial Imaging Group at the Media Research Laboratories, Massachusetts Institute of Technology, until his untimely death. His research was directed chiefly at realizing holographic video, and his invention of the rainbow hologram was an early by-product of this research.

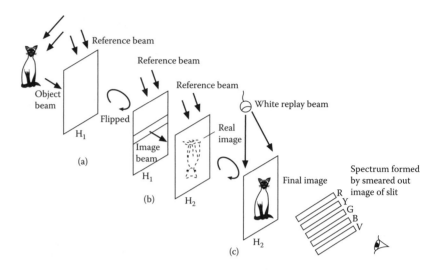

Figure 16.10 Making a rainbow hologram. (a) Conventional master hologram H_1 flipped and masked down to a horizontal slit. (b) Image-plane transfer hologram H_2 made. (c) When illuminated by a white spotlight, images of the slit form a spectrum, through which the viewer sees the final image in one or other spectral hue depending on vertical position.

you are viewing the image by only one wavelength at a time there is no color blur, and the image can be very deep.

Rainbow holograms can be very beautiful aesthetically, and offer a wide scope for creativity in terms of color and depth, as they permit the creation of multiple images in different planes and in different hues. Some practitioners have produced images of lasting artistic value using this process. Rainbow images are also the basis of all embossed security holograms.

Pulse Laser Holograms

Most lasers used for making holograms emit a continuous beam of from 3 to 35 mW for amateur work, depending on the image size, and up to 5 W for large professional holograms. One of the major headaches for holographers is the need to keep the subject matter stable to within a tenth of a wavelength during an exposure that may be anything from a few seconds to several minutes. This rules out portraiture and most live subjects, even flowers, with continuous-wave (CW) lasers.

The solution is to use a pulse laser. This is a laser that emits its light in a giant flash lasting only about 25 nanoseconds (ns). The total light energy in the flash is only a few joules, about the same as a small studio flash on half power, but the actual power, the rate of energy emission, runs into megawatts, so for safety's sake the subject illumination has to be well diffused. A holographic portrait studio is much like a photographic studio, except that all the "floods" are reflectors, because the light has to come from a single source. Until recently, all pulse lasers used an artificial ruby rod as lasing material, and the wavelength was 694 nm, very near the infrared. Complexions thus looked very waxy without full makeup. The introduction of green pulse lasers has improved matters greatly, and the recent advent of RGB pulse laser combinations is making full-color holographic portraiture a practical possibility.

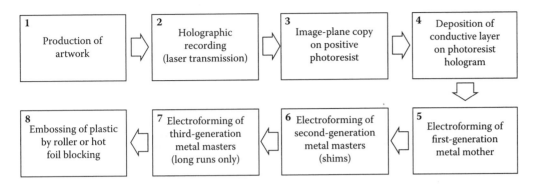

Figure 16.11 The production of an embossed hologram. All embossed holograms are basically rainbow holograms.

Embossed Holograms

The little shiny holograms you see on credit cards, banknotes, and other items as a security device are rainbow holograms backed by a mirror. They are produced mechanically by a process resembling that used for producing CDs. The master hologram is made in the usual way, but the transfer hologram (often containing more than one image) is made on positive photoresist (see above) instead of a photographic emulsion. The interference pattern thus consists of surface ridges (Figure 16.11). This intermediate master is then nickel-plated. This nickel "mother" is stripped and replicated in hard nickel, which can then be fixed to a hot roller and used to stamp out copies on thin plastics material with an aluminum foil backing. This in turn is backed with heat-sensitive adhesive, and can then be flush-mounted on documents or cards.

Holographic Stereograms

You probably have seen holograms containing two or more images, which switch suddenly from one to the other as you shift your viewpoint. There is no magic about making this type of hologram; it simply involves transferring two or more master holograms side by side in the same holder. As explained earlier, when you view a transfer hologram you are looking at it through the real image of the master hologram's aperture. So when you move across the field of view you see the first image through the image of the aperture of the first master, then the second image through the aperture image of the second master. If these two aperture images are close to the plane of your eyes, the transition is sudden. By making a series of master holograms and setting them up trimmed down to narrow vertical strips and butted together, then transferring the whole set in one exposure, you can produce a final image with a somewhat jerky sort of animation.

An extension of this idea is the making of a series of master holograms only a millimeter or so wide, on a single film masked with a slit, moving one slit width for each exposure. The individual holograms have no horizontal parallax (they are effectively vertical letterboxes), so the "objects" don't need to be three-dimensional. Usually they are prints made from photographs taken by a camera running along a rail, or from frames made by a movie camera. Drawn cartoons and computer animations can also be used. Figure 16.12 summarizes the whole process.

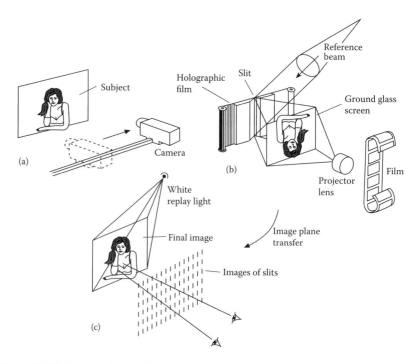

Figure 16.12 Process for making a holographic stereogram. (a) A series of photographs is taken by a camera moving along a rail. (b) The photographic images are used as objects to make adjacent narrow holograms on a film masked by a slit. The film moves one slit width between exposures. (c) The resultant intermediate hologram is transferred so that the image is across the plane of the final hologram and the image of the slits is in the plane of the viewer's eyes and is thus invisible. The initial image may also be a cine clip or a computer animation.

Holograms in Natural Colors

There are two ways of producing holograms in full natural color: three-color holography and holographic stereograms.

Three-Color Holography

If you use three lasers—one red, one green, and one blue—you can make a reflection hologram containing three independent sets of Bragg planes, replaying with white light to give an additive color image. The triangle enclosing the three wavelengths, as plotted on the CIE chromaticity diagram, needs to enclose as large an area of it as possible, in order to produce good, saturated colors. A good combination is a krypton-ion laser at 647 nm or a helium-neon laser at 633 nm (red), a solid-state laser at 532 nm (green), and a helium-cadmium laser at 442 nm (blue-violet). This combination produces good results for pastel or desaturated colors (Figure 16.13, color plate), but some saturated colors are inevitably somewhat falsified in hue. Recent research has shown that this falsification can amount to as much as 30 percent in the wavelength gaps, and that the addition of a fourth (yellow) laser beam reduces errors to no more than 5 percent, virtually undetectable by eye.

For example, using an RGB laser, a pure yellow object that reflected neither red nor green light would appear black. Fortunately, few natural colors are pure.

Color Holographic Stereograms

This is a way to achieve an image in true color without any falsifications in hue or saturation. Theoretically, you should get the best color reproduction when the

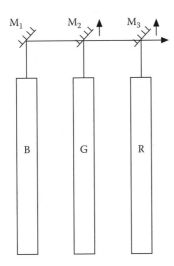

Figure 16.13 Making a natural-color hologram using three lasers. The mirrors M_2 and M_3 are removed in turn for three successive exposures.

taking filters cover the whole spectrum without gaps and with minimum overlap, and the printing filters are as narrow-band as possible. This method approaches that ideal. The original is a run of color transparencies, from which you need to make a set of red-, green-, and blue-separation positives in black and white. You now need to make a holographic stereogram master from each, after which you make a transfer rainbow hologram from all three, simultaneously. They need to be set up so that the "red" separation views as red, the "green" as green, and the "blue" as blue, all in register—not an easy task (most modern equipment uses computer control). But this process does give exceptionally good color rendering, although the viewing has to be from a very specific height or the hues will be incorrect. Figure 16.14 is a schematic diagram of the set-up.

Holographic Interferometry

All holograms are interferograms, of course: their images are produced by interference. It was noticed fairly early in their history that any very slight movement of the object during the exposure produced an image overlaid with stripes. This turned out to be a moiré pattern generated by the two sets of primary interference

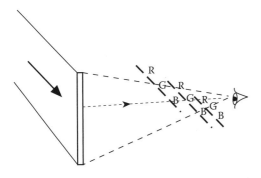

Figure 16.14 Natural-color rainbow stereogram. The original color photographs have black-and-white separations made, and these have to be aligned so that the spectra overlap in correct register (in practice the "windows" are close together).

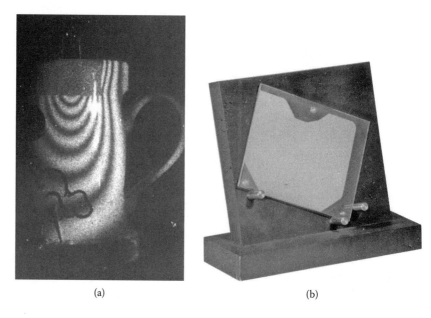

(a)   (b)

Figure 16.15 Real-time holographic interferogram. (a) The cup is positioned precisely in its own holographic image space. The fringes contour the distortion, here caused by a bulldog clip clamped to the rim. (b) The self-locating plateholder designed by Nils Abramson. (From Saxby G. *Practical Holography*, 3rd ed., CRC Press, Florida, 2002. With Permission.)

fringes from the two positions of the object. The first person to take these patterns seriously was Karl Stetson, who in 1965, with R. L. Powell, published a paper on the analysis of vibration via these secondary fringes. Holographic interferometry has now become an important tool in industrial and scientific research. The principle is simple: if an object is slightly distorted between two holographic exposures, the distortion will be contoured by dark and light fringes each contouring one half-wavelength of distortion. There are three basic kinds of technique.

Real-Time

You make a hologram of your test piece, using a holographic plate and a very accurate holder, and after processing you position the plate back in the holder. The test piece now coincides with its image, and any force on it will produce a strain that shows in the form of secondary fringes (Figure 16.15).

Double-Exposure

Here you make one exposure before applying the stress and another one after applying it, on the same plate or film. The strain is contoured by fringes (Figure 16.16).

By employing separate plates for the initial exposure and for several subsequent exposures, and using the special plateholder, you can examine the effect of progressive stressing by examining the plates in pairs. This method, called "sandwich holography," also enables you to see whether the distortions are inwards or outwards, in cases of doubt. To record dynamic stress (e.g., a blow on a crash helmet) a double-pulse laser is employed, with the two pulses a few milliseconds apart, timed just before and just after the blow.

The sandwich hologram was another of Nils Abramson's inspirations.

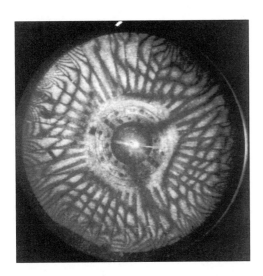

Figure 16.16 Double-exposure holographic interferogram of a Rolls-Royce RB211 fan rotating at speed, showing vibration patterns. This image was made using a modified 35-mm camera to make a focused-image hologram. The image was stabilized by an Abbe inverting prism rotating in the opposite direction to the fan. (From Saxby G. *Practical Holography*, 3rd ed., CRC Press, Florida, 2002. With Permission.)

Time-Average

When a component is in a state of stable vibration, it is momentarily stationary at the two extremes, and these two positions provide the secondary interference pattern. This lends itself to Denisyuk techniques, and the contour fringes are approximately one quarter-wavelength apart. The bright patches indicate nodes, where there is no movement. This was the system originally investigated by Stetson and Powell (Figure 16.17).

Holographic Optical Elements

I said earlier that you could think of a master hologram setup for making a transfer hologram as a kind of lens. This isn't just being whimsical. If you make a hologram of a single bright point, using a collimated reference beam, you can use a reversed replay beam to recreate that point. This hologram is indeed a lens in every sense, and the point image is at its principal focus. What is more, the hologram obeys all

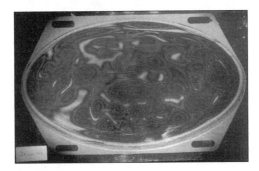

Figure 16.17 Time-average holographic interferogram of a loudspeaker fed with a pure sine wave at 5 kHz. The bright areas are the stationary modes; the antinodes are contoured by fringes. (From Saxby G. *Practical Holography*, 3rd ed., CRC Press, Florida, 2002. With Permission.)

247

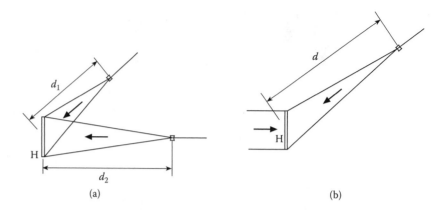

Figure 16.18 Setups for making (a) a holographic lens of focal length $d_1 d_2/(d_2 - d_1)$, (b) an optical mirror of focal length d.

the lens laws, and it is, truly, that otherwise mythical object, a "thin" lens. If you go through a similar process with a reflection hologram set-up, you will produce a holographic mirror that will focus light just like the mirror of a telescope. The layouts for making these two types of hologram are shown in Figure 16.18.

These holograms are called *holographic optical elements (HOEs)*, and they have brought about something of a revolution in optical equipment. Not only can you copy any shape of lens or mirror, but you can do it on a flat plate (or a curved one, if necessary). An equally important characteristic of HOEs is that you can incorporate more or less as many elements as you like into the same space, something quite out of the question with refractive devices. To take a simple example, suppose you make a hologram of three points at different distances and different angles. The resultant HOE is a holographic lens occupying a single space, but possessing three separate independent foci. No glass lens can do this. The new technology of optical computing relies heavily on HOE technology. So do head-up displays in aircraft, simulators, virtual reality sets, and more worldly items such as elevated rear lights for cars and focusing screen brighteners for studio cameras.

The main disadvantage of the HOE is its high dispersion, which means that it can often only be used for applications using lasers (such as bar code scanners), but in many applications the dispersion is actually useful, as in diffraction gratings. Because the dispersion characteristics are in the opposite sense to those of optical glass, an HOE can also be used as one element in an achromatic doublet.

As the interference patterns in HOEs are comparatively simple, modern micro-engraving techniques can copy them mechanically from computer programs. Such *diffractive optical elements (DOEs)* (this name now often subsumes HOEs) are more robust than HOEs, and do not deteriorate or require sealing as do silver halide or dichromate holograms. DOEs engraved on metal masters can be used to mass-produce copies in the same manner as embossed holograms. The originals can also be produced on photoresist using microlithographic techniques.

Computer-Generated Holograms

The interference pattern generated by a point source and a collimated reference beam is, in fact, the pattern of a zone plate. The intensity profile over the light and dark bands, though, is sinusoidal, not rectangular as in a Fresnel zone plate. It is termed a *Gabor zone plate*, after Denis Gabor, the father of holography.

When you make a hologram, you are actually making a Gabor zone plate for every point on the object, so when you shine the conjugate beam back through the hologram, you are focusing the light at every point that existed on the surface of the original object.

This is an alternative and valid way to describe the formation of a holographic image. I think the wavefront approach I have used is easier to grasp, though, particularly with complicated types of holograms.

ZONE PLATES

The simplest kind of DOE is the *Fresnel zone plate*. This is a transparent plate on which are drawn concentric circular opaque bands alternating with clear bands. The ratios of successive diameters of the circles are in proportion to the square roots of 1, 2, 3, 4, etc. (Figure 16.19).

When a collimated monochromatic light beam is incident perpendicularly on a zone plate, the clear rings all provide constructive interference at a point on the axis a distance r^2/λ from the plate, where r is the outside diameter of the innermost disc (it doesn't matter whether this disc is clear or opaque). The zone plate thus acts like a lens of focal length r^2/λ.

It can be shown that by engraving the zones with a slope, like a Fresnel lens, the opaque portion can be dispensed with, and the diffraction efficiency raised to nearly 100 percent.

As an example, with light of 500 nm wavelength, a zone plate with a central disc of 0.2-mm diameter would have a focal length of 80 mm. Such a device is well within modern micromachining capabilities.

Now, we can generate a zone plate from a single source by applying some simple calculations, so we should be able to generate other zone plates in the same way. We can indeed do this for a small number of points, but to do it for a whole continuous three-dimensional object is too complex a calculation for even a large computer. So we tackle the problem in a different way.

The Fourier-Transform Hologram

If you position an object at the front focal plane of a lens and illuminate it with coherent light, then hold a screen at the rear focal plane, you will see a diffraction pattern—a very small one for a large object, and a much bigger one for a very small object. It represents the *optical Fourier transform* of the object wavefront.

Fourier transforms are explained in more detail in Appendix 3.

If you set up a similar lens system on the far side of the pattern (and remove the screen), you will see an image of the object, the same size, but inverted. If you add a reference beam and make a hologram of the optical Fourier transform, you can use the hologram to reconstruct the image, via the second optical system.

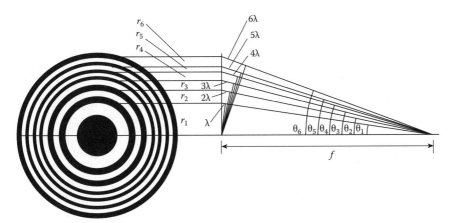

Figure 16.19 Principle of the Fresnel zone plate.

Now, the interference pattern of an ordinary hologram is too complicated for a computer to calculate (except for very simple objects), but a Fourier transform is a matter of straightforward maths. You can calculate the Fourier transform of a simple function such as a single square pulse on the back of an envelope. For a more complicated, three-dimensional object with shading you do need a computer, though. But the appropriate software will not only calculate the Fourier transform, it can also draw the interference pattern that characterizes the hologram, and feed the result into an electron-beam microlithography setup or a spatial light modulator. You can use the result as a hologram to create a three-dimensional image of an object that never existed, such as a car body design, a bridge, or a proposed new city center. The ability to achieve such a result is well within the reach of modern computers.

Dot Matrix Images

A dot matrix image is much like a halftone illustration in principle, being constructed of tiny dots small enough to be below the resolution of the eye. However, each dot is a tiny hologram, produced by two finely focused beams from a laser under computer control. The angle between the beams dictates the color of the dot as viewed, and this angle is varied under the control of the computer program to produce a colored image by diffraction. The images of the dots can be recorded on film, or on photoresist for mass production, e.g., for labels or gift-wrapping paper.

Digital Holography

A spatial light modulator (SLM) is basically a high-resolution version of an LCD.

Digital holography is a term that is much misused, being applied variously to computer-generated images, dot matrix holograms, and computer-controlled holographic printing. Strictly, it applies only to holograms that are generated and analyzed digitally, such as those making use of SLMs as recording media. Digital holography is severely limited in scope at present by the minimum pixel size of an SLM, which limits the imaging techniques to in-line and small-angle systems, but the ability of modern computer programs to manipulate Fourier transforms makes it possible to remove unwanted and spurious information from an in-line image. Once a holographic image has been recorded digitally, computer analysis of the recording can, in theory, yield an immense amount of information applicable particularly to microscopy and biological imaging, and digital holography is thus the subject of continuing research.

Digging Deeper

Literature on holography is limited to a fairly small selection of books, which cover a wide spectrum from the near frivolous to the inscrutable. If you want the full theory with no holds barred, the best book is P. Hariharan's *Optical Holography* (Cambridge University Press, 1996). The math isn't too difficult provided you can cope with complex numbers and exponentials. The most approachable text on applied holography is Nils Abramson's *The Making and Evaluation of Holograms* (Academic Press, 1981, temporarily out of print). If you want to try making your own holograms, there are several instructional manuals. Fred Unterseher has collaborated with Bob Schlesinger and Jeannine Hansen to produce *Holography Handbook* (revised edition, Ross Books, 1996), which is friendly and amusing, though some of the setups are less than ideal. If you want to build your own holographic studio/lab in your garage, Don McNair shows you how in *How to Make Holograms* (Tab Books, 1983, now out of print), which is excellent on concrete mixing, less so on making good holograms. My own *Practical Holography* (3rd edition, CRC Press, 2002) explains how to make every type of hologram in detail, with a full account of the principles of holography and a discussion of its applications. An earlier book of mine, *Manual of Practical Holography* (Focal Press, 1991, now out of print) is aimed at the less ambitious enthusiast. The most recent book on development in digital holography is *Digital Holography and Three-Dimensional Display* (Ed. T.-C. Poon, Springer, 2006), but be prepared for some pretty arcane maths. If you want to keep up with the latest developments, you will need to get hold of conference proceedings. The best collections come from the SPIE, which holds a conference on photonics every year, usually including a session on practical holography. The best way to get copies of these proceedings is via the Internet at www.spie.org. The mailing address is the International Society for Optical Engineering, P.O. Box 10, Bellingham, WA 98227, USA. The proceedings of the triennial Symposium on Display Holography at Lake Forest College, Illinois, are a goldmine. The first four were published by the College; you can get copies from Integraf, P.O. Box 586, 745 North Waukegan Road, Lake Forest, IL 60045, USA. The last symposium to be held at Lake Forest was in 1997, and since then these symposia have been held in a different country every third year, the most recent (at the time of writing) in China. Proceedings are published by the SPIE under its own volume numbers. You can also find much information on the Internet, but the Wikipedia entries on holography are not 100 percent reliable.

Chapter 17 Astronomical Imaging

Early History

It is far from clear who actually invented the telescope. At least four people seem to have done so independently, near the beginning of the seventeenth century. The first patent was granted in Holland to a German-born spectacle maker called Hans Lipperhey, though it seems that the possibility of using two lenses to obtain a magnified image of a distant object had been noticed much earlier. Like so many other groundbreaking scientific advances, it just seems that the time was right. What is certain is that the first person to turn a telescope to the heavens and make fundamental discoveries that confirmed Copernicus's heliocentric theories was Galileo. His telescope, apparently designed in ignorance of Lipperhey's methods, was to use a long-focus positive lens coupled with a short-focus negative lens, in a configuration that is still used in low-power opera glasses (Figure 17.1). He managed to achieve a magnification of 30 times. With this instrument he was able to observe the moons of Jupiter and the rings of Saturn.

It is difficult to produce a high magnification with the Galilean setup, as this is obtained by dividing the long focal length by the short one. A few years later Kepler designed an improved telescope using a convex lens as an eyepiece to examine the real image produced by the primary objective (Figure 17.2). This not only simplified the construction (it was difficult to fabricate concave lenses with high curvature) but also shortened the instrument, as the magnification was now the product of the two focal lengths. The fact that the image was inverted was unimportant in the context of astronomical observation.

At the time it was difficult to make satisfactory transmission objectives because of the residual aberrations of simple lenses, and the first telescope to use mirrors was designed by the Scottish mathematician James Gregory (1638–1675), using a paraboloidal primary mirror and an ellipsoidal concave secondary mirror positioned behind the primary focal point and reflecting the light through a circular aperture in the primary mirror to an eyepiece behind it (Figure 17.3).

His instrument (and possibly his eyesight) was not good enough to fully resolve the rings. It was Domenico Cassini (1625–1712) who identified these, along with the prominent gap in the rings that bears his name.

Johannes Kepler (1571–1630) was born in Württemberg, but produced his best work in Prague, working under the Danish astronomer Tycho Brahe, and inheriting the position of Imperial Mathematician on the death of the latter. Using Tycho's meticulous data, he showed that planetary orbits were elliptical, with the Sun at one focus, and that their radius vectors swept out equal areas in equal times. He also discovered many of the laws of optics, established the year of Christ's birth, explained the cause of tides and coined the word "satellite."

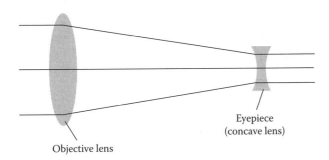

Objective lens

Eyepiece
(concave lens)

Figure 17.1 Optical arrangement of the Galilean telescope.

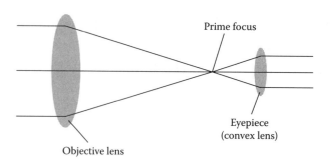

Figure 17.2 Optical arrangement of Kepler's telescope.

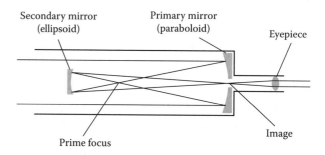

Figure 17.3 Optical arrangement of the Gregorian reflecting telescope.

The first working instrument based on Gregory's design was built by Sir Isaac Newton, who demonstrated it to the Royal Society in 1671. His version differed from Gregory's in that the secondary mirror was plane and inclined to the primary axis at 45° so that the light emerged from the side of the tube, from which the image was viewed by an eyepiece (Figure 17.4). The simplicity of the design made it popular, and it is still widely used in amateur telescopes.

Most large astronomical telescopes employ the Cassegrainian design (Figure 17.5). There are several versions of this system, differing only in detail. The *Ritchie-Chrétien* variant is differs from the basic Cassegrainian design only in that both mirrors are hyperboloidal. This arrangement minimizes coma and increases the useful field. An important variant of the Newtonian telescope is the *Coudé* system (Figure 17.6). This diverts the final part of the primary beam to a rotatable plane mirror, which means that the eyepiece can be kept in a fixed position irrespective of the telescope's orientation, simplifying the installation of spectrographs and other analytical equipment. Modern telescopes are usually of the Ritchie-Chrétien design,

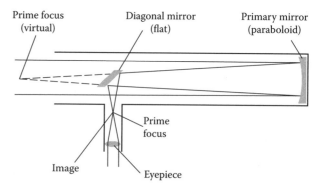

254

Figure 17.4 The Newtonian configuration.

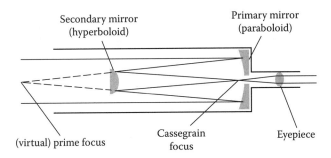

Figure 17.5 Optics of the basic Cassegrainian telescope.

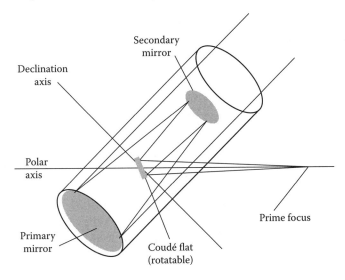

Figure 17.6 The Coudé modification to the Cassegrainian system, shown in an equatorially mounted system.

equipped with interchangeable secondary optical systems able to adapt the telescope to Newtonian, Cassegrainian, or Coudé geometry as needed.

The invention of the achromatic doublet in the mid-eighteenth century produced a considerable improvement in the optical qualities of refractor telescopes, and some large examples were built, culminating in 1895 in the 40-inch Yerkes telescope. Due to the weight, and the difficulty of producing such a large block of glass free from striations and bubbles, this represented a practical limit. This telescope, the largest refractor ever built, is still in use today. In the meantime reflector instruments were also increasing in size: the 100-inch (2.5 m) Hooker telescope at Mount Wilson was founded in 1904 and for many years its mirror remained the world's largest. Mirror technology continued to advance throughout the twentieth century, and today the largest monolithic mirrors are some 8 m in diameter, with apertures of around f/3.

The Schmidt Configuration

Mirrors do not suffer from chromatic aberration, and the use of paraboloidal and hyperboloidal surfaces for primary and secondary mirrors eliminates spherical aberration, though at the expense of introducing coma, which limits the useful field of view of large telescopes to a few square degrees. The Schmidt camera

Very little is known about Cassegrain (fl.1670), not even his first name. He appears to have been a professor at the College of Chartres. He submitted his paper describing his invention to the Academy of Sciences in Paris. About the same time he also submitted a paper on the acoustics of megaphones.

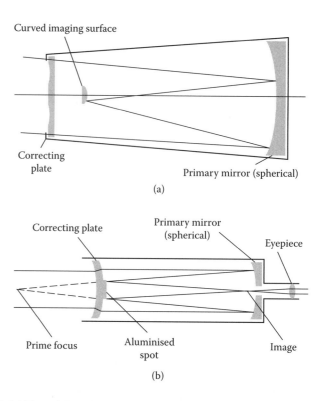

Figure 17.7 (a) The original Schmidt camera (with field flattener). (b) The Schmidt-Cassegrain catadioptric system, in a version designed by the Ukrainian astronomer Maksukov.

(Figure 17.7) was designed in 1930 to reduce coma and allow a wider angle for surveying purposes. It employed a spherical primary mirror, and in order to compensate for spherical aberration had a thin glass "corrector plate," thicker in the center and at the edges, mounted in front. Its field of view was about 15° at an aperture of $f/1.75$, and initially it had a strongly curved focal surface, which Schmidt later eliminated with a field flattening lens (Figure 17.7a).

Schmidt cameras are extensively used for sky survey, but as the prime focus is inside the tube and cannot be readily reached for visual examination of the image, several variations have been devised, in particular, a design that combines the Schmidt and Cassegrain principles (Figure 17.7b). This configuration has also found wide use as a catadioptric long-focus lens for general photography (see Chapter 5).

Mountings

Structural Supports

In modern telescopes the primary and secondary mirrors are held in their correct relative positions by what is still called the tube, though it is nowadays invariably an open structure. In large instruments it is not possible to avoid a certain amount of structural flexing; the important thing is that if either of the principal focusing elements do flex, they should stay in precisely the same spatial relationship. Most large telescopes are built using the *Serrurier truss*, which consists of struts arranged in triangular patterns (Figure 17.8).

Secondary mirror cell

Serrurier trusses

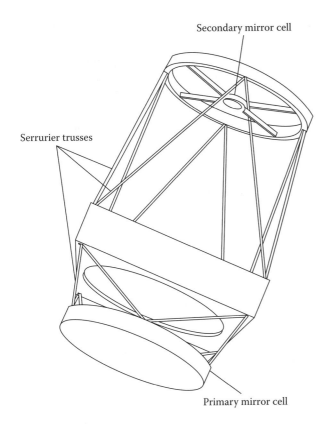

Primary mirror cell

Figure 17.8 A typical Serrurier truss mounting. (After Kitchen, C. R. *Astrophysical Techniques*, 5th ed., CRC Press, Florida, 2008.)

Steering Systems

There are two basic arrangements for pointing telescopes, both of which require a pair of orthogonal movements. In the *equatorial mounting* one of these movements, the *inclination axis*, is parallel to the Earth's rotational axis and the other, the *declination axis*, is perpendicular to it. Tracking is simple, requiring only a constant-velocity motor drive. The other arrangement is the *alt-azimuth mounting*. In this arrangement the two directions are vertical and horizontal with respect to the Earth's surface. This arrangement is simpler from a constructional point of view, but needs simultaneous adjustments in both directions with varying speeds; also, the image field rotates with the telescope's motion. However, now that computer control has become commonplace, almost all the telescopes built in the last 20 years use alt-azimuth mounting. Some telescopes cannot be moved at all, and employ a pair of plane mirrors, known as a *coelostat*, to direct the light. The first mirror is mounted on the polar axis and driven at half the sidereal rate, while the second mirror is mounted and driven so that it directs the light along the telescope axis, again under computer control.

Mirror Supports

The Rayleigh limit for surface imperfections in an optical mirror specifies a wavefront distortion of not more than $\lambda/4$. This implies that the mirror surface must be accurate to within $\lambda/8$. With a large mirror, its weight alone could cause a greater distortion than this, so the support system is important. Mirrors are nowadays usually made from a ceramic glass material called Zerodur,

There are few telescopes exactly on the Equator, where the alt-azimuth and equatorial mountings would be identical, but there has been an observatory at the South Pole since 1957. Apart from its being at an altitude of 2837 m, an astronomical telescope at this site would have a similar structural simplicity, at least in theory. Until recently there has been no such telescope. However, a new base, the Scott-Amundsen base, dedicated early in 2008, is now being equipped with a large instrument. Full information is available from the NSF Web site: www.nsf.gov.

257

The 8.2-m primary mirror of the ESO's Very Large Telescope has no fewer than 150 actuators to maintain its accurate shape.

which is very rigid and has an extremely small coefficient of thermal expansion. There are two approaches to weight reduction. One is to use a honeycomb construction for the glass backing; the other is to have a very thin mirror with numerous support points. Such a mirror may be monolithic, or it may be made up of individually adjustable hexagonal segments, the usual procedure for very large mirrors. All large mirrors, though, require some kind of active support structure in order to retain their correct surface shape no matter where they are pointing.

Atmospheric Effects

Ground-based observations face two problems. The first is the opacity of the atmosphere to much of the electromagnetic spectrum, in particular to the infrared (except for a few narrow "windows") and to almost all UV radiation with a wavelength of less than 320 nm. The second is atmospheric turbulence. Both these problems are avoided in space telescopes, but these have their own limitations (especially cost and weight). There have been a number of sorties into the upper atmosphere using aircraft and balloons, but the biggest telescopes remain doggedly earthbound. The difficulties are alleviated somewhat by locating them on the tops of high mountains, where they are above most of the more troublesome aspects of the atmosphere and away from major sources of light pollution. Even so, the effects of atmospheric turbulence still place serious limitations on their performance.

It can be shown that any telescope of more than 112 mm in diameter will have its images degraded by atmospheric turbulence.

One way of dealing with the turbulence problem is called "lucky imaging." A large number of short exposures are made, and those that are sharp are selected and combined. This method is capable of giving results close to the diffraction limit even for mirrors as large as 3 m diameter, but as something like 99 percent of the images have to be discarded, it is expensive in telescope time. This method has only become practicable in the digital era, as photographic emulsions designed for astronomy are far too slow for short exposures to register.

The second method, which needs to be built into the mirror support system, is to correct the distortions of the image in real time, using adaptive optics based on the servo principle.

Don't confuse "active" with "adaptive" optics. Although both are closed-loop systems depending on feedback, active optics operate on a timescale of seconds, whereas the adaptive system operates in fractions of a millisecond.

Sampling Systems

In order to provide the stimulus for the adaptive optics it is necessary to measure the distortion of the wavefront as it occurs. To do this a sensor has to monitor the light from a bright star close to the telescope axis, or, more commonly, an artificial star image produced by a powerful laser with a tightly collimated beam tuned to one of the sodium D-line frequencies. This is reflected by sodium atoms present in the upper atmosphere at about 90 km altitude, appearing as a starlike patch.

Sensor

The sensor commonly used consists of a mosaic of small lenses focused on a detector array. The degree of displacement and direction of the images from the optical centers of the detectors is used to generate the error signal, which is fed to the adaptive optics. These are piezo actuated, their operation being virtually instantaneous, and the results can come close to the diffraction limit even for very large objectives.

TELESCOPES IN SPACE

The most direct way of improving the resolution of a telescope is to put it above the atmosphere. We can get above more than half of it (i.e., the murkiest part) by siting the telescope at the top of a high mountain. Indeed, almost all of the world's highest accessible mountains are fairly fully occupied. We can go one better by putting the telescope into a high-flying aircraft or a balloon, and that has been done since the early 1960s. But we can do better still by putting the telescope into orbit. This carries the huge additional advantage that we have access to the entire electromagnetic spectrum. Both the mid-infrared and mid-ultraviolet are blocked by the Earth's atmosphere, and the orbiting Hubble Telescope can see both, from 150 nm to 2.5 μm, although it is designed to operate mainly in the visible spectrum. It is being supplanted by the more powerful James Webb Telescope, which is designed to cover the whole IR spectrum up to 10 μm and possibly beyond, with a 6.5-m beryllium mirror in 18 segments, folded for transport. (Beryllium is very light and hard and does not corrode easily.) An important difference from Hubble is that it operates from Lagrange Point L2 instead of being in Earth orbit, so it is effectively stationary. The Lagrange points are the five points where the gravitational pull of the Sun and the Earth balance out, and a body at any of these points could remain there indefinitely. However, two of these points are unstable. Point L2, the most stable, is in line with the Earth and Sun, and 1.5 million km farther out. Many small, specialized telescopes have already been launched on satellites, probing such areas as X- and γ-radiation from the Sun and other cosmic sources. Apart from the Sun, which is a major emitter of X-radiation, there are many X-ray sources in the cosmos, and there are numerous brief *gamma-ray bursts* from various points in the cosmos, which are thought to represent catastrophic stellar events. To tie these up with visible or radio sources it is necessary to find some way of focusing the radiation. The way this is done is explained in Chapter 19. One area that is still to be investigated is so-called *T-rays* (terahertz radiation).

Types of Detector

In the optical region, i.e., visible light plus infrared and ultraviolet radiation, there are two main groups of detectors: photon detectors and thermal detectors. The former detect the energy directly, whereas the latter monitor the increase in temperature through the absorption of energy, and are useful mainly for IR detection. The longest wavelengths are in the radio spectrum, and are detected by electronic means that are able to record phase as well as intensity. This is something that can be achieved with visible light only by means of interferometry.

Photon Detectors

Until the early nineteenth century the only photon detector available was the human eye, but from the 1840s photography gradually took over, with the advantage that long exposures on emulsions specially developed for the purpose allowed the image to build up, sometimes over several nights. Although the old visual and photographic records are still of incalculable value to the science of astronomy, as image detectors they have become totally outclassed by semiconductor sensors, and even photography is rarely used nowadays. Charge-coupled device (CCD) arrays were first applied to astronomical measurements in the 1970s. Their method of operation was described in Chapter 11. To recap, photons incident on a CCD produce electron-hole pairs, which are trapped in potential cells and accumulate until they are read out and recorded. The sensitivity of a CCD extends roughly from 400 to 1100 nm, with a peak in quantum efficiency at about 750 nm. If they are kept at liquid nitrogen temperatures, CCD pixels can continue to accumulate charge in terms of single photons for several hours. Various patterns have been evolved to reduce the dead space between pixels: the possible loss of information is minimized by dithering the image, i.e., shifting the optical image by a fraction of a pixel over a series of exposures. In terms of the time taken to access a given amount of information, a CCD array is between 20 and 50 times as fast as a photographic emulsion, and does not suffer from reciprocity failure.

Photomultipliers

A photomultiplier is a vacuum tube device containing a photocathode and a cascade of electrodes of progressively increasing positive potential, which provide considerable amplification. They are now little used for visible light, their main application being the detection of single photons where a very rapid response is required, and as detectors in the far IR region.

Superconducting Tunnel Junction Detectors

These depend for their operation on the behavior of the *Josephson junction*, a device that consists of two superconducting layers separated by a layer of dielectric material a few nanometers thick, through which electrons can tunnel, creating a current that can be controlled by a magnetic field. The device has great potential, as it responds to radiation of all wavelengths from IR to X-rays and has very high resolving power. At present its operation is confined to liquid helium temperatures.

Photoconductive Cells

Group 3 elements have three outermost electrons and Group 5 have five. Compounds of type 3/5 form crystals with alternate atoms, which behave in a similar manner to Group 4 semiconductors such as silicon.

These exhibit a change in conductivity when illuminated. They are used mainly for photon detection in the IR region, the detector material being determined by the frequency band it is intended to detect (mostly Group 3/5 compounds such as gallium arsenide and gallium nitride in various combinations with other elements).

Bolometers

These are devices that change their resistivity in response to heating by IR illumination. Each cell forms part of a Wheatstone bridge, which records varying degrees of imbalance depending on the amount of radiation absorbed. Bolometers are slow in action and are suitable only for the longer IR wavelengths.

UV Detectors

All the standard detectors operate in the UV band, and for this purpose are usually fitted with a filter excluding visible light, though in some instances fluorescent screens are added to increase the sensitivity in the far UV where this is necessary.

X-Ray and γ-Ray Detectors

These are described in Chapter 19.

Detectors for Radio

These are dealt with below, under the headings for the equipment concerned.

Solar Telescopes

This is one area where visual observation has not yet become obsolete, though it is necessary to take considerable precautions. Not only the eye, but also the telescope itself, can be at risk from overheating. If you want to use your own amateur telescope you must fit a filter designed specifically for solar observation, or project the focused image onto a white card. Never use a filter over the eyepiece, as it will overheat and may shatter. You should also restrict the telescope aperture to 75 mm or less. Professional solar telescopes use a simple lens as objective, to minimize internal reflections, with a restricted aperture and a monochromatic filter. They also have a facility for fitting a *coronagraph*, an occulting disc that simulates a total eclipse (Figure 17.9). Elaborate baffles are included to eliminate all internal reflections, and a mountain location (or a balloon) is mandatory.

If you want to look at the Sun yourself, the best option is to buy a hydrogen-α filter. This is a monochromatic interference filter centered on the hydrogen-α line (deep red), which will reveal the granular structure of the Sun's surface as well as the detail in sunspots.

Infrared and Terahertz Astronomy

Infrared (IR) Radiation

In order to make observations in the IR regions, it is necessary to be above most of the water vapor in the atmosphere. Many of the early investigations of the IR regions were carried out with telescopes carried by aircraft or balloons, but the advent of satellites has simplified the process.

Another serious problem is background radiation. Everything above absolute zero radiates electromagnetic energy. The sky is full of random IR radiation, and this has to be shielded out as far as possible. A further source is the equipment itself, so all IR detectors have to be cooled: in the near-IR (0.7–5 μm) liquid nitrogen (77K) suffices, but for the mid-IR (5–30 μm) liquid helium (4K) is needed, and for the far-IR, temperatures have to be below 1K. These temperatures are standard on space telescopes that operate in the IR region, such as the new James Webb telescope.

Standard reflector telescopes are perfectly suitable for IR observations, though refractors may need special IR-transmitting glasses, and even (for the far-IR) lenses made of germanium or silicon. The main detectors are photoconductors

Infrared windows. The Earth's atmosphere is largely opaque to IR radiation, which is strongly absorbed by water vapor. There are narrow "windows" at 1.25, 1.6, 2.2, 3.6, 5.0, and 21 μm, and a broader one between 8 and 11 μm. Most of the water vapor in the atmosphere exists below 10,000 m.

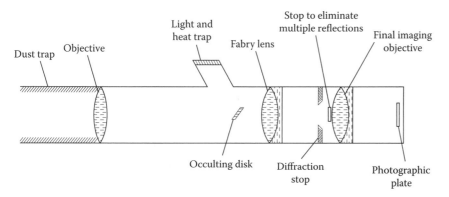

Figure 17.9 Typical layout of a coronagraph camera. The Fabry lens and stop is a device for eliminating diffraction by the edges of the objective. (After Kitchen, C. R. *Astrophysical Techniques*, 5th ed., CRC Press, Florida, 2008.)

for the near-IR and bolometers for the longer wavelengths, but for faint sources arrays of photomultipliers may be appropriate. The IR images of astronomical objects may differ considerably from their visible-light images, and can tell us a great deal about, for example, star formation, which begins at low temperatures. IR spectrographs are also important for objects that are so far distant that the red shift has moved the Fraunhofer lines beyond the visible spectrum into the IR.

Terahertz Radiation (T-Rays)

Terahertz radiation (known as "T-rays" by analogy with X-rays) is radiation with wavelengths lying between 1 mm and 100 µm, or frequencies of 300 GHz to 3 THz. Research in this region has been wanting, owing to the lack of any suitable receiving device, but new interest has been aroused by the recent development of devices that can produce T-radiation, by pushing the upper limits of oscillators and the lower limits of lasers, and by the need to design detectors (see Chapter 19). Space telescopes are already being designed to take advantage of the information offered by this once arcane region of the electromagnetic spectrum, though to date little has emerged.

Radio Telescopes

Radio emission begins where IR and T-radiation leave off, with a band of wavelengths going from a few millimeters to about 20 meters, the limit of atmospheric penetration (longer waves are blocked by the ionosphere). In some ways, a radio telescope is just an optical telescope writ large. It employs a paraboloidal reflector and obeys the usual laws of geometric optics and diffraction. As with optical telescopes, the reflecting surface has to be smooth to within one-eighth of a wavelength, but as we are talking of wavelengths that are of the order of a million times as great as those of visible light, we can get away with surface irregularities of millimeters or even centimeters. In order to have any kind of directionality, however, the dish needs to be large in comparison with the wavelength, which often means diameters of many meters.

The first radio dish to discover an astronomical source did so almost by chance. In 1931 the Bell Telephone Company commissioned Karl Jansky, one of their engineers, to construct a device that would identify sources of static that might interfere with their proposed radio telephone links. Jansky came up with a 30-meter dish that soon found a powerful radio source in the constellation Sagittarius, the direction of the center of the Galaxy.

With lessons learned from radar during the Second World War, radio astronomy began to take off in the 1940s and 1950s. The 76-m Lovell telescope at Jodrell Bank in Cheshire (Figure 17.10) was one of the first to be built. Soon afterwards an enormous bowl 305 m in diameter was excavated in a conveniently shaped depression just north of Arecibo in Puerto Rico, and lined with perforated aluminium plates. It is still the world's largest full dish. Being by its nature immovable, the receiving antenna has to be moved instead (it has some 40° of horizontal movement), so the bowl itself is spherical rather than paraboloidal, to avoid coma and keep other aberrations constant over all the receiver angles. Even larger dishes are planned, but, as we shall see, arrays of smaller dishes can give an equal amount of information, and are steerable. Some radio telescopes are mounted equatorially, but most are built on alt-azimuth mountings. The dishes are usually constructed of mesh, to reduce wind resistance.

Figure 17.10 The first radio telescope in Britain, at Jodrell Bank, near Manchester. I took this photograph in 1952, when the genre was still something of a novelty.

Receivers

The usual type of receiver is a simple dipole antenna situated at the focus of the paraboloid; for the shorter wavelengths a horn antenna is used. This is simply a tapered waveguide matched to the wavelength range of the incoming signal.

Arrays

With a single receiver the signal is necessarily one-dimensional, so it is allowed to track across the receiver with the motion of the Earth, and the telescope itself can be tracked in other directions to give a full picture of the source. The signal can also be analyzed spectroscopically, and Fourier methods can be applied to the image to deconvolve the aperture function and remove some of the noise. As the signal can be very faint and the noise obtrusive, it is usually necessary to make observations lasting many hours in order to be able to separate out the signal (which is organized) from the noise (which is random).

Instead of moving the dish (or the receiving antenna) a simpler method would be to have an array of antennas at the focus, receiving a two-dimensional image in the same way as the CCD array of a light telescope. But we can do better than that. Unlike detectors of light waves, detectors of radio signals can record the phase of a signal directly, so it is possible to link a number of smaller radio telescopes to receive a signal over a large area, and to synchronize the phase of the signal to obtain a full picture. The resolution is determined by the diameter of the array, so it can be very high indeed, often better than the equivalent optical signal. One of the first arrays was built in 1954 near Sydney, Australia, under the auspices of the Commonwealth Scientific and Industrial Organization (CSIRO), by Bernard Mills, and called after him a *Mills Cross*. Each arm of the cross was 450 m long and consisted of 250 half-wave antennas backed by mesh reflectors. The beam could be steered by adjusting the phasing of the elements in each arm. That observatory is now closed, but arrays on the Mills Cross principle have been assembled in a number of other countries.

As this mounting makes the receiver difficult to reach, a secondary reflector may be installed to place the receiver in a more convenient position, or the dish may be an off-axis segment (like the TV dish antennas installed on the sides of houses).

The aperture function is an Airy pattern, plus any diffraction patterns resulting from the spider mounting of the receiver. Deconvolution is simply the process of removing these distractions from the genuine information, i.e., the true image.

263

By linking the arrays in different countries together we can have a "dish" as wide as we like—in fact, as large as the Earth's diameter. The snag is that the records need to be synchronized to within a few nanoseconds, and the losses and delays in long lines and intermediate amplifiers would ruin the coherence of the recorded wavefront. In order to avoid this, the observations are recorded with precise timing from an atomic clock, and assembled into an image at a central point. In this way resolution well in excess of any optical image is possible (Figure 17.11).

Interferometry

There is another string to the radio astronomer's bow, namely *interferometry*, the making of measurements through the interference of electromagnetic waves. At present there is no optical telescope with sufficient resolving power to be able to show us the disc of any star. Even the supergiant star Betelgeuse (it forms the top left-hand corner of the constellation Orion) is beyond the resolving power of our best telescopes. But if we can't resolve the disc of Betelgeuse, how is it that we know its angular diameter? The answer is by means of interferometry. You came across the phenomena of interference and diffraction briefly in Chapter 1, and saw some applications in Chapter 16. Here it is used as a method of measurement. It depends on the fact that when two coherent wavefronts cross in space their amplitudes add up algebraically—that is, if the peaks and troughs of the waves coincide ("in phase") the resultant is a large amplitude, and if the peaks of one coincide with the troughs of the other and vice versa ("in antiphase"), they more or less cancel out. The method of measurement associated with interference, which became known as interferometry, was worked out by A. A. Michelson in 1920.

If a point source is imaged by two separated slits the combined image will appear as a band of alternate bright and dark fringes (bright where the two wavefronts are in phase, dark where they are in antiphase). Michelson's insight was to fit two slits to the optics of a large telescope. As the angular diameter of the star was so small the slits needed to be well separated, so he used a periscope arrangement to separate

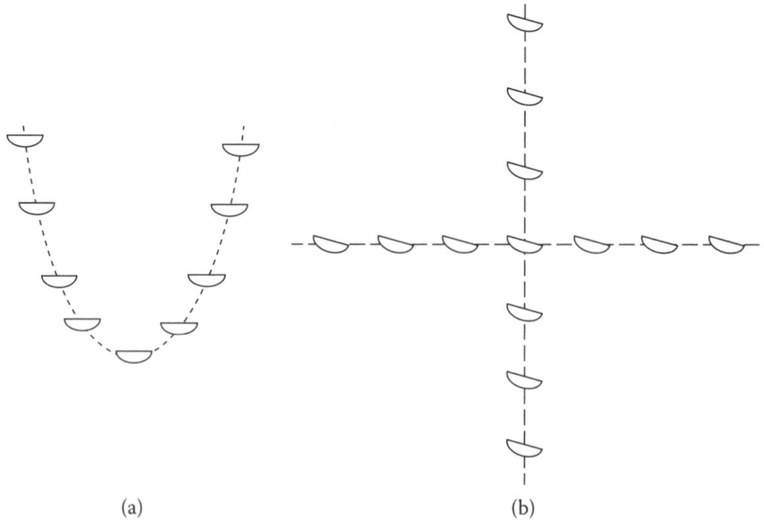

Albert Abraham Michelson (1852–1931) was the same Michelson who, along with Edward William Morley (1838–1923), had carried out the famous interferometric experiment of 1887, which confirmed the nonexistence of a "luminiferous ether." I used to reassure students having trouble with their optics experiments with the fact that at least one Nobel Prize for Physics had been handed out to a team for achieving a null result.

(a)   (b)

Figure 17.11 Two kinds of telescope array. (a) Dishes arranged to form a single parabola. (b) Mills cross layout; in this configuration all the dishes are steerable, and the received phases are matched by delay lines.

them beyond the diameter of the telescope objective. Michelson fitted the system (illustrated schematically in Figure 17.12) to the 100-inch Mt. Wilson telescope, and confirmed its worth by successfully measuring the diameter of Betelgeuse. The diameters of Antares and other giant stars were established soon afterwards, using the same method. Since then hundreds of stars have had their diameters measured interferometrically, and many binary systems identified.

How to Measure the Diameter of a Star

The largest of our telescopes, even using a finegrain photographic plate, still can't resolve the disc of any star, even a supergiant like Betelgeuse. But Michelson succeeded in measuring its diameter in 1920, using his stellar interferometer. The principle is shown in Figure 17.12. This is how it works:

If you illuminate a double slit with a monochromatic point source, and focus the image of the slits with a lens, you will see a series of interference fringes. Their spacing is in inverse proportion to the spacing of the slits, the formula being $\sin \theta = \lambda/d$, where d is the slit spacing, λ is the wavelength of the light, and θ is the angle (in radians) subtended by adjacent fringes at the focal plane. (For large distances and small angles we can use the approximation $\sin \theta \approx \theta$.) Now a binary star will produce a separate set of fringes for each star. As the slit separation is adjusted, there will be a spacing at which the two sets of fringes will overlap and exactly fill in the other set, so the fringes will vanish (or at least their contrast will fall to a minimum). This will happen when $\theta = \lambda/2d$, and this gives the angular separation of the binary pair. The same applies to a single star. Here you need to consider the star as being made up of two halves. In this case the fringes will disappear when $\theta = \lambda/d$. Michelson's stellar interferometer had a maximum spacing of 20 feet (6 m). He examined the image of Betelgeuse at the Cassegrain focus with a microscope, and from his measurements calculated the angular diameter of the star to be 0.047 of a second of arc. When this figure was slotted into the known distance of the star (from previous parallax measurements) its diameter was calculated to be 140 million miles (2.24×10^6 km), about the same size as the orbit of Mars. (For comparison, Venus at its largest subtends about 1 minute of arc, i.e., just about the resolution limit of a particularly acute human eye.)

With modern interferometric practice we can synchronize more widely spaced sources to obtain higher and higher image resolution. The largest stellar interferometer is the giant Keck telescope on Mt. Mauna Kea in Hawaii, which has two 10-m diameter mirrors spaced up to 85 m apart, giving a resolution of some 5 milliseconds of arc. Thousands of stellar diameters are now known. More than 700 exoplanets have been identified from the small drop in light intensity as they transit their stars. Most are Jupiter-sized, but the Kepler space telescope, which follows the Earth's orbit, has so far found more than 170 Earth-sized planets. Optical interferometry is difficult to carry out because it requires both wavefronts to be coherent, and with white light only one fringe is discernible, so the distances of the two receivers have to be less than a wavelength in optical path difference. This is much easier to achieve with radio telescopes because of the much longer wavelength, and because the wavefront phase can be recorded directly. Consequently interferometry can be carried out with almost any combination of radio telescopes—which is why we have so much information about radio sources and their structure.

To get a parallax measurement on a star image you have to take two readings of its position six months apart, when the Earth is at the extremes of its orbit. The distance is reckoned in parsecs (the reciprocals of seconds of parallax), 1 parsec being 3.26 light years (a light-year is approximately 9.5×10^6 km). There are 60 seconds to a minute of arc, and 60 minutes to a degree. These somewhat odd divisions, like our time divisions into hours, minutes, and seconds, are legacies of the ancient Babylonians, the earliest astronomers, who happened to count in alternate tens and sixes.

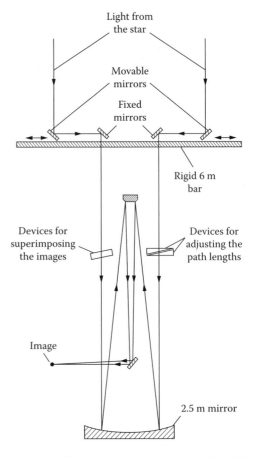

Figure 17.12 Principle of the Michelson stellar interferometer. (After Kitchen, C. R. *Astrophysical Techniques*, 5th ed., CRC Press, Florida, 2008.)

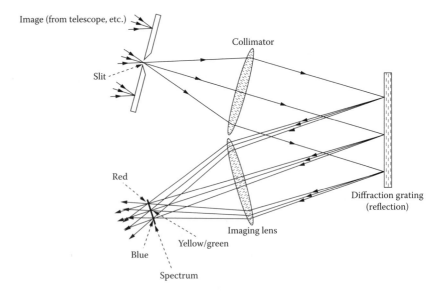

Figure 17.13 Schematic of an astronomical spectroscope. This layout, using a reflection grating, is the one most commonly used by astronomers. (After Kitchen, C. R. *Astrophysical Techniques*, 5th ed., CRC Press, Florida, 2008.)

Spectroscopy

Spectroscopy is a huge subject, and has filled many books and many astronomers' lives; it has dominated astronomy for more than a century, and has numerous applications in ground-based work. The main purpose of astronomical spectroscopy is to find out what the Universe is made of.

It was, of course, Newton who discovered the spectrum of sunlight using a prism, but it was Fraunhofer who mapped the dark lines in the spectrum that were given his name.

Fraunhofer discovered that the lines corresponded exactly to the bright lines produced when certain elements and compounds are heated in a Bunsen flame and examined through a spectroscope. It was from these lines that the gas helium was first identified in the Sun's atmosphere, before it was found on Earth. The Fraunhofer lines are difficult to see with a prism, as it is not easy to obtain high enough resolution. You need a diffraction grating, which gives greater and more uniform dispersion.

A type of spectrometer popular with astronomers is shown in Figure 17.13. The telescope is centered on the image of the star being examined, and the spectrometer is inserted in the optical path as shown.

Two important results emerge from a spectroscopic analysis. The first is the position and intensity of the various lines, which tell us about the chemical make-up of the object. The second emerges from the position of the whole range of lines. In general these will be shifted somewhat towards the red end of the spectrum as compared with those seen in the spectrum of the Sun, this "red shift" being greater the farther away the object. This is caused by the Doppler effect, and indicates that the source is retreating; the amount of red shift can be used to estimate the speed of this.

Apart from prism and diffraction grating spectrometers there are several other devices, which operate by scanning through the spectrum rather than displaying it as a whole, and give better resolution as a result. One such is the Fabry-Pérot etalon (Figure 17.14). This consists of two optically flat partially silvered parallel surfaces perpendicular to the slit. It operates by multiple reflections, which single

As this book is about imaging, most of spectrometry lies outside its remit, but in the context of astronomy it is too important to leave out altogether.

Joseph von Fraunhofer (1787–1826) was a German chemist and optician who developed new methods of glassmaking and produced the finest telescopes of his time, as well as inventing the diffraction grating. He was not the first to discover the lines that bear his name (that was the English chemist William Hyde Wollaston, in 1802), but he was the first to explain them.

Modern gratings are often made holographically, and can be very good indeed. I once built a spectroscope from two cardboard tubes, a spectacle lens, and a homemade holographic grating, with two razor blades to make the slit, and the grating quality was high enough to separate the two sodium-D lines at 589.0 and 589.6 nm, a feat beyond the capabilities of a glass prism.

A few sources, such as the Andromeda galaxy, show a *blue* shift, indicating that it is approaching our Galaxy, not retreating. But don't worry: it is going to be a very long time before the collision occurs.

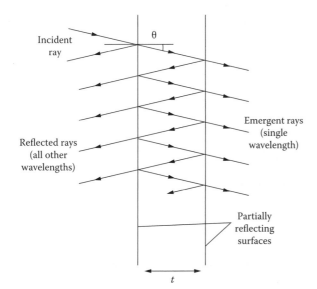

Figure 17.14 Principle of the Fabry-Pérot etalon. The path difference between the emergent rays determines the wavelength that is transmitted, and this can be varied by changing either the separation of the faces or the angle of incidence.

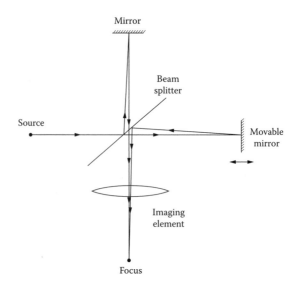

Figure 17.15 Layout of a Michelson interferometer.

out the one wavelength that resonates, i.e., fits into the space by a whole number of half-wavelengths, so that the emergent waves are all in phase. The etalon thus acts as a monochromator. Adjustment of the spacing between the surfaces changes the wavelength, and the whole spectrum can be scanned at a resolution up to a hundred times that available from a diffraction grating.

The Michelson interferometer (Figure 17.15) is a somewhat different device (not to be confused with Michelson's stellar interferometer). When correctly adjusted this produces a set of uniformly spaced vertical fringes from a monochromatic source. However, if the source is not monochromatic, the fringes will overlap, and a complex diffraction pattern will appear. This is in fact equivalent to the Fourier transform of the power spectrum, and can be analyzed by an appropriate computer program.

This has been a very brief and somewhat superficial account of spectroscopy, but in a book about imaging, this subject is only marginally relevant. If you want to learn more, you will need to consult the references below.

Digging Deeper

There is a vast amount of literature on astronomy and astrophysics. If your interest lies in the beautiful images astronomy can produce you can't do better than *Hubble: The Mirror on the Universe* (Robin Kerrod, David & Charles, 2003). It is frustratingly short on technical information about the telescope, though. An excellent textbook at a readable level is *Introductory Astronomy and Astrophysics* (4th editon, Eds. Michael Zeilik and Stephen A. Gregory, Saunders Golden Sunburst Series, 1997). The most recent book is *Astrophysical Techniques* (5th edition, C. R. Kitchin, CRC Press, 2008), which is pretty comprehensive at undergraduate level, though some of the explanations are not as clear as they might be. The most comprehensive of all is *An Introduction to Modern Astrophysics* (2nd edition, Bradley W. Carroll and Dale Osborne, 2006), a massive (1400-odd pages) and very readable book. The weekly magazine *New Scientist* is a must for accounts of the latest developments in astrophysics. There is also a plethora of Web sites available, not all of them totally reliable.

Chapter 18 Macrography, Micrography, and Microimaging

It is perhaps unfortunate that four quite different imaging techniques should possess very similar names. This has led to a certain amount of confusion, sometimes extending even to the authors and editors of technical manuals. So let's begin by settling the nomenclature.

- *Macrophotography* is the technique of producing giant images, called *photomurals.*
- *Photomacrography* is the technique of producing images that are larger than the subject matter using conventional camera techniques, often called simply *macrography.*
- *Microphotography* is the technique of producing tiny images, now usually called *microimaging.*
- *Photomicrography* is the technique of recording images produced by a microscope, often called simply *micrography.*

In order to avoid any possible confusion I shall be using the alternative terms above throughout, as they are clear and unambiguous. We are not concerned with photomurals, as their technique belongs in a working manual; but as microimaging is, practically speaking, a form of reverse micrography, it is included at the end of the chapter.

Macrography

Strictly, the term "macrography" covers only imaging techniques where the magnification is greater than unity. Many modern zoom lenses are claimed to have a "macro" facility, although their maximum magnification may be as low as 0.3. A true macro lens needs to be specially corrected for ultraclose-up work, as general-purpose lenses are designed to give their best performance at subject distances of several meters. When used for close-up work such lenses, however good the original aberration corrections may have been, show spherical aberration and curvature of field, both of which aberrations can be satisfactorily corrected only for specific image conjugates. So-called zoom macro lenses mitigate this by having a special setting for close-ups, which alters the relationships of the floating elements to provide some compensation. This usually works best when the lens is set to a longer focal length. Lenses that are designed for operation at around 1:1 magnification, such as process lenses, are usually symmetrical or quasi-symmetrical, giving a flat field and zero distortion at these magnifications.

You can use an ordinary camera lens with extension tubes or a bellows extension for a magnification around unity without losing noticeable image quality, but once you start to focus on images two or three times natural size the usual image conjugate positions are reversed. Under such conditions you generally obtain a better result by reversing the lens in its mount. The conditions governing depth of field and depth of focus will also be reversed, i.e., there will be little depth of field but considerable depth of focus. You are therefore better off focusing the camera by adjusting its back rather than the lens, or, if this is not possible, moving the entire camera. In order to obtain a reasonable depth of field you will need a small stop, and that brings further consequences. The effective f-number will be increased, since it is obtained by dividing the exit pupil diameter into the image distance v rather than the focal length f. As we saw in Chapter 3, this is given by the formula $v = f(1 + m)$, where f is the focal length and m the magnification, so the effective f/no is equal to the marked f/no $\times (1 + m)$.

The result of this is that the required exposure increases as (magnification plus 1) squared. Thus a magnification of 1:1 requires $(1 + 1)^2 = 4$ times the indicated exposure; a magnification of 2:1 requires $(2 + 1)^2 = 9$ times the indicated exposure, and so on. Through-the-lens (TTL) metering will take care of this, but when the extra exposure is combined with a small aperture you may be looking at exposure durations of the order of a minute, so if you are working with film you may experience reciprocity problems (and your subject may move during the exposure). If you are using artificial light there is also the risk of cooking your subject matter. In such a case you should consider using an optical fiber system or electronic flash. Under these circumstances the easiest way to calculate depth of field is from the magnification. It is given by the formula

$$D = 2cN (1 + m)/m^2$$

where c is the diameter of the acceptable disc of confusion and N is the f/no. For a full frame 35-mm format you can take c as approximately 0.02 mm.

Most newcomers to ultraclose-up work select their shortest focal length lens, probably because it will produce the largest image with their existing extension tubes. This is not as logical as it may seem. For one thing, at a long extension such a lens will be covering a much larger field than the actual format, and will therefore not be operating at its best. In addition, this arrangement puts the lens very close to your subject, which is often awkward, and unnecessarily exaggerates the perspective. Given the choice of a 50-mm or a 200-mm macro lens, you would be well advised to choose the latter. It will put you farther from the subject (important if it is, say, a live butterfly), and the perspective will be greatly improved. Don't expect an increase in depth of field, though. As you can see from the formula, focal length doesn't come into the equation (Figure 18.1).

(a)   (b)

Figure 18.1 Comparison of (a) short-focus and (b) long-focus macro lenses. Note the difference in the backgrounds.

RESOLUTION CRITERION

Because of the long lens extension, the formula for the diameter d of the Airy disc becomes

$$d = 2.44\lambda N(1 + m)$$

where N is the marked f/no. The resolution R is half this figure, and for light of wavelength 550 nm it is given by

$$R = 1.22 \times 550 \times 10^{-9} N(1 + m)$$

$$= 671 \times 10^{-9} N(1 + M)$$

For unit magnification (M = 1) and a resolution of 50 lp/mm on 35-mm format, we have

$$N = 10^6/(1.342 \times 5 \times 10^4) \approx 15$$

So the diffraction-limited aperture is about f/15 for unit magnification, which gives you a depth of field of only about 2.5 mm. Stopping down below this will result in a fall-off in image sharpness as diffraction takes over. For a magnification of ×4 the diffraction-limited aperture is f/6, giving a depth of field of just about a millimeter. However, unsharpness due to diffraction is less obtrusive than unsharpness due to lack of focus, so you may have to compromise and use a smaller stop.

Micrography

Microscopy is a big subject, big enough to have gathered a huge literature, but micrography itself is simply photography through a microscope. However, there are several different kinds of images in both light and electron microscopy, so we need to begin by examining the basic principles of microscopy.

A light microscope has the same basic layout as a refracting (Kepler) telescope, i.e., the magnified real image produced by one lens is examined by a second magnifying eyepiece lens (Figure 18.2).

The main difference is that the objective is set to a close (macro) focus instead of at infinity, and has a short focal length, typically 2–10 mm, much shorter than that of the eyepiece. For viewing, the microscope is focused so that the virtual image is formed by the eyepiece at a normal close focus for the eye, about 250 mm. For photography it is necessary instead to form a real image in the focal plane of the camera. The standard viewing eyepiece is not well suited for this purpose, as it is corrected for virtual images only. Special projection eyepieces are available for photography. If you want to use an SLR camera you need to have an adaptor that replaces the camera lens and attaches the camera to the eyepiece. You can then focus via the camera viewfinder.

Microscope Optics

Because of their specialized function, microscope objectives and eyepieces are designed differently from standard photographic lens systems.

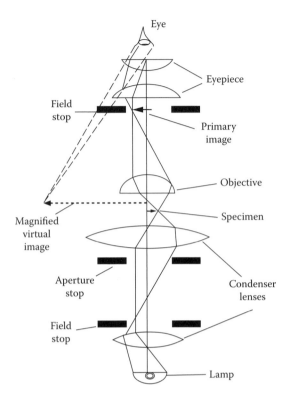

Figure 18.2 Principle of the optical microscope. The primary (magnified) image is further magnified by the eyepiece to give a virtual image for visual examination, or a real image for photography. The substage illumination optics is also shown schematically.

Objectives

The objective is basically an extreme version of a macro lens. Objectives are classified by their magnification rather than their focal length, as the u- and v-distances are constrained by the structure of the instrument. They are usually available only in a series of standard magnifications, in a range from ×5 to ×100, those with the highest magnifications usually being *oil-immersion objectives*. When using these, the space between the objective and the microscope slide is filled with cedarwood oil, a fluid that has the same refractive index as the glass of the microscope slide. An important optical fact about microscope objectives is that they are so well corrected that their resolution is diffraction limited at full aperture, and a stop is therefore unnecessary.

Numerical Aperture

In the mid-nineteenth century Ernst Abbe worked out the mathematics of microscopy. Using diffraction theory, he showed that the performance of a microscope objective depends on its resolving power and not on its magnification. In particular, the detail resolvable is proportional to the sine of the half-angle of acceptance of the objective. Abbe coined the term *numerical aperture (NA)* for this angle:

$$NA = n \sin \theta$$

where n is the refractive index of the intervening medium (1 for air) and θ is the half-angle of acceptance. You can get an approximately equivalent f/no by dividing 0.4 by the NA.

If you fill the intervening space with cedarwood oil, you will increase the NA by 50 percent. This is why the most powerful objectives are of the oil-immersion type.

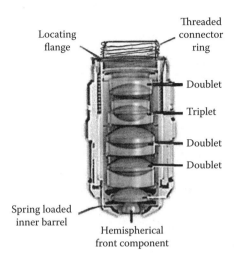

Figure 18.3 A typical medium-power achromatic microscope objective.

From these premises Abbe derived a formula for the resolution R:

$R = \lambda/2n \sin \theta$

where R and λ are expressed in the same units. Abbe calculated from this, and the resolution of the human eye, that the maximum useful magnification of a microscope objective was about $1000 \times NA$, and that any further increase would be "empty magnification," i.e., would not show any more detail. Figure 18.3 shows a cross-section of a typical microscope objective.

Eyepieces

The objective focuses a real image near the upper end of the microscope tube, and this is examined with the *eyepiece*, which is basically a magnifying glass producing a magnified virtual image of the real image produced by the objective at a distance from the eye suitable for normal close vision (about 25 cm). Two of the better-known versions of eyepieces are shown schematically in Figure 18.4.

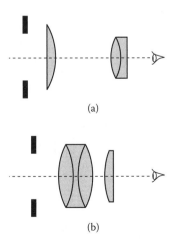

Figure 18.4 Two popular eyepiece designs: (a) Kellner Achromat; (b) Abbe Orthoscopic. Both these designs are also used in astronomical telescopes for visual observation.

Overall Magnification

Rather than doing calculations using focal lengths and conjugate distances, microscopists use formulae involving magnifications, which are much simpler. For example, a 4-mm objective is rated at ×40; a 25-mm eyepiece is rated at ×10. The combined magnification is ×400. If instead you were to do the same calculation using focal lengths, you would start with the effective focal length of the system, which is $f_o f_e / l$, where f_o and f_e are the focal lengths of the objective and eyepiece, respectively, and l is the length of the tube (usually about 160 mm). Taking the near distance of clear vision as 250 mm also gives us $m = 250 l / f_o f_e = $ ×400 for the figures above. As you see, you get the same answer, but more laboriously and no more accurately, as these figures depend ultimately on the operator's eyesight. If you need an exact figure (which you often do) you have to replace the slide with a *stage micrometer*, a tiny scale graduated in hundredths of a millimeter.

The depth of field is always very small, around 10 μm. The exact figure depends on a number of factors, including the refractive index of the medium the specimen is mounted in. To a first approximation

$$\text{Depth of Field} = \lambda/(\text{NA})^2$$

where λ is expressed in μm.

Note that microscopists almost always refer to depth of field as "depth of focus."]

For photography, the microscope needs to bring a real image out beyond the eyepiece, and it is advisable to fit a *projection eyepiece*, which will produce a real image without refocusing. The more sophisticated microscopes have a special camera port separated from the viewing eyepiece by a beamsplitter or a movable mirror. Exposure estimation is tricky. Even through-the-lens metering is unreliable, and wide bracketing is in order, especially with color transparency material. Keep a full record of all exposures, including meter indications, whether correct or incorrect.

Illumination Systems

The main purpose of the illumination optics is to provide uniform illumination of the specimen, with no light wasted. This is not as easy as you might think, and sometimes the substage optics is more complicated than the viewing optics, including as it does condensers, apertures and field stops, filters (both color and spatial), and polarizers. The standard bright-field illumination is called *Köhler illumination*, and is illustrated in Figure 18.5. It is a two-stage system in which an auxiliary condenser produces an image in the aperture of a substage condenser, which itself forms an image of the auxiliary condenser on the object. The two stops control the field area and the numerical aperture, respectively.

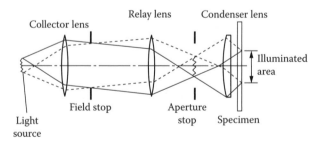

Figure 18.5 Schematic of Köhler illumination. This system minimizes flare and ghosting, and is essential for image recording.

Dark-Field Illumination

Many specimens are hard to see under conventional lighting, because they are nearly (or even totally) transparent and may in addition be unsuitable for staining. Two important modifications to the lighting arrangement are designed to deal with this problem. The first is *dark-field illumination*. The substage optical system contains an opaque annular stop that prevents direct light from reaching the objective, while still illuminating the specimen from the sides (Figure 18.6).

Phase Contrast Illumination

A more subtle arrangement is *phase contrast*. In this system, developed by Frits Zernike, the annular stop is positioned such that both direct light and light transmitted by the specimen reach the objective, but a quarter-wave retarding annulus is installed on (or in) the objective so that the direct light has a phase retardation. The varying refractive index of the specimen produces varying degrees of interference in the primary image, making the detail of the specimen visible (Figure 18.7).

Frits Zernike (1888–1966) was a Dutch microscopist and an early proponent of the Fourier model for image formation. For his invention of the phase contrast microscope he received the Nobel Prize for Physics in 1953.

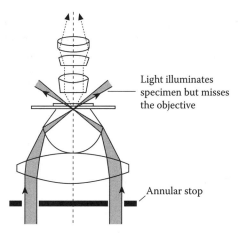

Figure 18.6 Dark field illumination.

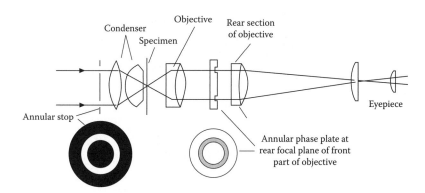

Figure 18.7 Phase contrast microscopy. The annular stop permits separate concentric beams to illuminate the specimen and the background. An annular phase plate at the rear focal point of the front section of the objective delays the phase of the "ring" illumination so that it interferes with the object beam and turns the phase differences into amplitude differences, rendering a transparent specimen visible.

275

Polarized Illumination

A third, rather different way is to position the specimen between two crossed polarizing filters. This is particularly useful in petrological microscopy (studies of rock specimens). Transparent sections of crystalline rock are often optically active, that is, they rotate the plane of polarization, and you can learn a good deal by rotating the upper polarizing filter (the *analyzer*) and observing the color and brightness changes.

Confocal Microscopy

For deep specimens where out-of-focus detail would degrade the image, *confocal microscopy* is used. Telecentric lenses were described in Chapter 5, where it was explained that only rays entering the lens parallel to the optic axis are accepted, as there is a field stop (effectively a pinhole) at the principal focus. In a confocal system for microscopy two telecentric systems are mounted back-to-back, with the second system focused on the field stop (Figure 18.8). The out-of-focus light is greatly attenuated by the stop, the depth of field is much reduced, and both illumination and detection are limited to the plane of sharpness. Refocusing on a different object plane does not alter the magnification, as the only light that can enter the system is parallel to the axis. This type of optics is standard in scanning microscopes, which may also be equipped with a device to vibrate the stage up and down during the exposure to increase the effective depth of field.

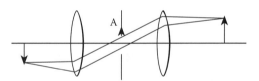

Figure 18.8 Confocal system of two telecentric lens systems back-to-back, as used in scanning microscopy.

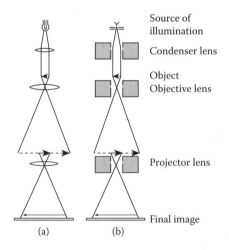

Figure 18.9 (a) Light microscope. (b) Equivalent two-stage transmission electron microscope.

Electron Microscopy

There are two main types of electron microscopes, the *transmission electron microscope (TEM)* and the *scanning electron microscope (SEM)*. In the TEM (Figure18.9), the "lens" components are similar in optical principle to those in a light microscope, but are specially shaped electromagnetic coils. The "light beam" is a beam of electrons, and the image is formed either on a display tube or directly onto a photographic emulsion. The equivalent focal length is about 1 mm and the effective aperture around $f/50$, but the wavelength of the electron beam is so short that resolution is limited by low NA and residual aberrations, not by diffraction. Resolution can be much better than 1 nm, with magnifications up to ×500,000. It sounds marvelous, but there's a snag: you can't just mount the specimen and insert it into the microscope. Not only would the actual specimen be destroyed by the beam of electrons, but in addition it is transparent to such a beam. So the specimen has first to be "shadowed" by a very thin sputtered film of some heavy element (usually gold), and this, of course, may both hide some structural details and create unwanted artifacts.

The SEM (Figure 18.10) uses a different principle. A very narrow diameter (<10 nm) beam of low-energy electrons is scanned in raster form across the specimen, and the scattered electrons are collected by a system that distinguishes between forward and backscatter and secondary emission as well as detecting direction, so various image aspects can be presented to a display screen. There is a large depth of field, and images can have both perspective and stereoscopic depth, at magnifications from about ×10 to ×50,000. The resolution can again be counted in nanometers. There is little or no need to prepare specimens, provided they can stand up to the high vacuum that is needed. As the image is computed, all the usual kinds of enhancement are possible, and magnification can be controlled between both these extremes with a few keyboard strokes.

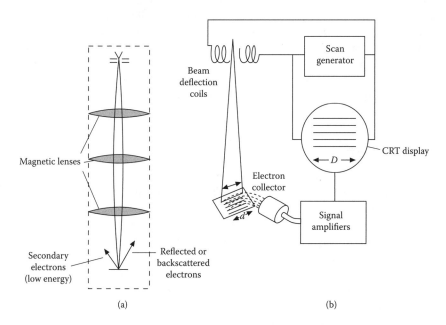

Figure 18.10 Principle of scanning electron microscope. (a) shows the beam focused and scanned by magnetic lenses (shown schematically); in (b) the scattered and secondary electrons are collected and counted separately, and the computed image is displayed on a monitor screen. The magnification D/d can be varied over wide limits at the display output.

There are other, more specialized types of electron microscope, such as the scanning tunneling microscope, capable of imaging single atoms, but for information on these you will need to go to the references in Digging Deeper.

Microimaging

The techniques of microimaging evolved almost as early as photography itself. In the mid-nineteenth century silver halide materials were very slow, but they were also very fine-grained. In the 1860s jewelers would set tiny images behind lenses as dress ornaments, and during the Franco-Prussian War the citizens of beleaguered Paris sent microfilmed messages by pigeon post, each bird carrying enough messages to fill a smallish book. By the end of the First World War spies were sending "microdot" messages concealed under full stops in otherwise innocuous typewritten documents. These were made by reversed microscope optics, and were read using ordinary microscopes. The limit of resolution for fine-grain photography is usually around 1–2 μm, but because of loss of contrast owing to the shape of the MTF, words and drawings needed (and still need) a two-stage reduction. Although microdots may have gone out of fashion in Smiley's twilit world, microimaging techniques are still useful for document recording, and were at one time an important part of library records in the form of microfiches; these were usually made on diazo material, which is grainless. The growing popularity and convenience of computer records on disc has led to the phasing out of most microfiche applications.

One area in microimaging that is continuously expanding, however, is microlithography, which is important in the making of electronic microchips. Lenses need to be specially designed for this purpose. As the field is small (about 10 mm diameter) the aberrations can be (and must be) very well corrected, in a quasi-confocal configuration. Chromatic aberration is irrelevant, as the images are produced by exposure to single wavelengths in the UV region. The light-sensitive material is usually a positive photoresist, a substance that becomes soluble when it is exposed to short-wave radiation. The component with its developed mask is then etched or doped as appropriate. As chips become smaller and more complex, UV lasers, X-ray beams, and electron beams are being increasingly used, in scanning mode. Electron beam lithography can generate patterns only a few nanometers in size.

Digging Deeper

Macrography has always been something of a Cinderella for technical writers. The Kodak Workshop Series, which is regularly updated, contains *Close-Up Photography*, and Kodak's earlier Technical Publications Series No. N–16, *Close-Up Photography and Photomacrography*, by Henry Louis Gibson, is a gem, if you can find a copy. Two good books have appeared fairly recently: *The Art of Close-Up Photography*, by Joseph Meehan (Fountain Press, 1994), which is a guide to the photographer who wants to produce beautiful pictures of tiny things, and *Scientific PhotoMACROgraphy* (sic), by Brian Bracegirdle (Bios, 1995), which concentrates on the technology side. In micrography, however, one is spoiled for choice. In the same series as the previous book is *Scientific PhotoMICROgraphy*, also by Brian Bracegirdle with S. Bradbury (Bios, 1995). The series is called The Royal Microscopical Society's Microscopy Handbooks (there are more than 30 of them), and between them they tell you everything you ever wanted to know about microscopy. T. Wilson has written the definitive *Confocal Microscopy* (Academic

Press, 1990), and a more recent book, from Bios again, is *Confocal Laser Scanning Microscopy*, by C. J. R. Sheppard and D. Shotton (1997). Ian Watt says the last word on *The Principles and Practice of Electron Microscopy* (2nd edition, Cambridge University Press, 1997). And if you are interested in the history of microscopy, you can sample *Microscopy from the Very Beginning*, by H. Kapitze (Carl Zeiss, Jena, undated), and see some astonishing images from the earliest photographers in *Brought to Light: Photography and the Invisible, 1840–1900*, Ed. Corey Keller (Yale University Press, 2008).

Chapter 19 Imaging the Invisible

In Chapter 1 I explained that the range of frequencies in the electromagnetic spectrum spans some 23 octaves, of which we directly perceive less than a single octave. But every part of this vast spectrum is capable of conveying information in the form of images. We have now looked in detail at the methods of recording the visible part. In the case of astronomy, we have gone further, into the realms of cosmic radiation covering the whole electromagnetic spectrum. It is time to look at the way we capture and use images from the invisible radiation we find within the living world.

Radio Images

We are all familiar with the ability of radio waves of wavelengths ranging between 2 and 2000 meters to carry signals that can be decoded into sound and visible images, but the radiation itself can contain information too, in particular the radio sources in outer space, which I discussed in Chapter 17. The trouble with these longer wavelengths is that radio waves are subject to the same rules of diffraction as all radiation, and to minimize this effect and produce adequate images the receiver has to be many wavelengths in diameter. In Chapter 17 we looked at the ingenious ways in which this can be achieved without the need for kilometer-wide dishes. However, at centimeter wavelengths the radiation is better behaved, and 3-cm microwave sources are widely used in physics teaching to demonstrate such phenomena as diffraction, interference, reflection, and refraction, on a scale large enough for whole classes of students to appreciate. And when we reach wavelengths of the order of a millimeter (the T-rays), with suitable equipment we can produce viewable two- and even three-dimensional images.

Terahertz Imaging

The terahertz (T-ray) band of electromagnetic radiation lies between 300 GHz and 3 THz, corresponding to a wavelength range of 1 mm to 100 μm.

A GHz is 10^9 Hz; a THz is 10^{12} Hz.

This band was first described in the 1890s, but as at that time it was beyond the upper limits of electronic oscillators and below those of available infrared sources, it was not explored until comparatively recently. Although astronomers were aware that the universe was filled with terahertz radiation, probably carrying precious information about its structure, they were frustratingly unable to obtain access to it because of the opacity of the Earth's atmosphere to these wavelengths. As we saw in Chapter 17, the latest generation of space telescopes is now ready to remedy this. At ground level, too, it has now become possible to build semiconductor lasers that can produce coherent beams of THz radiation in both continuous-wave and pulse mode. THz-radiation is strongly reflected by metal surfaces, so purpose-built THz cameras use mirror optics. Because of the long wavelength and consequent diffraction, the mirrors need to be at least 50 cm in diameter. Images are typically built up by scanning.

Applications

Radiation at these wavelengths penetrates paper, cardboard, and textiles, so that terahertz imaging has potential applications in airport security scanning, as it can readily detect concealed weapons, and can also be employed in the inspection of packages. This aspect is at present under active development. As the radiation penetrates plaster, it also promises to allow art historians to see murals hidden beneath coats of plaster or paint in old buildings. Unlike X-rays, the radiation is non-ionizing and does not damage tissues, and as it can penetrate several millimeters below the skin it promises to be valuable in the examination of subcutaneous tissue for tumors and other lesions. Some frequencies will penetrate tooth enamel, and could be used for three-dimensional imaging of teeth, offering more information than current X-ray technology. There are also considerable possibilities in spectroscopy, in a region so far almost completely unexplored.

Infrared Imaging

In the early part of the twentieth century new dyes were discovered that could sensitize silver halide emulsions to infrared (IR) radiation in the region of 700–950 nm. This region is called the *optical infrared*, as the radiation can be focused by conventional camera lenses. The important characteristics of IR radiation for photography are:

- It is invisible, so you can use it to shoot photographs in total darkness. The uses here are self-evident. Almost half of the radiant output of an electronic flash happens to be in the optical IR, so by covering a standard electronic flash head with a "black" IR filter you can take covert photographs in darkness.
- It penetrates atmospheric haze, human skin, and some plastics materials and textiles. Black-and-white IR photographic emulsions have been in use for long-distance aerial photography since the First World War, because of the ability of the radiation to penetrate haze. The presence of atmospheric haze causes low contrast owing to Rayleigh scatter, which is inversely proportional to the fourth power of the wavelength of light. As the wavelength of optical IR radiation is more than twice that of blue light, it penetrates haze about 20 times more effectively.
- It is reflected differentially by living matter, and very strongly by broad-leaved plants.

It doesn't penetrate fog, though, because fog consists of water drops, which cause refraction. (Nor nylon dresses, despite early optimistic press claims.)

Perhaps the most important property of IR radiation in this respect is that it is absorbed and reflected in a manner very different from visible radiation. Tarmac surfaces and water absorb it completely, and chlorophyll reflects it strongly. A black-and-white landscape taken on monochrome IR material (Figure 19.1) shows roads as jet black and deciduous trees as snowy white. Pictorial photographers have made much use of these qualities.

However, CCD chips in digital cameras are usually protected by an IR-absorbing filter, so if you want to take infrared images with a digital camera, you will probably need to have it modified by a specialist.

Infrared False Color

This tripack material was, like Ektachrome, developed by Eastman Kodak during the Second World War for aerial reconnaissance, though it arrived too late to be useful

Figure 19.1 (a) Visible spectrum and (b) infrared images of an ornamental garden. The yellow flowers in the foreground reflect IR radiation strongly and appear very light, as does the shrubby foliage, whereas the painted labels and the rocks absorb IR and appear unnaturally dark. (Photographs by Adrian Davies.)

in that conflict. It has subsequently more than earned its keep in ecological, agricultural, and demographic surveys. It is a false-color system, with three emulsion layers sensitive, respectively, to IR, red, and green. Blue light is excluded by a yellow (minus blue) filter on the camera lens. The film is processed to a positive transparency. The IR emulsion layer then gives a positive image in cyan, the red emulsion in magenta, and the green emulsion in yellow. The result is that in the final transparency IR records as red, red records as green, and green records as blue. Blue, of course, records as black. If you think of infrared as a fourth primary, along with red, green, and blue, you can draw the four primaries as a color quadrant, with a shift round of one quadrant between the subject hue and the photographic image hue (Figure 19.2). This will enable you to work out exactly how any color (including IR) will appear in the final image.

In ecological terms, the most important qualities detected by these hue shifts are concerned with live vegetation. Healthy broad-leaved plants record as magenta ([IR + green] → [red + blue, i.e., magenta]), and dead or dying vegetation as

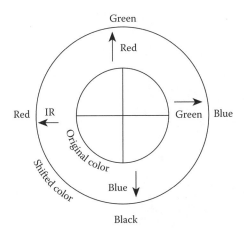

Figure 19.2 Color shifts in false color infrared images.

greyish green (yellow → cyan, brown → grey-green). These differences show in the image before anything is visible to the unaided eye. Silt and other watery contaminations show up light, in sharp contrast to the black images of clean seas and rivers. Freshly ploughed fields show blue-grey, roads are black, and towns a bluish patchwork. An example is shown in Figure 19.3 (color plate). You can buy infrared color roll film, and creative photographers have occasionally exploited its anomalous characteristics, sometimes using filters other than plain yellow.

Thermal Imaging

Thermal imaging, or *thermography*, is involved with the far infrared region of the electromagnetic spectrum, between 1 and 20 micrometers. There are two main methods of recording: direct imaging and scanning. The direct imaging or *focal plane array* camera is a digital camera equipped with a CCD array of platinum silicide (PtSi), which is sensitive to wavelengths in the 3–5 μm band, and a lens of amorphous silicon, which is transparent to IR radiation (refractive index = 3.45) or fluorite (1.38). The system needs to be cooled with liquid nitrogen to eliminate noise from stray radiation. As there are no moving parts the framing rate can be very high, up to 1000 pps, allowing the photography of high-speed events. Germanium optics (refractive index = 4.1 to IR) allows a further extended response up to 12 μm.

Figure 19.3 (See color insert following page 154.) Infrared false color aerial photograph.

In the *scanning imager* there is a single detector. Two orthogonal pivoted mirrors or prisms provide a scan of the scene. The system operates at television speeds. Figure 19.4 shows one practical combination of mirrors.

The *pyroelectric vidicon camera* is a portable maid-of-all-work camera used by firefighters and rescue workers to find trapped persons, and for general surveillance at night. It is a conventional vidicon camera except that the sensitive surface is a thin sheet of pyroelectric material such as lithium tantalate ($LiTaO_3$) or PZT (see margin note below), which both have flat spectral responses throughout the far infrared. A surface charge proportional to change in temperature builds up (Figure 19.5). The system is slow, with a refresh rate of several seconds, but it is cheap and light, and does not require special cooling.

Several types of *photoelectronic* or *quantum detectors* are employed in specialized cameras such as those used in satellites for astronomical observations at wavelengths that are absorbed by the Earth's atmosphere. As explained in Chapter 17, these require cooling to liquid nitrogen or even liquid helium temperatures.

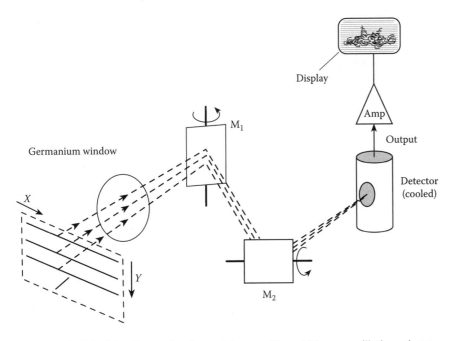

Figure 19.4 Principle of scanning thermal imager. M_1 and M_2 are oscillating mirrors.

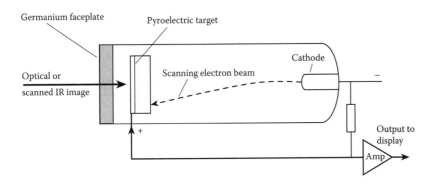

Figure 19.5 Pyroelectric vidicon camera.

Image Conversion and Intensification

Image Converters

These turn optical images (or scans) produced by near IR, UV, or X-radiation into visible light. The image is either focused or scanned onto a layer of phosphor on a faceplate, which is coupled via a fiber optics disc to a photocathode, part of an *image intensifier tube* (Figure 19.6). The photons that form the initial image produce electrons at the photocathode, and these are accelerated and focused to form a much brighter image at the phosphor screen, which is at a potential of 15–20 kV. This image can then be read by a CCD chip or recorded on film.

Even light in the visible spectrum may be too dim for us to be able to see by, and again an image intensifier is the solution. In the "starlight" camera, designed for photographic or visual observations in very dim light (hazy starlight), up to three intensifier tubes can be cascaded, though the resultant image has low resolution and is noisy. Image intensifying low-light cameras are generally fitted with wide-aperture (*f*/1.5) catadioptric lenses, to gather as many as possible of the meager supply of photons.

An alternative (or additional) amplifier stage can be provided by a *micro-channel plate intensifier*. This is a flat plate of lead glass about 1-mm thick, bored with up to 10 million tiny channels a few tens of micrometers in diameter (Figure 19.7). A potential of around 1 kV accelerates the photoelectrons through the channels,

Infrared image converters are standard battlefield equipment. Their peaceful uses include the now familiar TV sequences of badgers and other nocturnal animals. For these the camera is usually equipped with its own IR illumination.

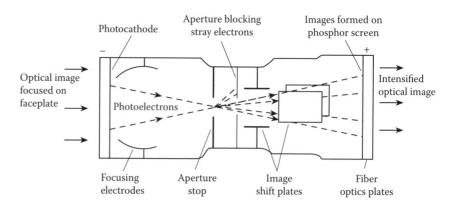

Figure 19.6 Image converter/intensifier tube (schematic).

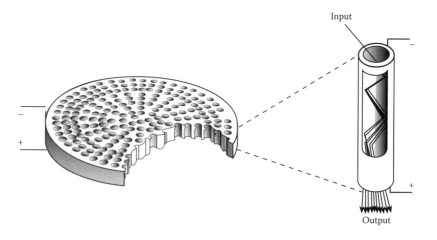

Figure 19.7 Microchannel plate image intensifier.

and during their travel they gain further photoelectrons in an avalanche process before striking a phosphor screen. This type of device saves space, and is now used in most handheld low light and IR conversion equipment. It is also used in satellite imaging systems to amplify celestial X-ray bursts.

Ultraviolet and Fluorescence Imaging

Direct UV Imaging

If human vision were to extend into the near ultraviolet, say to 300 nm, things would often look quite different from the way they do now. Some animals, including many insects, can see color in the ultraviolet. The study of insect vision is important to both botanists and entomologists. To bees, for example, many flowers look quite different from the way they look to us. From UV photography we can get a good idea of what a bee sees. Figure 19.8 shows how a daisy flower looks to us, and how it might look to a bee. (Incidentally, bees can see the polarization of skylight, too.) An important application of UV imaging, along with holography, IR, and X-rays, is the examination of old artworks, both to establish authenticity and for conservation purposes. Among other disciplines, microscopy, astronomy, and camouflage detection also make use of UV imaging.

The UV spectrum is usually partitioned into regions in terms of its biological effects. UV–A (315–400 nm) is the radiation that gives you a suntan. It is transmitted by ordinary window glass and by Wood's glass ("black") filters. Modern optical glasses transmit UV radiation only weakly, and if you want to take your own UV photographs with a Wood's glass filter (Wratten 18A) on the camera lens, you should if possible choose a lens made before 1950. This region of the UV is the one that is most often used for the purposes I have outlined above. However, silver halide emulsions are of low sensitivity in this region, as gelatin is not very transparent to UV radiation. Even when using pre-1950 lenses you should downgrade your film by a factor of about 10.

UV–B (280–315 nm) is the region that can cause serious sunburn and can also damage your eyes. UV–C (100–280 nm) is lethal to most life forms, and is used for sterilization and bactericidal purposes. It is radiated by the sun, but fortunately

The β-cone cells of the eye (see Chapter 3) are in fact sensitive to the near-UV, but the eye lens blocks this radiation below about 380 nm. Since undergoing a cataract operation some time ago involving lens replacement, I can now see the beam of a 355-nm laser. (It just looks violet, by the way.)

Be warned that some films actually have an anti-UV coating, and digital camera sensors come with an anti-UV screen.

(a)   (b)

Figure 19.8 Flower as seen by (a) white light and (b) UV illumination. (Photographs courtesy of Adrian Davies.)

for us it does not penetrate the atmosphere. General-purpose films are almost insensitive to these regions: to obtain images on film you require special equipment and materials. Gelatin is totally opaque to this radiation, and you need to use a *Schumann emulsion*, which has very little gelatin, and the silver halide crystals protrude from its surface. You also need special lenses made from fluorite or quartz. UV–D is ionizing radiation, overlapping with soft X-rays. These very short wavelengths, from 10 nm to 100 nm, are sometimes called *Grenz rays*, and can be used like X-rays to show the inner details of, for example, the interior of botanical specimens.

Light Sources for UV Imaging

Many gas-discharge sources have a UV content, and can be used for UV imaging with a Wood's glass filter. The best sources are high-pressure mercury-xenon lamps, which have lines down to 265 nm, plus a continuum.

UV Fluorescence Photography

Again, there are plenty of applications. Many substances glow in the visible spectrum when irradiated with UV radiation of wavelength 280–350 nm. In some biological research work fluorescent tags are attached to organisms so that they can be tracked by UV illumination. Fluorescent labeling of goods and documents is widespread. Many chemical substances and body fluids can be identified by their fluorescence, so the uses in forensic investigation are legion. Another important use is in nondestructive testing, where the application of a fluorescent dye dissolved in penetrating oil can reveal microcracks that would otherwise be invisible. To make fluorescence photographs using UV in this region (which is the one most used for medical and scientific purposes) you don't need any special camera equipment, but you do need total darkness and a Wood's filter over the light source. You also need a UV-absorbing filter on the camera lens, or your image may be compromised by unwanted direct UV imagery. A xenon discharge lamp is a useful portable source. To isolate specific substances you may need a spectral footprint, and this means making a number of exposures using a series of narrow-band filters on the camera in addition to the UV-absorbing filter. In this type of investigation photography is often coupled with spectroscopy.

Endoscopy

An *endoscope* is an optical device for looking inside the body by way of its orifices, or through a surgical incision. Endoscopes are also used for examining the interiors of, for example, gun barrels, pipework, and, on a larger scale, sewers. Clinical endoscopes are usually specialized, and have names such as *laryngoscope* (for the throat) and *otoscope* (for the ear canal). The rigid kind may use periscope optics, which consist essentially of two telescopes back to back (Figure 19.9a). A thinner tube employs relay lenses (Figure 19.9b). A third type has an extended GRIN rod, which acts as a set of relay lenses (Figure 19.9c). If the endoscope needs to be flexible, the light guides can be a coherent bundle of optical fibers.

In these endoscopes the camera is outside the body, and you can consider the optical system an extreme form of retrofocus lens. Illumination is usually straight down the endoscope through its optical system, the light being fed in through a beamsplitter. Recently, the advent of very small CCD arrays has made it possible to fit a tiny television camera on the end of the probe, with the image information

A coherent fiber bundle is one in which the fibers remain in the same spatial relationship throughout the length of the device. The image is thus transferred from one end of the device to the other without becoming scrambled. Note that the use of the term "coherent" in this context has nothing to do with coherent light.

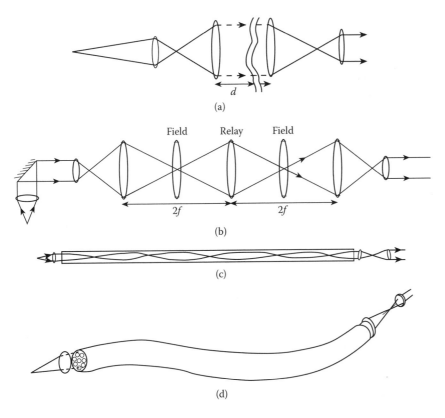

Figure 19.9 Optical systems for endoscopes: (a) periscope; (b) relay lenses; (c) GRIN relay lens; (d) coherent fiber bundle.

conveyed out by cable. The resolution is better than that given by an optical fiber relay system. The cable is carried in a flexible tube or catheter, which also contains a camera steering cable and an optical fiber for illumination, and even the instruments for microsurgery.

There are a number of other cameras that are specialized for clinical examinations. One such is the *fundus camera*, used for photographing the retina of the eye. This has a lens system that compensates for the focusing power of the eye, enabling the viewer to see the retina directly. As in modern microscopes, the camera port is separate from the viewing port.

Radiography

Radiography and its clinical applications represent a huge area of medical technology, though the principles of radiographic imaging are, in general, fairly simple.

Origins

Near the end of the nineteenth century Wilhelm Röntgen was investigating fluorescence in Crookes tubes. He discovered that a mysterious radiation from the tube at the lowest pressure was causing barium platinocyanide crystals to fluoresce, although there was a cardboard barrier between the crystals and the tube; they even fluoresced when they were in an adjacent room. He called this radiation "X-rays," and published his findings in 1895. X-rays were first used in clinical practice the following year.

Wilhelm Konrad Röntgen (1845–1923) became Director of the Physical Institute of Würzburg in 1888. He made important contributions to other branches of physics, but his discovery of X-rays (which are still called "Röntgen rays" in Germany), and of their possible applications, earned him what was the very first Nobel Prize for Physics, in 1901. Like du Hauron and Maddox (see earlier notes) he refused to make any financial gain from his discovery, and died in poverty.

Generation of X-Rays

X-rays are generated when a beam of high-energy electrons in a vacuum tube strikes a material of high atomic number (in practice usually tungsten). The kinetic energy lost by the electrons reappears in the form of very high frequency photons (wavelength around 0.1 nm). The present-day form of X-ray tube has been developed from a design by Coolidge in 1913. The electrons are emitted by a hot cathode and focused by an electromagnetic coil to a fine beam, striking a tungsten anode specially shaped and angled to produce what is as nearly as possible a point source of X-radiation. This has a continuous spectrum with a few characteristic spectral lines due to the anode material. Higher (kilo)voltages produce shorter wavelengths; higher anode currents produce increased beam intensity. Figure 19.10 is a schematic diagram of a modern Coolidge tube. The entire tube is lead-shielded except for a window with a filter of lead or copper compounds to absorb the longer wavelengths (soft radiation) and reduce scatter.

A radiograph is a shadowgraph, as X-radiation is not refracted or reflected (at least, not in the usual sense), but different materials selectively absorb it. Its clinical importance stems from the fact that bone has a transmittance to X-rays that is only one-seventh of that of soft tissue. As the beam of X-rays is divergent, it can give an enlarged image if the subject is positioned fairly close to the X-ray source and some distance away from the sensitive material. A specially designed tube with a very small source size (~100 nm diameter) can even obtain magnifications of over ×1000, with resolution at least as good as that of a light microscope, as at these very short wavelengths diffraction effects are negligible.

Some of the radiation is scattered within the subject matter, so the sensitive material is usually wrapped in thin lead foil to reduce the effect (which resembles flare in a conventional photograph). When the subject matter is both dense and deep, as for example when imaging the cranium, a device called a *Potter-Bucky grid* is positioned between the subject matter and the detector to reduce the effect of scattered radiation. It takes the form of a venetian blind constructed of lead strips interspersed with polymer spacers, and it is oscillated laterally during the exposure to avoid bar lines appearing on the image.

Detection of X-Rays

Although Röntgen discovered X-rays through their action on a photographic plate, silver halide emulsions are not very sensitive to X-radiation—for which

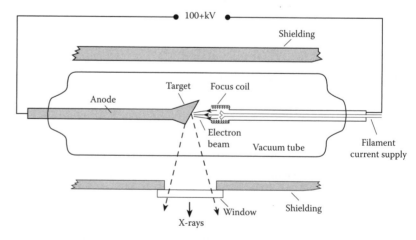

Figure 19.10 Modern Coolidge X-ray tube (schematic).

photojournalists who use film and travel frequently by air should be grateful. The sensitivity falls, too, as the wavelength decreases, since shorter wavelengths, being more energetic, are also more penetrating, so most of the X-ray photons pass straight through the emulsion. As you might expect, absorption increases with increasing silver halide crystal size. Photographic films for X-ray detection are therefore coated with a thick coarse-grain emulsion on both sides. They have a high contrast to X-radiation (the maximum density is a whopping 6), but to visible radiation the contrast is low.

With plain film the necessary exposure duration would be so long that a patient might possibly receive a dangerous dose of radiation, so fluorescent *intensifying screens* are used, in contact with both sides of the film. Traditional phosphors have been lead sulphide, lead barium sulphate, or calcium tungstate, but compounds of lanthanides such as gadolinium are more efficient and are now standard. In most modern medical X-ray apparatus film has been displaced by digital detection systems.

By using a phosphor screen on its own, it is possible to view the image in real time. Where observation of movement is necessary (as in assessing blood flow or lung action), the fluorescence image is picked up by an image intensifier system. Where movement is irrelevant (as in mass screening), the exposure is as brief as possible, and the image is held by a storage tube (a recording tube that holds an image for as long as is required). A similar system operates at airport X-ray machines.

The element selenium is also sensitive to X-radiation. A selenium-coated metal plate forms the sensitive surface for *xeroradiography*. The image is transferred to a transparent sheet or to paper for viewing. The edge enhancement that can be a nuisance in photocopying is an advantage here. Its main use is in mammography, where it gives quick results cheaply, but like silver halide X-radiography, it is being supplanted by digital techniques.

Gamma Radiography

The year after Röntgen published his discoveries, Henri Becquerel, inspired by Röntgen's work, discovered natural radioactivity in uranium compounds. He was able to show that much of the emission consisted of streams of electrons, and concluded, correctly, that uranium was decaying into other elements.

The emissions are now known to be of three types: alpha (α) particles, which are helium nuclei; beta (β) particles, which are fast electrons; and gamma (γ) radiation, which is X-radiation with a range of wavelengths from 10^{-1} to 10^{-3} nm. This radiation will penetrate up to 100 mm of steel, and this makes radioactive substances suitable sources for radiography of metal structures as a part of non-destructive testing (NDT) procedures. The source, which is usually an artificial radioisotope, is contained in a lead casket with a small aperture. This is positioned a short distance from the structure under investigation, the film being fixed to the other side of the structure. As the process is very inefficient, the exposure duration may often be several hours. In order to calibrate the source, a step tablet is used. This operates like the Kodak step tablet discussed in Chapter 10, but is a good deal more robust, being a set of metal steps machined from a single steel block. Gamma radiography is a valuable adjunct to UV and ultrasonic tests in the detection of cracks and other faults in structures such as aircraft wings. Radioisotopes are also used as γ-ray sources placed inside metal pipes with the detector wrapped round the outside, to check the quality of seam welds.

Antoine Henri Becquerel (1852–1908) was a professor at the École Polytechnique in Paris. He discovered, more or less by accident, that compounds of uranium fogged photographic plates even when they were wrapped in opaque material. He investigated the nature of the radiation, and in 1903 was awarded the Nobel Prize for Physics jointly with the Curies, who had succeeded in isolating polonium and radium from uranium ores.

A *radioisotope* is an element that has a nucleus containing too many neutrons for stability. Atoms of the isotope spontaneously decay into more stable forms, usually emitting neutrons and γ-radiation in the process. Radioisotopes occur naturally in small amounts (radium itself is one), but usually they are created as by-products of nuclear reactions; they can also be produced in particle accelerators. Each radioisotope has a characteristic *half-life*, that is, the time it takes for half its atoms to decay. This time may be anything from less than a microsecond to more than a million years.

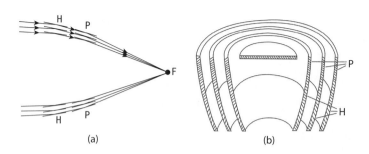

Figure 19.11 Grazing incidence mirror system for imaging an X-ray source: (a) schematic of mirror layout; (b) a nested grazing incidence X-ray telescope objective. (After Kitchen, *op.cit.*)

Units of Measurement

Like photometry, radiography was in the past lumbered with a surfeit of unrelated units of measurement, but again the adoption of SI units has whittled the number down to a manageable minimum. Although not entirely abandoned, non-SI units such as the curie, rad, rem, and röntgen are no longer *persona grata* with the ISO. The becquerel (Bq) replaces the curie as the unit of radioactivity (1 Bq = 1 s^{-1}); the gray (Gy) replaces the rad as the unit of absorbed radiation (1 Gy = 1 m^{2}s^{-2}); the sievert (Sv) replaces the rem and röntgen as the unit dose equivalent (1 Sv = 1 m^{2}s^{-2}). The energy associated with wavelength is often quoted in electron volts (1 eV = 1.6021 × 10^{-19} J). All the other terms are now obsolete.

X-Ray Optics

I mentioned earlier that X-rays are not reflected or refracted in the way light waves are. Nevertheless, there are some ingenious ways of obtaining a focused X-ray image. The simplest way is to use a pinhole camera lined with lead sheet. This was the earliest type of camera used for identifying celestial X-ray sources. X-rays can in fact be reflected, though only from metals and at grazing incidence. One form of X-ray telescope, based on this principle, uses a nest of narrow hyperboloids and paraboloids of revolution (Figure 19.11). (This principle is also used for X-ray microscopy.) In a different approach, using microlithographic or holographic methods, it is possible to construct a zone plate that will focus a beam of X-rays to produce an image. It is also possible to construct Bragg mirrors to make a reflecting objective by coating metal mirrors with alternate half-wave layers of tungsten and carbon, only 2–3 atoms thick, utilizing the very small refractive indexes (for X-rays) of these elements (Figure 19.12)

In a completely different area of research, diffracted beams of X-rays are used to deduce the molecular structure of crystals (Figure 19.13).

Tomography and Scanning Systems

Tomography evolved as a method of obtaining clear images of internal organs that lay within bony structures such as the skull and ribcage. The X-ray source was moved across the subject during the exposure, and the film, with a Potter-Bucky grid, was moved in the opposite direction, both movements being centered on the organ to be imaged. The specific organ to be examined would have been perfused with a radiopaque dye (Figure 19.14).

Energy is inversely proportional to wavelength. A photon with a wavelength of 1 nm has an energy of 1.234 MeV.

It was Rosalind Franklin's brilliance at X-ray crystallography that enabled Crick and Watson to work out the structure of DNA, for which they received the Nobel Prize for Physiology or Medicine along with Maurice Wilkins in 1962. Franklin had died four years previously, aged only 38, and by the Nobel rules was thus not eligible.

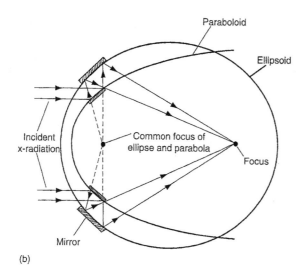

Figure 19.12 Reflection X-ray focusing objective making use of Bragg diffraction. (After Kitchen, *op.cit.*)

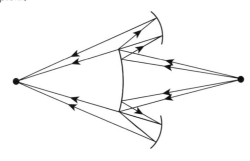

Figure 19.13 X-ray diffraction image of a crystal. By comparing patterns at varying angles of incidence with known diffraction patterns, a three-dimensional picture of the crystal structure can be deduced.

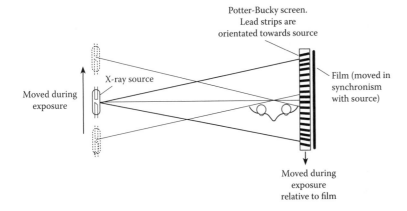

Figure 19.14 An early X-ray tomography system.

A modification of this method appears in one piece of modern equipment. If you have recently had a dental X-ray, you may have been treated to a panoramic view of your teeth. For this type of radiograph the film is mounted in an elliptical holder with one focus at the back of your head and the other focus at the front. The X-ray source follows the outline of the rear half of the ellipse while a slit traverses the film at the front, in synchronism (Figure 19.15).

293

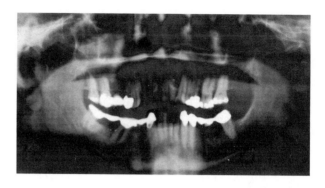

Figure 19.15 Panoramic X-ray scan of a complete dentition, showing tooth roots (pale grey) and bridges and crowns (white).

X-Ray Body Scanning

Once it had become possible to scan round a patient's body with an X-ray machine and a detector diametrically opposite one another, and to feed the acquired data into a computer, 360° *computerized axial tomography* (CAT) became a reality. Each "slice" produces a cross-section of the patient's body, stored for display and examination. The earliest systems were similar to that of the dental panoramic X-ray, with a single source and detector swinging through the full angle. The most recent design has no moving parts. The X-ray emitters are installed in parallel rows in a semicircle below the patient, and the detectors are in similar rows above. An electron beam directed by magnetic coils scans the emitters, and scanning is complete in under a second (compared with several minutes for the original process) (Figure 19.16).

Radioisotope Scanning

X-ray tomography also lends itself to autoradiographic scans. The organ for examination is perfused with an appropriate radioactive tracer (e.g., radioiodine for the thyroid gland). A gamma camera, which is basically just a pinhole camera equipped with a scintillation plate instead of a film, maps the emissions. The scintillations are recorded as the camera moves round, and the computer builds up a three-dimensional picture of the organ. The monitor can display a picture in near real time, and can show movement too.

Positron Emission Tomography (PET) Scanning

This is a similar device, but records γ-rays emitted when positrons emitted by appropriate tracers interact with electrons present in body chemicals.

A positron is an antielectron, i.e., an electron that bears a positive instead of a negative charge. When a positron encounters an electron they mutually annihilate, emitting a burst of γ-radiation.

The existence of the positron, a fundamental particle of antimatter, was first predicted by Paul Dirac (1904–1984). (see p.316.)

Nuclear Magnetic Resonance Imaging (NMRI)

The protons in the nucleus of an atom possess *spin*, giving them a magnetic polarity that is usually random (Figure 19.17a). In the presence of a powerful magnetic field the polarities line up. However, the spins have a wobble, a random precession like a spinning top (Figure 19.17b). An applied radio frequency (RF) pulse of the same frequency as the precessions aligns them so that the spinning protons all precess together. The whole magnetic field thus rotates in phase (Figure 19.17c). When the RF pulse ceases the spins relax into their former precessions, emitting energy. The strength, frequency, bandwidth, decay patterns, and directions of the signals are recorded: they locate and identify the tissues concerned, each tissue having its own signature. Watery and fatty tissues are imaged, but bone is not, nor is blood. Contrast media can be injected to highlight specific structures.

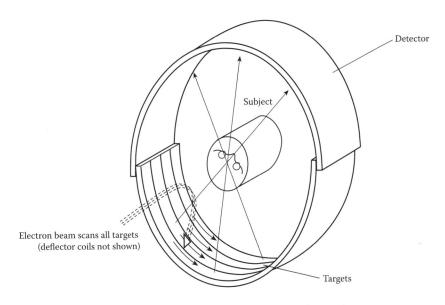

Figure 19.16 A modern CAT scanning system with no moving parts.

The magnetic field required is of the order of 1–1.5 teslas. This is an enormously strong field, over 100 times that of an ordinary bar magnet. As you can imagine, the sorting out of the information is complicated, and needs a good deal of computer power. The slices are pixellated into three-dimensional elements (*voxels*), and the final image is comparatively easily interpreted, especially as it can be compared with, and superimposed on, PET slices, which show chiefly biochemical function, and CAT slices, which show bone structure.

Analysis of Scanning Outputs

The actual outputs of these scanning systems are complicated, especially in the case of NMRI scans, and are well beyond the scope of this book (and most others). There are two main methods of processing the data as they accumulate. The first is an iterative method, whereby a crude "guesswork" map is successively reprofiled until it fits all the data. The second is the Fourier approach, which has become increasingly common since the introduction of fast Fourier transform software into modern computer technology. Each set of data is analyzed into its sinusoidal components. When these have all been specified in terms of amplitude, direction, (spatial) frequency, and phase, the final map is obtained by a reverse Fourier transform. (The Fourier approach to imaging is explained in Appendix 3.)

Ultrasonic Imaging

In its own way, this is also a form of scanning, but it is different from the methods above in several fundamental ways. *Ultrasound* consists of longitudinal compression waves of the same type as sound waves, but with a frequency higher than the human ear can detect. Strictly, ultrasound applies to any frequency higher than about 20 kHz (some two octaves above the highest note of a piano), but for imaging purposes 20 kHz is much too low a frequency, as the corresponding wavelength is about 18 mm, and because of diffraction any object smaller than this would not be seen by the beam at all. At frequencies of around 1 MHz the wavelength is of the order of 100 μm, the effects of diffraction are much reduced, and a broad source gives a tight, well-collimated beam. Higher frequencies are greatly attenuated in

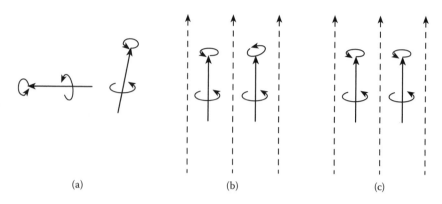

(a)   (b)   (c)

Figure 19.17 Principle of NRMI scanning system. (a) Proton spin gives a random magnetic polarity. (b) A powerful magnetic field aligns all the proton spins, but not the phases of their precessions. (c) An applied RF pulse of the appropriate frequency puts all precessions into phase. When the pulse ceases the precessions return to their former phases, giving off energy that identifies the tissue concerned. A computer analyzes these data and constructs and displays the image.

PZT, a ceramic material, is often used as the driver in small speakers (tweeters) and for the generation of "beeps" in electronic equipment. It is also used in optical equipment for making very small positional adjustments.

fluids and air. Ultrasonic imaging operates by the analysis of echoes. The three main applications are sonar, clinical scanning procedures, and nondestructive testing for hidden flaws in structures.

The ultrasound generator is a *piezoelectric transducer*, usually made of lead zirconate titanate (PZT). A piezoelectric transducer is a substance that changes its dimensions in response to an electric field and, conversely, produces an electric field when it is strained.

To generate the ultrasound, a radio frequency (RF) voltage of the required frequency stimulates a specially shaped PZT transducer. As the device also operates in reverse, it can be used to detect echoes from its own emissions. By alternating short pulses of RF stimulation with pauses to "listen" for echoes, the incoming signal can be amplified and compared with the outgoing signal, and the result analyzed.

Sonar

The Doppler effect refers to the raising of the pitch of an echo when an object is approaching, and the lowering of pitch when it is retreating. Knowing the speed of sound and the change in frequency makes it possible to calculate how fast the object is moving towards you or away from you. The Doppler effect also causes the redshift in the spectra of celestial bodies as discussed in Chapter 17.

In investigations of the sea floor, a transducer directed downwards has its echo pulses displayed in real time in terms of return time (and hence distance) on a monitor screen. This gives an immediate reading of the depth of the water, and in addition an indication of the nature of the sea floor (rocks, mud, coral, etc.). Shoals of fish show up as intermediate blips. A lateral oscillation of the transducer beam, when combined with the vessel's forward movement, creates a two-dimensional map of the sea floor. Forward- and side-looking sonar sweeps can also detect submerged rocks and (with luck) mines. The movement of any submarines that are within range can also be analyzed by making use of the change in frequency of the returning beam (the Doppler effect). Sonar is illustrated in Figure 19.18.

Non-destructive Testing (NDT)

Ultrasound systems for NDT operate on similar principles, though here the aim is usually to detect air gaps in objects that should be solid. The transducer is coupled to the structure's surface by a film of grease. Any hidden cracks or inclusions give their presence away by their echoes. It is necessary to use higher frequencies, as sound travels very fast in metals, and has a correspondingly longer wavelength.

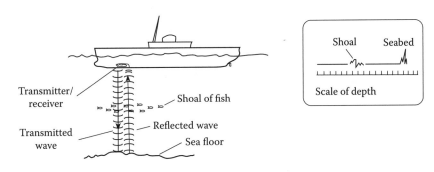

Figure 19.18 Principle of sonar depth measurement.

Clinical Ultrasonic Imaging

For clinical purposes such as brain, kidney, and fetus scans, the transducer is coupled to the skin surface by a film of jelly, and moved over the whole area while its movement is tracked and, as in sonar mapping, a picture is slowly built up. Moving organs such as the heart, and blood flow in veins (not forgetting fetuses) can be monitored by means of the Doppler effect (see above). In this mode the signal is continuous and the frequency changes can be compared, as in Doppler sonar.

Schlieren Photography

We cannot normally see variations in atmospheric density, except in cases such as the heat shimmer familiar in long-focus TV shots of the African plains. Schlieren is a German word meaning "striations." The term schlieren photography was coined by Albert Töpler, who first described the technique in 1906. It is based on an idea by Foucault, who originally devised a knife-edge technique for examining astronomical telescope mirrors for defects in figuring.

The general principle of schlieren optical systems is that if you focus a collimated beam of light, then position a knife-edge precisely at the focal point, the slightest disturbance in the medium the beam has passed through results in either a cutoff of part of the beam or its enhancement. Now suppose you pass your collimated beam of light through the test chamber of a wind tunnel containing a test piece, say, an aerofoil section, and that after the knife-edge you position a focusing lens and a screen, or a camera (Figure 19.19). The shadow of the test piece will be there, and will be in focus, but the rarefied air above the aerofoil, and the over-pressure on its underside, will have diverted the beam. In both cases the density gradient will show up as a darkening or lightening of the image. The most dramatic results are with shockwaves generated at transonic and supersonic speeds (Figure 19.19), though in ballistics research simple spark photography also gives a bright image. However, the schlieren method is sensitive enough to show the patterns of rising air from a candle flame or even from a human hand, and is easier to analyze.

In practice it is difficult to make good lenses large enough for anything but the smallest test volumes. Optical mirrors are the answer. Unfortunately, a pair of spherical mirrors introduces a double dose of spherical aberration and an unacceptable amount of coma. So, in spite of the extra expense, it pays to use parabolic mirrors designed for off-axis beams (Figure 19.20). For maximum resolution the light source needs to be focused to a slit.

Jean Bernard Léon Foucault (1819–1868) was a physicist at the Paris Observatory. He took the first detailed photographs of the Sun's surface in 1845, and demonstrated the rotation of the Earth in 1852 with "Foucault's pendulum." He followed up this line of reasoning by inventing the gyroscope. He determined the speed of light in air and water to an accuracy of better than 1 percent, and made many contributions to telescope engineering. His observations led to the invention of spectroscopy.

297

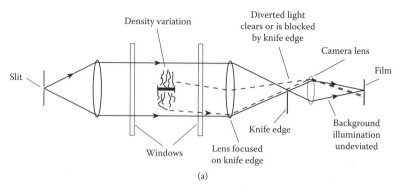

(a)

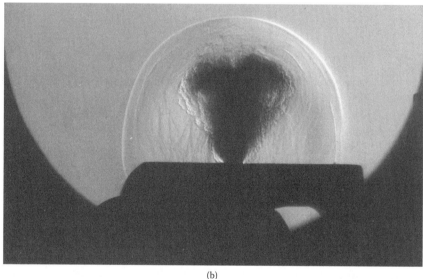

(b)

Figure 19.19 (a) Principle of schlieren photography. (b) Schlieren photograph of discharge of a .22-calibre starting pistol, made using a 0.2-μs flash. (Photograph by Ron Graham.)

We can obtain more informative results by substituting color filter strips for the knife-edge, for example, red above, green in the middle, and blue below. The image will then be basically green, with a high-to-low density gradient giving a yellow band (red + green) and a low-to-high gradient giving a cyan band (blue + green). Figure 19.21 (color plate) illustrates this.

A variant of the method uses interferometry. The beam is split into two by a beam-splitter, and the two beams, one of which has been through the test space, are recombined on the far side by a beam combiner (Figure 19.22). It is necessary to use a single wavelength in order to be able to see interference patterns, so laser light is mandatory. The principle is that of the *Mach-Zehnder interferometer*. If the air in the test volume is undisturbed you will see a set of straight fringes, which you can adjust to be vertical or horizontal as necessary by making small adjustments to the parallelism of the mirrors. Density variations are contoured by the disturbances in the fringe pattern. This system is more sensitive than the schlieren configuration, and has the advantage of direct reading, as each displacement of a whole fringe width depicts a change in the optical path of one wavelength. The interference method thus gives a direct reading of density variations, whereas the schlieren method shows the first derivative of the density, i.e., the density gradient,

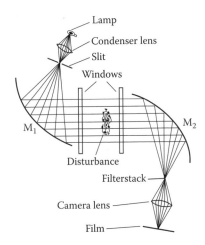

Figure 19.20 Schlieren optical configuration employing off-axis paraboloidal mirrors.

and is more difficult to analyze quantitatively. The spark method shows the second derivative and, being more recalcitrant to analysis, is usually employed only for showing the general form of a shockwave.

Digging Deeper

There is a great deal of literature on most of the subjects in this chapter, but the disciplines are still developing rapidly, and many of these recommendations are already out of print. Rather than producing new editions, the tendency seems to be to update the older ones by publishing papers in the specialist journals, until the whole business falls so far behind that some enterprising author writes a completely new book, and then the whole process begins again. Perhaps by the time you read

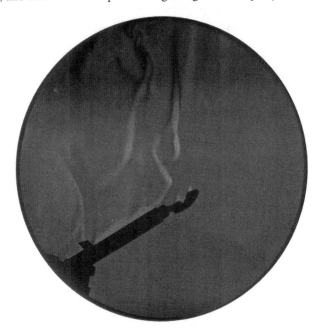

Figure 19.21 (See color insert following page 154.) Visualization of air currents using color filters instead of a knife-edge.

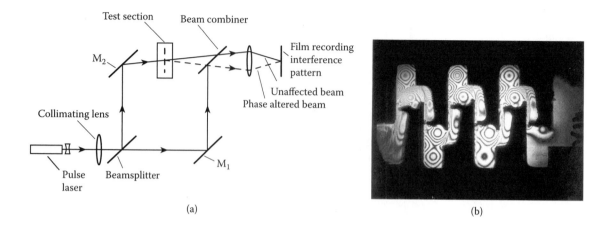

Figure 19.22 Schlieren interferometry. (a) Basic Mach-Zehnder interferometer setup. (b) Visualization of the gas flow pattern in a labyrinth seal. (Interferogram by Ric Parker, courtesy of Rolls Royce Plc.)

this book some of the works recommended here will have been superseded, so I have tried to keep to the ones whose information is likely to stand the test of time.

There is a fair amount of literature on UV photography. If you can find a copy, Eastman Kodak's *Ultraviolet and Fluorescence Photography* (Technical Publication M–27, 1972) is full of sound practical advice. *Photography for the Scientist* (Ed. R. Morton, Academic Press, 1984) has a chapter that deals with the theoretical background in some detail. B. Duffey's *Ultraviolet Radiation in Medicine* (Institute of Physics, 1982) gives a comprehensive review of the applications of UV in general. Sidney Ray's *Scientific Photography and Applied Imaging* (Focal Press, 1999) has a chapter that gives comprehensive data on light sources, filters, and lenses, with a very full bibliography.

With the rapid growth of keyhole surgery techniques, endoscopy is constantly evolving, and for up-to-date information you need to consult the appropriate medical journals such as *Biological Photography*. There is an excellent book, *Biomedical Photography* (Ed. I. Walter, Butterworth-Heinemann, 1992), which contains a chapter on photoendoscopy by L. Morris. It also contains a full account of ophthalmic photography by L. Merin. For an in-depth treatment of TV endoscopy, try *Electronic Videoendoscopy* (Ed. K. Harubini, Harwood Academic Publishers, 1993). Radiography, as you might expect, has spawned a vast literature; I can only point to some excellent reference works I have come across. The standard work, covering the whole field, is *Radiological Imaging* (2 volumes) by H. Barrett and W. Swindell (Academic Press, 1981). It does need some updating, however. A more recent book, concentrating on the theoretical side, is the multiauthor *X-Ray Science and Technology* (Ed. A. Michelle and C. Buckley, Institute of Physics, 1993). The principles of tomography are covered in Barrett and Swindell (see above), and various authors cope with the analysis of scanning data in *Computer Methods: The Fundamentals of Digital Nuclear Medicine* (Ed. D. Liebermann, Mosby, 1977), which needs less updating than you might expect, but could certainly do with a new edition. Probably the best book on ultrasonic imaging is *Modern Acoustical Imaging*, (Ed. H. Lee and G. Wade, IEEE, New York, 1995).

The most complete account of thermal imaging is *Applications of Thermal Imaging* (Ed. B. Burney, T. Williams, and C. Jones, Institute of Physics, 1988), but there is a collection of recent papers on the subject published by SPIE Publications (1999) called *Practical Applications of Infrared Thermal Sensing and Imaging*

Equipment, which is well up to date. If you are interested in the birth pangs of this fascinating area of technology you should try to get hold of a back copy of the *Journal of Photographic Science* (1961, *9*, pp. 375–9), where there is an article by A. Bouwers et al. "Low Brightness Photography with Electro-Optical Intensifiers." Dr. Bouwers of De Oude Delft was a pioneer of image intensification, and almost all modern developments in the technology stem from his insights into both wide-aperture optics and electronic light amplification.

Schlieren photography is an old technique, and apart from the use of lasers for schlieren and holographic interferometry, little new has developed recently. The best summaries of the technique are to be found in two Focal Press books with Sidney Ray as author or editor: Chapter 17, "Flow Visualisation," by Peter Fuller, in *High Speed Photography and Photonics* (Ed. S. F. Ray, Focal Press, 1997), and Chapter 24 in Ray's *Scientific Photography and Applied Imaging* (see above). If you want to go back to beginnings and can read German, call up a copy of Töpler's original paper, "Beobachtung nach einer Neuen Optischen Methode," *Ostwalds Klassiker derExacten Wissenschaften* (Leipzig, 1906, Nr. 157).

Appendix 1: Logarithms: What They Are, What They Do

Logarithms were the brainchild of John Napier, who invented them as a way of simplifying the lengthy calculations involved in astronomy, his chief interest. He coined the name "logarithm" from two Greek words meaning "ratio" and "number."

The ubiquity of the pocket calculator has made the use of logarithm tables redundant for calculations (the slide-rule, too, into the bargain), but the concept of the logarithm (usually abbreviated to "log") is still important, particularly where perceptual processes are concerned. Its basic thesis is that any positive number can be expressed as a power (usually of 10), and that this power (the *logarithm* of the number) has a unique relationship with the number itself.

Consider two series of numbers:

1   2   3   4   5 ...

10 100 1000 10,000 100,000 ...

Do you see the connection? The most obvious one is that the upper row represents the number of zeros in the lower row. You can see a closer connection if you write the lower row

10^1 10^2 10^3 10^4 10^5 ...

The upper row is called the *logarithm* of the lower one, usually abbreviated simply to *log* (strictly, *log base 10* or log_{10}). The series can continue for as long as you like: you can see that the next number in the upper row will be 6, and in the lower row 1,000,000 (i.e., 10^6).

But what *kinds* of series are these? Well, the upper row is plainly a "counting" or *linear* series: to obtain the next number you simply add 1 to its predecessor. The lower row is a "times" or *power* series: to obtain the next number you multiply its predecessor by 10.

You've seen how we can continue both series upwards. But what happens if we try to count downwards? The answer is that you continue both series in the same way: you subtract 1 from the upper row, and divide the number in the lower row by 10, giving the continuation downwards:

... –3 –2 –1 0 1 ...

... 1/1000 1/100 1/10 1 10 ...

John Napier (1550–1617) was a Scottish aristocrat and a brilliant scientific amateur. He was fiercely Protestant, and published a long treatise denouncing Catholicism. He invented a water pump for coal mines, and researched the use of fertilizers in agriculture. He produced the first calculating device, using numbered rods ("Napier's bones"). He was also the first person to use decimal point notation.

By analogy, the lower numbers can be written ... 10^1, 10^0, 10^{-1}, 10^{-2}, 10^{-3} ... . This is perfectly logical, since for all values of x and y, $a^x \times a^y = a^{(x+y)}$ and $a^x \div a^y = a^{(x-y)}$. Thus $10 \div 10 = 10^{(1-1)} = 10^0$, and $1 \div 10 = 10^{(0-1)} = 10^{-1}$.

The next question is: do the numbers *in between* those in the lower row have log equivalents, too? The answer is yes, they do. For example, consider the number $10^{0.5}$. From the same rule of adding of powers, $10^{0.5} \times 10^{0.5} = 10^{(0.5+0.5)} = 10^1 = 10$. This means that $10^{0.5}$ must be $\sqrt{10}$, which is approximately 3.2. Using similar arguments, you can work out the log of *any* positive number. As the log of 0 is $-\infty$ (minus infinity), there are no logs of negative numbers.

You can also show that $10^{1.5}$ is 32, $10^{2.5}$ is 320, and so on. This indicates that the number behind the decimal point of the log (called the *mantissa*) tells us what the figures are in the original, and the number in front of the decimal point (called the *characteristic*) tells us its order of magnitude. A log value with a characteristic of, say, 2 tells us that the number lies between 100 (10^2) and 1000 (10^3), and a log value with a characteristic of 0 corresponds to a number between 1 (10^0) and 10 (10^1). When we count downwards from 1 (log 1 = 0.0) we write the characteristic as -1, -2, etc., but keep the mantissa the same. So the log of 0.32 is written $\bar{1}.5$, the log of 0.032 is written $\bar{2}.5$, and so on. (By the way, pocket calculators don't do this; they actually indicate the full negative value. This makes no difference to its calculations.)

To the image-maker, the most important logs are the log of 10 and the log of 2. The log of 10 we already know: it is 1. The log of 2 we can find by looking at the log of 32, which you will remember is 1.5, that is, 32 is $10^{1.5}$. Now, $32 = 2^5$, so 2 is $\sqrt[5]{32}$, which is $10^{(1.5 \div 5)}$, i.e., $10^{0.3}$. So the log of 2 is 0.3. We can now write down the series

No	1	2	4	8	[10]	16	32	64	[100]	128	256	512	1000	2000
Log	0.0	0.3	0.6	0.9	[1.0]	1.2	1.5	1.8	[2.0]	2.1	2.4	2.7	3.0	3.3 etc.

Leaving out the figures in square brackets, which I have included only as an indication of where they fit in, this is a doubling-up series, and it matches the series of log values we use in plotting characteristic curves.

Logarithmic Scales

Some graphs use scales in which equal intervals along the scale represent, not equal arithmetic increments, but equal multiples. These are called *logarithmic* scales. They are often convenient for plotting purposes where there is a very large range of values, and the small values are also important in themselves. When there is a power-law relationship between the two variables, for example, $y^3 = x^2$, the use of log paper gives a straight line instead of a curve. When there is an exponential relationship such as $y = a^x$, you can obtain a straight line by using log–linear paper, that is, paper in which one axis is logarithmic and the other is arithmetic.

Logs Base 2

These underlie the light-value scales found on most exposure meters. Each increment on the scale represents a doubling of the stimulus:

No	1	2	4	8	16	32	64
Log	0	1	2	3	4	5	6

The small discrepancy at 1000, which you would expect to be 1024, is because the logarithm of 2 is not exactly 0.3 but actually 0.301—but an error of one-third of a percent isn't very important.

When we speak colloquially of (say) giving three stops (= 8 times) more (or less) exposure, we are actually using logs base 2 (written log_2).

To convert logs base 10 to logs base 2, all you have to do is divide the log by 0.3 (strictly, 0.301); and to convert logs base 2 to logs base 10 you multiply by 0.3.

Logs to Other Bases

You can in fact have logs to any base you like, but the only base other than 10 and 2 that you are likely to come across is e. This is one of those curious constants like π that seem to crop up in all sorts of odd places in mathematics. Its value is 2.71828 ... , and like π, it goes on forever. It is derived from the equation $y = e^x$, which when plotted gives a curve having a slope that is equal to its own y-value, an *exponential* curve, and it has a number of properties that make it particularly useful in advanced mathematics. To convert logs base e to base 10 you divide the figure by 2.303, and to convert the other way you multiply by the same figure. Logs base e are properly called *natural logarithms*; they are also sometimes called *Napierian logarithms*, although Napier himself had nothing to do with this constant. (Logarithms base 10 are sometimes called *common logarithms*.) In textbooks Napierian logarithms are abbreviated *ln* rather than log_e, and the expression e^x is often shown as *exp (x)*.

Napier's original logarithms were to the base $1/e$, though Napier himself knew nothing of e. (He had originally arrived at the concept of a logarithm by a geometrical process.) It was Napier's friend Henry Briggs who suggested the base 10, and he perhaps deserves equal credit with Napier for the modern concept of a logarithm.

Appendix 2: How a Hologram Works

Although we mostly talk of sine waves, I am going to use cosines here. There is no difference in the outcome, but it simplifies the picture, as a cosine function is symmetrical about the origin, whereas a sine function is not. The only bit of trigonometrical manipulation you need to know is the relationship

$$\cos X \cos Y \equiv \tfrac{1}{2} \cos (X + Y) + \tfrac{1}{2} \cos (X - Y)$$

Any traveling wave that is (co)sinusoidal when viewed as it passes a fixed point can be described by an equation of the form

$$a = A \cos (\omega t)$$

where a is the instantaneous amplitude at a time t. A is the maximum amplitude, and ω (omega) the angular frequency, which is related to the wavelength λ (lamda) by the relationship

$$\omega = 2\pi/\lambda \text{ (radians)}$$

If the waveform is not centered on the origin we need to put in a "phase" term φ (phi) that tells us how many radians it is away from a wave that *is* centered on the origin. If we have two waves of the same frequency ω, which we can call U_1 and U_2, we can write down their equations as

$$U_1 = A_1 \cos (\omega t + \varphi_1)$$

and

$$U_2 = A_2 \cos (\omega t + \varphi_2)$$

Now U_1 and U_2 are amplitudes, and the time-averaged intensity = $\tfrac{1}{2}$(amplitude)^2; the intensities I_1 and I_2 are given by $\langle \tfrac{1}{2}U_1^2 \rangle$ and $\langle \tfrac{1}{2}U_2^2 \rangle$, respectively, where the angle brackets mean that the intensities are time-averaged. Now, if U_1 and U_2 were not mutually coherent, their combined intensity would be merely $\tfrac{1}{2}(U_1^2 + U_2^2)$. But as the beams *are* mutually coherent, their combined intensity I is $\tfrac{1}{2}(U_1 + U_2)^2$, that is

$$I = \tfrac{1}{2}(U_1^2 + U_2^2 + 2U_1 U_2)$$

Writing out the equations in full gives

$$I = \tfrac{1}{2}[A_1^2\cos^2 (\omega t + \varphi_1) + A_2^2\cos^2 (\omega t + \varphi_2) + 2A_1 A_2 \cos (\varphi t + \varphi_1) \cos (\varphi t + \varphi_2)]$$

Using the identity $\cos X \cos y \equiv \tfrac{1}{2} \cos (X + Y) + \tfrac{1}{2} \cos (X - Y)$ for the third term gives

$$I = \tfrac{1}{2}[A_1^2 \cos^2 (\omega t + \varphi_1) + A_2^2 \cos^2 (\omega t + \varphi_2) + A_1 A_2 \cos (2\omega t + \varphi_1 + \varphi_2) + A_1 A_2 \cos (\varphi_1 - \varphi_2)]$$

Notice that the final term does not contain the symbol t, and is thus independent of time.

We now have to time-average this unwieldy expression. This simplifies it quite a lot, because the time-average of $\cos^2 X$ is simply ½, and the time-average of $\cos X$ is 0. We are left with

$$<I> = \tfrac{1}{4}(A_1^2 + A_2^2) + \tfrac{1}{2}A_1A_2 \cos (\varphi_1 - \varphi_2)$$

As there is no t term present, the intensity must be constant at this point in space. But how does it vary in space (i.e., with respect to the plane of the emulsion)? Well, the term φ_1 refers to the reference beam and, as this is undisturbed, φ_1 is either constant (if the beam is perpendicular to the plane of the emulsion) or varies linearly across the plane (if the beam is at some other angle). The term φ_2 refers to the object beam, and varies irregularly across the plane. So the positions of the fringes are coded in the phase relationships $\varphi_1 - \varphi_2$, and their intensities are coded in their amplitude variations A_1 and A_2. The emulsion records the positions and intensities of these fringes.

How do we replay the image? We simply direct the beam U_1 onto the hologram. We obtain the equation of the emergent wave, which we can call U_3 by multiplying U_1 at that point by the film transmittance function at that point. This is proportional to $<I>$. So U_3 will be proportional to $U_1 \times I$.

Writing out the values for these functions,

$$U_3 = \tfrac{1}{4}U_1(A_1^2 + A_2^2) + \tfrac{1}{2}U_1A_1A_2 \cos (\varphi_1 - \varphi_2)$$

Replacing U_1 in the final term by its actual value $A_1 \cos (\omega t + \varphi_1)$ gives

$$U_3 = \tfrac{1}{4}U_1(A_1^2 + A_2^2) + \tfrac{1}{2}A_1^2A_2 \cos (\omega t + \varphi_1) \cos (\varphi_1 - \varphi_2)$$

Using the identity $\cos X \cos Y \equiv \tfrac{1}{2} \cos (X - Y) + \tfrac{1}{2} \cos (X + Y)$ for the last term, we have

$$U_3 = \tfrac{1}{4}U_1(A_1^2 + A_2^2) + \tfrac{1}{2}A_1^2A_2\cos (\omega t + \varphi_2) + \tfrac{1}{2}A_1^2A_2 \cos (\omega t + 2\varphi_1 - \varphi_2)$$
$$= U_1 \times \text{Constant} + U_2 \times \text{Constant} + \text{Modified } U_2 \times \text{Constant}$$

The first term is thus the reference/replay beam; the second is the object/image beam. The third term is another image beam turned through an angle $2\varphi_1$ and reversed in phase (φ_2). This last term represents the "spurious" pseudoscopic real image in the (-1) order of diffraction.

Note: Most textbooks use the exponential form $a = A\ e^{i(\omega t + \varphi)}$ as the basic wave equation, which enables a more rigorous analysis: $[i = \sqrt{(-1)}]$. However, if you are unfamiliar (or uneasy) with complex algebra, the trigonometrical approach I have used is a good deal easier to follow.

Standing Waves

The previous section used the equation $a = A \cos (\omega t)$ to describe the variation of amplitude with time at a fixed point in space. But in a full description of a traveling wave we must also consider the way the amplitude varies in space at a given time. This is, of course, also a (co)sinusoidal function. In this case we write the equation as $a = A \cos (kx)$, where x represents distance, and $k = 2\pi/T$ (radians), where T is the period of the wave. We can combine these into a single equation by writing

$$a = A \cos (\omega t - kx)$$

Conventionally, this describes a wave moving from left to right. If the wave moves from right to left the distance variable is reversed, and we have

$a = A \cos (\omega t + kx)$

If two such traveling waves occupy the same space, we can add their equations to obtain the resultant disturbance:

$a = A [\cos (\omega t - kx) + \cos (\omega t + kx)]$

Using the identity $\cos (X - Y) + \cos (X + Y) = 2 \cos X \cos Y$ gives

$a = 2A \cos (\omega t) \cos (kx)$

Distance x is now linked to time t, and both the instantaneous amplitude a and the peak amplitude A are linked to distance. This equation represents a (co)sinusoidal waveform that has an amplitude varying from $+2A$ to $-2A$ with time, but does not move in space. The nodes are separated by one half-wavelength.

Appendix 3: The Fourier Model
for Image Formation

Of all the models there are for the behavior of light, the one that gives the best and most satisfying description of the formation of an image is the Fourier model. In particular, it provides a simple and elegant explanation of two aspects of image formation, namely the limits of resolution of an optical system and the optical transfer function. In this necessarily brief account I shall try to give an intuitive feel for the Fourier approach. You have already seen something of it in Chapter 6, where I suggested that every scene could be considered as made up of a large number of sinusoidal patterns of luminance with different orientations, amplitudes, and spatial frequencies. The converse, that you can make up any sort of pattern, repeated or not, from a set of sinusoidal components, is also part of the Fourier model, and is perhaps easier to grasp.

To see how this works, we can take an example that may be familiar to you if you have had some training in electronics engineering: building up a square wave from sinusoidal components. For our purpose it is preferable to use cosines instead of sines, as a cosine function is symmetrical about the origin, i.e., $\cos (-x) \equiv \cos (+x)$. Let us consider first a simple cosine function with spatial frequency q cycles per millimeter and amplitude A (Figure A3.1).

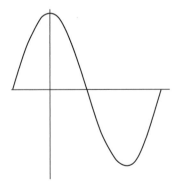

Figure A3.1

Now add a second cosine function with spatial frequency $3q$ and amplitude $A/3$, with a phase at the origin that is opposite to the fundamental cosine function (Figure A3.2).

The pattern, as you see, is no longer a simple cosine pattern, but it is still a periodic waveform, and still symmetrical about the origin. And it is more flat-topped than the fundamental cosine. Now let us add a further cosine function, this time of frequency $5q$ and amplitudes $A/5$, in antiphase to the second function (Figure A3.3).

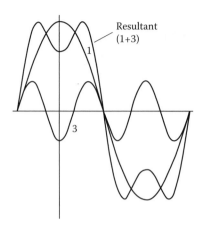

Figure A3.2

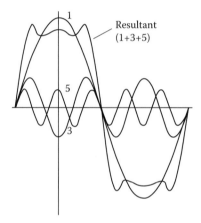

Figure A3.3

You can see that the resultant function is still periodic and symmetrical, and is becoming progressively more flat-topped. Now if we add all the odd multiples of the fundamental spatial frequency as far as the fifteenth, the result is as in Figure A3.4.

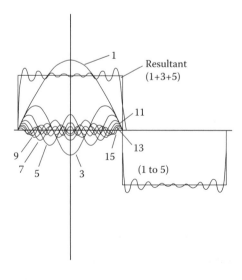

Figure A3.4

The combined function is now looking much more like a square wave. If you continue to add in the odd multiples of frequency (in electronics terminology, the odd harmonics) all the way to infinity, you will get a true square wave, or, to be mathematically correct, a rectangular function. This is *Fourier synthesis*.

Fourier methods also prove the converse: that a rectangular function does actually consist of those cosine components, and with those phase relationships. The proof is mathematical, not intuitive, of course, but I hope you can see that it makes sense. It applies not only to rectangular functions, but also to all kinds of repetitive functions: sawtooth, triangular, trapezoidal, and so on. This converse is called *Fourier analysis*.

There is a graphical method of depicting the spatial frequencies and amplitudes of the components of our rectangular function. It is called a *Fourier spectrum*, and is a graph of amplitude against spatial frequency for all the components of the function. Because of the symmetry of the cosine function, I have split the amplitudes on either side of the origin, but have kept the phases correct (Figure A3.5).

A music synthesizer does exactly this sort of thing, putting together new tone flavors by adding harmonics to the fundamental frequency in varying proportions.

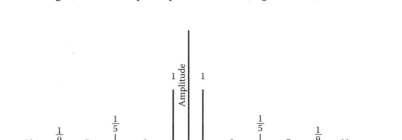

Figure A3.5

Note: The reason for including a zero-frequency component is that when dealing with light energy, the total always has to be greater than zero: there is no such thing as negative energy. By analogy with electronics, this is often called "the d.c. component."

In contrast to this complicated graph, a pure cosine function has a Fourier spectrum that consists of only a single spatial frequency, so we can represent its frequency spectrum by just two lines (Figure A3.6).

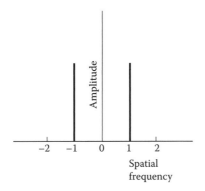

Figure A3.6

313

Now, if you pass a narrow parallel beam of light through a cosine grating, that is, one in which the transmittance varies (co)sinusoidally with distance, you get just two diffracted spots, one on each side of the zero-order spot (the d.c. component) (Figure A3.7).

• ● •

Figure A3.7

But if you pass it through a rectangular grating you get a whole row of spots (Figure A3.8).

•　　•　●　●　•　　•

Figure A3.8

You can see what is happening here: the position of the spots is giving us the spatial frequency plot of the grating function.

The mathematical statement of the amplitude and frequency spectrum of a function of this type is in the form of a series of sine and cosine terms, called the Fourier series for that function. The diffraction pattern of a grating is the optical equivalent of the Fourier series that describes the grating function.

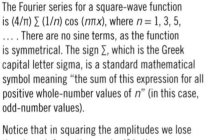

The Fourier series for a square-wave function is $(4/\pi) \sum (1/n) \cos (n\pi x)$, where $n = 1, 3, 5, \ldots$. There are no sine terms, as the function is symmetrical. The sign $\sum$, which is the Greek capital letter sigma, is a standard mathematical symbol meaning "the sum of this expression for all positive whole-number values of n" (in this case, odd-number values).

Notice that in squaring the amplitudes we lose the phase information, as $(-x)^2$ is the same as $(+x)^2$. Of course, the positive and negative phases do still exist.

In the pattern for the square grating you can see clearly that it is made up of discrete regularly spaced spatial frequencies. You can work back to the values of those frequencies by applying the formula $\lambda = d \sin \theta$ from Chapter 1, and solving it for d using the measurements of the spot distances. If the fundamental pitch of the grating is w, then $d = w, 3w, 5w, \ldots$, and if you measure the light intensities, you will find them to be $E, E/9, E/25, \ldots$, which are the squares of the amplitudes.

Figure A3.5 is the frequency plot for a true square function, that is, one where the width of the slits is the same as the width of the spaces between them. If we double the spacing between the slits, the frequencies in the spectrum will become twice as close together, because if you double d in the diffraction formula, you have to halve $\sin \theta$ (Figure A3.9).

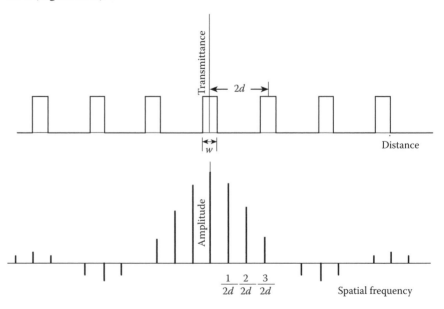

Figure A3.9

If you double the spacing again, the frequencies again become twice as close together (Figures A3.10 and A3.11).

As we keep increasing the spacing between the slits the frequencies will get closer and closer together, until in the limit, when the adjacent slits have moved away to infinity, we are left with a single slit, and the frequency spectrum has become a continuum (Figure A3.12).

The equation of the envelope of this spectrum is called the *Fourier transform* of the single rectangular function. This represents the amplitude distribution in the

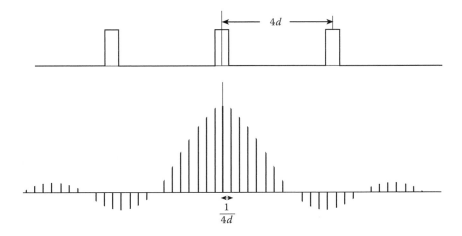

Figure A3.10

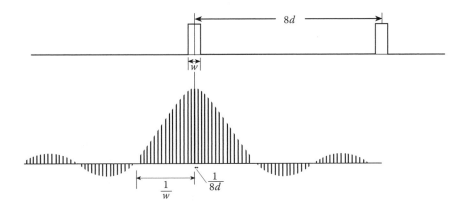

Figure A3.11

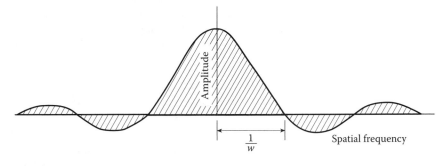

Figure A3.12

In the mathematical description, the sigma (Σ) symbol has turned into an integral (∫) symbol. Notice that the shape of the envelope has remained unchanged throughout. This is because we have kept the width of the pulse (or slit) constant.

The Dirac delta function is obtained by taking a rectangular pulse of unit area and making it progressively narrower, keeping the area the same. In the limit the pulse becomes infinitely narrow and infinite in amplitude, still with unit area. Paul Dirac (1902–1984) was one of the giants of modern physics. He played a crucial part in the development of quantum electrodynamics. He predicted the existence of antiparticles and integrated the work of Louis de Broglie and Erwin Schrödinger on wave mechanics with Werner Heisenberg's model of the electron and with special relativity. He shared the 1933 Nobel Prize for Physics with Schrödinger.

diffraction pattern of a single slit. The distance from the origin to the first zero is inversely proportional to the pulse width, and this is exactly what you see in the diffraction pattern as you change the slit width.

This last point has a further important implication. If we make the pulse narrower, the first zero moves farther away from the origin, until in the limit the pulse becomes infinitely narrow and the first zero moves away to infinity. The Fourier transform of an infinitely narrow pulse (called a Dirac delta function) contains all frequencies in equal amount (Figure A3.13).

Figure A3.13

Now look at this the other way round. If our slit (or pulse) is infinitely wide, its (spatial) frequency is zero. There is no diffraction in an infinitely wide slit! So this works in reverse (Figure A3.14).

Figure A3.14

No calculation of this type ever seems to be complete without some constant like 2/π turning up to make a nuisance of itself.

Without a lens you still get a diffraction pattern, but the electromagnetic field is not an exact Fourier transform.

This converse principle proves to be true for all Fourier transforms. If you take a Fourier transform of a Fourier transform, you get back to the original function, give or take the odd constant.

The second rule of Fourier transforms is also implicit. If you add several functions (in the same way as we added components to make a square wave) and take the combined Fourier transform, the result is the same as if you took the Fourier transforms of the functions separately and then added them together. They don't all have to be in one dimension, either; they add in the same way. A two-dimensional function has a two-dimensional Fourier transform.

There is one practical point to make. In order to produce an exact optical Fourier transform, you need a lens. The object must be situated at its front principal focal point. The optical Fourier transform is formed at the rear principal focal point. The diffracted waves from the object (which are plane waves) are focused on the screen at the rear principal focus, as in Figure A3.15.

The outermost waves are those that have been diffracted through the largest angle, and therefore represent the highest spatial frequencies present in the object, as we have already seen. But this simple fact is of fundamental importance in imaging optics, because *the relative size of the lens aperture determines the highest spatial frequency the lens can transmit*. This represents the diffraction-limited resolving power of the lens. You can calculate this figure from the *f*/no of the lens, or (as Abbe did for microscope objectives), from its NA.

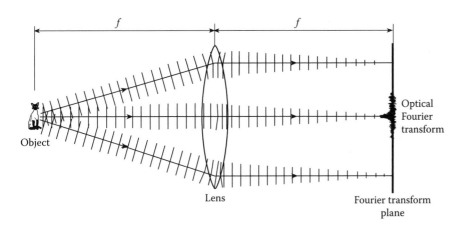

Figure A3.15

But what about the optical image? You can't get an image when the object is at the front principal focus—at least, the image, such as it is, is at infinity and infinitely large. But remember the rule about the Fourier transform of a Fourier transform: it goes back to the original function. So all we need to do is to put a similar lens system with its front principal focus at the Fourier transform plane, and we obtain a reconstruction of the object field in its rear principal focal plane (Figure A3.16).

If you go through the maths you can prove that this is true, and even show that the image is orthoscopic and inverted.

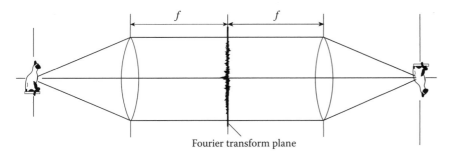

Figure A3.16

By blocking off (or attenuating) parts of the diffraction pattern, you can modify the final image. It isn't difficult, for example, to eliminate the raster lines of the image of a TV screen or the halftone dots of a newspaper photograph, as the diffraction patterns of these are characteristic patterns of spots that you can easily see and block off. You can even remove the blur from an image showing camera shake, though this is less easy.

The Fourier model evolved gradually, from work on diffraction theory by Fresnel and, later, Abbe. But it was only in the 1960s, with the arrival of the laser, that it became possible actually to see the diffraction pattern at the rear focal plane. (With white light it *is* there, but is smeared out by dispersion.)

One final, very important point. It is not difficult to show that the optical transfer function (OTF) of a lens system is the Fourier transform of its line spread function (LSF), and vice versa. The earliest derivations of OTFs in the early 1950s were in fact obtained from LSFs by (in those days highly protracted) computer calculations.

In this appendix I have only been able to give a flavor of the Fourier model. It is a fascinating approach that goes a good deal deeper than I have been able to indicate, and gives many insights into the whole subject of imaging. You can find a more comprehensive nonmathematical exposition in an appendix to my book *Practical Holography* (3rd edition, CRC Press, formerly IoP Pubs, 2004). The *Manual of Photography* (10th edition, Focal Press, 2010) has a whole chapter relating the Fourier model to digital imaging and image modification, an aspect I haven't had space for here. For a full treatment of the subject, including all the maths, the standard text is E. G. Steward's *Fourier Optics: An Introduction* (2nd edition, Ellis Horwood, 1989).

Appendix 4: The Meaning of pH

The pH of a solution is a measure of its acidity or alkalinity. A neutral solution has a pH of 7. Higher figures are associated with alkalinity, lower figures with acidity. Every aqueous solution contains hydrogen ions (H^+) and hydroxyl ions (OH^-). In neutral solutions and pure water there are exactly 10^{-7} moles per liter (mol/ℓ) of each.

The product of hydrogen and hydroxyl ion concentrations in an aqueous solution is always 10^{-14}. If a solution is made alkaline by adding a substance that increases the OH^- concentration by a factor of, say, 100, increasing it to 10^{-5} mol/ℓ, the H^+ concentration falls to 10^{-9} mol/ℓ. We say that the pH has changed from 7 to 9. Conversely, if the H^+ concentration is increased by a factor of 100, bringing it up to 10^{-5} mol/ℓ, we say that the pH has changed to 5. (The OH^- concentration will have fallen to 10^{-9} mol/ℓ.)

Each step on the pH scale means a change in the hydrogen-ion concentration by a factor of 10, so a solution with a pH of 10.5 is 10 times as alkaline as one with a pH of 9.5. It is a logarithmic scale (see Appendix 1), so that each positive pH increment of 0.3 represents a doubling of the OH^- concentration and a halving of the H^+ concentration. The pH values found in photographic processing solutions range from about 10.5 for the fiercest developers to about 5.5 for acid-fixing baths and 4.5 for acid bleach baths.

Laboratory pH meters are expensive and need frequent recalibration. Soil pH meters from garden shops are not usually accurate for values outside pH 6–8 and are unsuitable for testing photographic solutions. You are better off with pH indicator papers, which are strips of absorbent paper impregnated with a dye that changes color at specific pH values. These can be "universal," changing color through a whole spectrum as pH values vary between 2 and 13 (you match the color against a printed color chart) or "narrow-band," changing color over a given small pH range (say pH 8–9). You can get them from educational suppliers.

The *mole* (mol) is the fundamental SI unit of quantity of substance. It is defined as the amount of substance that contains as many entities as there are atoms in 12 grams of carbon-12. The "entities" are usually atoms, ions, or molecules, though in fact they could be anything—grains of sand, say, even people! The actual number of atoms, etc., in a mole is 6.023×10^{23}, Avogadro's number. A *molar solution* is a solution that contains 1 mole per liter of dissolved substance.

Index

Dufaycolor films, 27
Duffieux, P. M., 85
DVDs, 208–209
Dye-bleach process, 106, 148

E

Eastman Research Institute, 105
Eastman, George, 116
Edge spread function, 89
Edison, Thomas, 207
Einstein, Albert, 2
Ektachrome, 106, 282–283
Electromagnetic radiation, 2–3
Electromagnetic spectrum, 4
Electromagnetism, 2
Electron beam lithgraphy, 278
Electron microscopy, 277–278
Electronic flash, 26
Electronic information storage
    charge-coupled device (CCD), 163, 164
    field-effect transitor, 162–163
    junction transistor, 162
    overview, 161
    semiconductor diodes, 161, 162
Electrostatic copying. *See* Xerography
Endoscopy, 288–289
Entrance pupil, 49, 51
Escher, Maurits, 45
Exit pupil, 51
Eye
    anatomy of, 34
    aqueous humor, 34
    color perception of, 39–40, 41
    constancy of position, 42
    cornea, 34
    evolution of, 38, 39
    iris, 34
    long-sighted defect, 34
    optical system of, 33
    retina (*see* Retina)
    sensitivity range, 36–37
    short-sighted defect, 34

F

Fabry-Pérot etalon, 267
Faraday, Michael, 3
Fechner, Gustav Theodor, 37
Ferromagnetism, 200
Filters, light
    color print (CP), 29
    color separation, 28–29
    contrast, 27–28
    emulsion balancing, 27
    infrared (IR), 29
    light balancing, 29
    narrow-band, 28–29

neutral density (NP) filters, 29
    polarizing (*see* Polarizing filters)
    special effects (SFX), 29
    ultraviolet (UV), 29
Fisheye lenses, 67, 68, 69
Fizeau's fringes, 10
Fizeau, Armand Hippolyte Louis, 10
Flashes
    duration, 133
    electronic, 127
    flash heads, 128, 129
    flashbulbs, 26, 127
    guide numbers, 129
    night shots, for, 129
    ripple firing, 141
    synchronization of, 127, 128–129
Fluorescent lamps, 26
Focal length, 16
Focusing screens, 119
Forward-motion compensation, 130–131
Foucalt, Jean Bernard Léon, 297
Foucalt, Louis, 10
Fourier analysis, 86
Fourier model for image formation, 311–318
Fourier synthesis, 86
Fourier, Jean-Baptiste Joseph, 10
Fourier-transform holograms, 249–250
Frame finders, 119–120
Frame grabbers, 170
Franklin, Rosalind, 292
Frequencies
    electromagnetic radiation, as part of, 3
    image making, as part of, 4
Fresnel lens, 16
Fresnel zone plate, 248, 249
Fresnel, Augustin, 16
Fringes, 7

G

Gabor zone plate, 248, 249
Gabor, Denis, 234, 235, 248
Galileo, 253
Gamma radiography, 291
Gauss points, 50
Gauss, Karl Friedrich, 50
Gaussian optics, 50–51
Ghost images, 59
Godowsky, Leopold, 105
Goerz Hypergon lens, 67
Graded refractive index lenses, 34, 231
Granularity, 90
Grating condition, 10
Gray scale cards, 101
Gregory, James, 253
GRIN lenses, 77
Guide numbers, 129

325